IMAGE BASED
MEASUREMENT SYSTEMS

IMAGE BASED MEASUREMENT SYSTEMS

OBJECT RECOGNITION AND PARAMETER ESTIMATION

Ferdinand van der Heijden

University of Twente,
The Netherlands

JOHN WILEY & SONS
Chichester • New York • Brisbane • Toronto • Singapore

Copyright © 1994 by John Wiley & Sons Ltd.
Baffins Lane, Chichester
West Sussex PO19 1UD, England
National Chichester (01243) 779777
International +44 1243 779777

All rights reserved.

No part of this book may be reproduced by any means,
or transmitted, or translated into a machine language
without the written permission of the publisher.

Other Wiley Editorial Offices

John Wiley & Sons, Inc., 605 Third Avenue,
New York, NY 10158-0012, USA

Jacaranda Wiley Ltd, 33 Park Road, Milton,
Queensland 4064, Australia

John Wiley & Sons (Canada) Ltd, 22 Worcester Road,
Rexdale, Ontario M9W 1L1, Canada

John Wiley & Sons (SEA) Pte Ltd, 37 Jalan Pemimpin #05-04,
Block B, Union Industrial Building, Singapore 2057

British Library Cataloguing in Publication Data

A catalogue record for this book is available from the British Library

ISBN 0 471 950629

Camera copy supplied by the author using MS Word.
Printed and bound in Great Britain by Bookcraft (Bath) Ltd,

TABLE OF CONTENTS

Preface ix

Chapter 1: introduction 1
    1.1   scene, image and modality 1
    1.2   why use images? 3
    1.3   computer vision, image processing and computer graphics 3
    1.4   applications of image based measurement systems 4
    1.5   organisation of the book 6
    1.6   literature 6

Chapter 2: image formation 9
    2.1   interaction between light and matter 9
            2.1.1   radiometry and photometry 9
            2.1.2   reflection models 15
    2.2   projection 16
            2.2.1   world and camera co-ordinates 16
            2.2.2   pinhole camera model 17
            2.2.3   homogeneous co-ordinates 18
            2.2.4   point spread function 19
            2.2.5   2-dimensional convolution 21
    2.3   Fourier analysis 23
            2.3.1   harmonic functions 23
            2.3.2   the Fourier transform 25
    2.4   image conditioning 29
            2.4.1   front or rear illumination 29
            2.4.2   specular or diffuse illumination 30
            2.4.3   structured light 30
    references 32

Chapter 3: image models 33
    3.1   first order statistics 34
    3.2   second order statistics 37

                3.2.1   stochastic processes and 2-dimensional filtering 40
                3.2.2   autocorrelation and power spectrum 41
   3.3   discontinuities 43
references 48

Chapter 4: image acquisition 49
   4.1   the ccd-camera 50
   4.2   spatial sampling and reconstruction 53
                4.2.1   impulse modulation 53
                4.2.2   reconstruction 55
                4.2.3   area sampling and presampling filtering 58
                4.2.4   2-dimensional sampled stochastic processes 60
                4.2.5   finiteness of the image plane 62
                4.2.6   discrete Fourier transform 63
   4.3   amplitude discretisation 64
references 67

Chapter 5: image operations 69
   5.1   pixel-to-pixel operations 69
                5.1.1   monadic operations 70
                5.1.2   dyadic operations 76
   5.2   linear operations 77
                5.2.1   convolution 79
                5.2.2   orthogonal transforms 86
                5.2.3   differential operators 94
   5.3   correlation techniques 103
   5.4   morphological operations 107
                5.4.1   definitions 109
                5.4.2   basic operations 111
                5.4.3   morphological operations 117
                5.4.4   applications 124
                5.4.5   grey scale morphology 127
references 129

Chapter 6: statistical pattern classification and parameter estimation 131
   6.1   decision making and classification 134
                6.1.1   Bayes classification 136
                6.1.2   uniform cost function and minimum error rate 138
                6.1.3   Gaussian random vectors 140
                6.1.4   the 2-class case 144
                6.1.5   rejection 149
   6.2   parameter estimation 151
                6.2.1   Bayes estimation 152
                6.2.2   minimum variance estimation 153
                6.2.3   map estimation 155
                6.2.4   maximum likelihood estimation 158
                6.2.5   least squares fitting and ml-estimation 158
                6.2.6   simultaneous estimation and classification 160

       6.2.7   properties of estimators...161
       6.2.8   estimation of dynamically changing parameters................163
6.3   supervised learning...167
       6.3.1   parametric learning 168
       6.3.2   Parzen estimation...173
       6.3.3   nearest neighbor classification...175
       6.3.4   linear discriminant functions...179
       6.3.5   generalised linear discriminant functions 182
6.4   performance measures..186
       6.4.1   interclass and intraclass distance 186
       6.4.2   Chernoff-Bhattacharyya distance..190
6.5   feature selection and extraction..192
       6.5.1   performance and dimensionality..193
       6.5.2   feature selection 195
       6.5.3   linear feature extraction 198
references 206

Chapter 7: image analysis...207
7.1   image segmentation..212
       7.1.1   pixel classification 212
       7.1.2   region based segmentation...231
       7.1.3   edge based segmentation...238
7.2   region properties...253
       7.2.1   region parameters...254
       7.2.2   contours..260
       7.2.3   relational description..270
7.3   object recognition...272
       7.3.1   object models 274
       7.3.2   from object model to image model..277
       7.3.3   matching techniques...280
7.4   estimation of 3-dimensional body parameters 288
references 293

Appendix A topics selected from linear algebra and matrix theory 295
A.1   linear spaces 295
A.2   metric spaces 299
A.3   orthonormal systems and Fourier series..300
A.4   linear operators...302
A.5   vectors and matrices 305
A.6   trace and determinant 308
A.7   differentiation of vectors and matrices..309
A.8   diagonalisation of self-adjoint matrices 311
references 312

Appendix B probability theory and stochastic processes 313
B.1   probability theory and random variables...313
B.2   bivariate random variables 312
B.3   random vectors 319

	B.3.1	decorrelation	321
B.4		stochastic processes	322
	B.4.1	stationarity and power spectrum	323
	B.4.2	average, expectation and ergodicity	324
	B.4.3	time-invariant linear systems	326
references			327

Bibliography 329

Permission source notes 331

Index 333

PREFACE

"Measurement science" is a science that concerns the objective, meaningful and empirical assignment of symbols or numbers to properties of physical objects or events. As such, measurement science has an interdisciplinary nature. It integrates many other disciplines including sensor design, estimation theory, pattern classification, system identification and artificial intelligence. In the title of this book the term "image" refers to 2-dimensional representations of the "objects under measurement". Often, these images are formed optically, but we can also think of other types of image formation techniques (as in tactile sensory systems and ultra sonic imaging systems).

In many universities and technical colleges an increasing amount of time is now given in courses on "Computer Vision", "Statistical Pattern Classification" and "Estimation and Detection Theory". This book is intended for courses for undergraduate students in the third or fourth academic year. It should also be useful to beginning graduate students, practical engineers and scientists.

Although many textbooks have appeared with excellent treatments of the topics mentioned above there has not been any suitable book covering the integration of these disciplines. The present book is an attempt to meet the need for a text in which the treatment of image based measurements is founded on the integration of techniques. In accordance with that the book emphasises the importance of mathematical modeling of the underlying physical processes (e.g. image formation).

Chapter 6 has been added for the readers that are not familiar with statistical estimation theory and pattern classification. The chapter is an introduction to these two subjects. In fact, it can be read independently from the remaining text. As such, the book is hoped to be suitable for an introductory course on "computer vision", but also for a course on "statistical estimation and classification".

The significance of the theoretical part is underlined from time to time by a number of examples describing practical systems. Furthermore, listings of (pseudo) C code are added here and there in order to suggest how functional designs can be transformed into realisations.

Ferdinand van der Heijden
Enschede, The Netherlands

1
INTRODUCTION

The importance of images in the development of human civilisation is evident. Even in the ancient times images have played a significant role. Today, image technologies develop rapidly, and we may expect that in future the importance of images will increase progressively. Apart from spiritual or artistic connotations, the significance of images is especially found in information systems. Here, the application of images is roughly twofold. In the first place, images are used to improve or assist human visual perception. Thermography, for instance, helps us to perceive heat sources that are hard to observe with the naked eye.

In visual information systems images are the physical carriers of information. Images can easily be brought in a form suitable for further processing. Therefore, the second application of images is in automated information systems, i.e. *image based measurement systems*. A rudimentary example is an infrared detector equipped with only one sensor. These detectors are used, for instance, to open doors automatically as soon as a person wants to enter a building. Its output is a *symbol* with two possible states: one indicating that there is a heat source in its vicinity; the other that there isn't.

An example of a more complex image based measurement system is the so-called *optical character reader* (OCR), a system that reads the text on documents and transforms it into (ASCII-)files. Here, the output consists of a *sequence of symbols*, each symbol having a number of possible states.

Other systems are used to measure parameters describing objects and events, e.g. position, size, velocity, etc. The output are *numbers* that describe the object or the event.

1.1   SCENE, IMAGE AND MODALITY

In dealing with images, the ultimate area of interest is always an object or a collection of objects existing in the real 3-dimensional world. Examples are manifold. In photography: the face of a human, a building, a landscape, etc. In remote sensing: earthly terrain, urbanisation. In medical diagnosis systems: biological organs and tissues inside the human body. In robot vision and quality control: the products (or parts of them) at a production line in a factory.

2 IMAGE BASED MEASUREMENT SYSTEMS

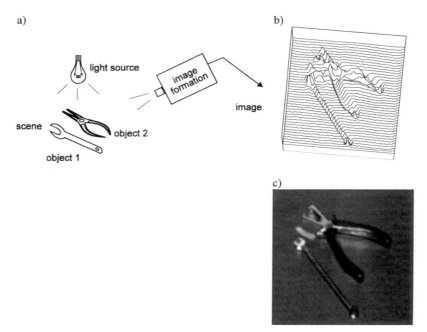

Figure 1.1 Scene and image
a) Image formation
b) 2-dimensional "light" distribution
c) Image displayed with grey tones

Together with their surroundings, these objects constitute a scene. Often, an image is formed by means of visible light. The light emerges from a light source and interacts with the surfaces of the objects. Then, part of the light is captured by an image formation system, which finally yields a 2-dimensional distribution of the light defined on the image plane. It is this distribution of light that we call the *image* of the scene. An illustration is presented in figure 1.1.

In the situation described above, visible light is the carrier of information about the scene. It is called the *modality* of the image. Visible light is not the only manner to form an image. Other parts of the electro-magnetic spectrum may be exploited: microwaves (radar), infrared (thermography), X-rays and gamma rays. Other physical entities that can serve as a modality are acoustic waves (ultra-sonic imaging systems), and magnetic fields (NMR imaging).

Using visible light, the main interaction between the modality and the scene is the *reflection* of light at the surfaces of the objects. Other possibilities are *absorption* and *scattering* inside the objects (as in X-ray imaging and gamma ray-imaging). Furthermore, one can distinguish *passive* imaging systems from the *active* one. In active measurements, the energy of the modality is provided by an external source (e.g. a light bulb, an X-ray source). In passive measurements, the objects of interests themselves supply the imaging system with energy (as in thermography).

A *digital image* is a digital representation of the image introduced above. It is a matrix of numbers that represents a sampled version of the 2-dimensional

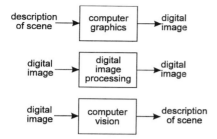

Figure 1.2
Relations between computer vision, digital image processing, and computer graphics

distribution resulting from the image formation. The advantage of having such a representation is that it can be manipulated by digital computer equipment.

1.2   WHY USE IMAGES?

Since, ultimately, only the scene and its objects are of interest, the question can be raised why (and when) one should use images. The answers are manifold:

- For archiving and retrieval. Images can be stored and retrieved on/from magnetic or optical materials. As such, images are a tool to memorise the details of a scene. Examples: video tapes, X-ray images in medical diagnosis systems.
- To bridge long distances. An image can be coded such that it is suitable for transportation. Examples: television, fax, radar, satellite images.
- To perceive the interior of non-transparent bodies in a non-destructive manner. Example: X-ray imaging.
- To perceive details of objects which are too small for human vision. Example: microscopy.
- To improve human vision by exploiting other modalities than visible light. Example: infrared in thermography.
- In automated measurement systems: to capture the multi-dimensional nature of scenes.
- In automated measurement systems: to prevent any mechanical contact between the objects of the scene and the sensors of the measurement system.

1.3   COMPUTER VISION, IMAGE PROCESSING AND COMPUTER GRAPHICS

Computer vision, image processing and computer graphics are all processes that deal with images. The difference between these processes is in the representation of the input and output data. This is shown in figure 1.2.

The input of a computer graphic system consists of a list that describes a scene. The purpose is to transform this list into a (digital) image which could have been formed if this scene would really exist. As such, a computer graphics system simulates the image formation.

4 IMAGE BASED MEASUREMENT SYSTEMS

a) 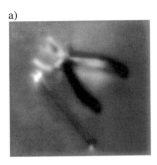   b)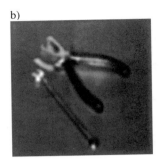

Figure 1.3 Image restoration: a) Blurred, observed image b) Restored image

In computer vision it is the other way round. Starting point is an image captured from an existing scene. The purpose is to gain information about the scene.

Digital image processing acts on images and results in images. It transforms one digital image into another. Image processing can be classified according to:
- image coding
- image enhancement
- image restoration
- image feature extraction.

Image coding is useful for storage and transportation of images. The goal of *image enhancement* is to facilitate human vision by enhancing specific image features. For example, in certain medical diagnosis systems, X-ray images are displayed on monitors (cathode-ray tubes). In order to adapt this presentation to human perception sometimes contrast manipulations and image filtering are applied. *Image restoration* is an operation that tries to correct distortions from the image formation system, for instance, image blurring and noise. An example is shown in figure 1.3. The goal of *image feature extraction* is to transform the image into another image from which measurements can be derived. This, of course, is useful in computer vision.

1.4 APPLICATIONS OF IMAGE BASED MEASUREMENT SYSTEMS

Image based measurement is a process that assigns meaningful numbers and symbols to objects in an imaged scene (see figure 1.4). This information may concern:

- A simple detection of an object.
  Is there a certain object in the scene?
- Classification of object(s).
  To what class does an object belong?
- Parameter estimation.
  What is the size, position and orientation of an object?
- Shape analysis.
  What is the shape and geometry of the objects?

- Scene description.
  What are the objects in the scene, and how are they related to each other?

Image based measurement systems find wide applications in various kinds of scientific areas. But also in various industrial, medical and agricultural environments the applications are rapidly growing.

The first scientific discipline that discovered the power of computer vision was astronomy. In the sixties, pictures of the moon were processed by computers. Initially, these techniques were developed to improve the quality of the images (i.e. image restoration). Later, scientists found out that similar techniques could be applied to take over the manual, time consuming procedures needed to extract the desired information from these images. Other scientific disciplines that use computer vision nowadays include:

- photogrammetry
- particle physics
- biology
- medical science
- geology and oceanology
- science of materials

In the industrial and agricultural area, the main application of image based measurement is quality control. In many production lines, for instance in the food industry, the quality of products must be guaranteed. Manual inspection is laborious and unreliable, and thus expensive. However, the quality control often involves a visual task. Therefore, in these branches, the application of computer vision is fruitful.

Robot vision involves visual tasks that are meant for real-time control of production systems and autonomous vehicle guidance (i.e. navigation, collision avoidance). In these examples, the reason to use image based systems is its accuracy, flexibility, reliability, and its cost efficiency. It must be noted, though, that in this area, computer vision is still in an immature phase. The reason is, that it involves 3-dimensional vision. In the present state of art, this is partly an unsolved problem.

Other situations in which image based measurements are applied is when it is impossible or not desirable to use human vision. An example of the first category is under-water inspection (e.g. under-water vehicle guidance). Examples of the second category occur in heavy polluted or hazardous environments (e.g. nuclear power plants).

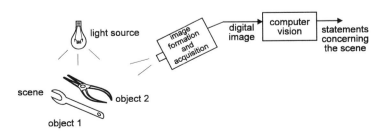

Figure 1.4 Image based measurement

1.5 ORGANISATION OF THE BOOK

This book contains seven chapters. Chapter 2 discusses the image formation process. As an example, the interaction of visible light with surfaces of matter is looked upon. Together with the pinhole camera, this modality is a well-known model of an image formation system. Chapter 3 deals with the mathematical modeling of images. These models are useful in order to develop image operators applied in computer vision systems. Chapter 4 applies to image acquisition, i.e. the transformation of the continuous image into a digital representation. The device most widely used for that purpose is the CCD-camera. Fundamental image operations, e.g. image filtering, are discussed in Chapter 5. Chapter 6 deals with statistical pattern classification and parameter estimation. This, and the topics from previous chapters, find application in Chapter 7, which addresses the problem of image analysis, i.e. transformation of the images into meaningful numbers and symbols.

The objective of the book is as follows. After studying the book, the reader is capable of recognising a measurement problem that can be solved by computer vision. Furthermore, he is able to design this system at a functional level. For that purpose, the reader has a broad overview of all kind of digital image processing techniques. He is able to access relevant literature. Listings of (pseudo) C-code are added here and there in order to suggest how functional designs can be transformed into realisations.

Pattern classification and parameter estimation is an important discipline which is applied in computer vision extensively. Therefore, the book also includes a chapter containing an introduction to this discipline. Chapter 6 is written such that it can be studied independently from the other chapters.

It is understood that the reader is familiar with the fundamentals of "probability theory and stochastic processes" and concepts from "linear algebra". Two appendices are added so as to help readers who have lost some knowledge in these areas. Furthermore, the appendices also explain the mathematical notations used in this book.

1.6 LITERATURE

There is a very wide range of literature covering the fields of image formation, image processing, parameter estimation, classification, computer vision, etc. The bibliography included at the end of the book (hopefully) provides an introduction to this extensive literature. Apart from that, each chapter contains a list of references where the reader can find the background of some topics discussed in the chapter.

Most titles listed in the bibliography require detailed mathematical background knowledge. Especially for the readers who are not much familiar with mathematical statistics and analysis the following list of titles has been added. These textbooks avoid a profound treatment.

Introductory textbooks

Bank, S., *Signal Processing, Image Processing and Pattern Recognition*, Prentice Hall, New York, 1990.

Fairhurst, M., *Computer Vision for Robotic Systems - An Introduction,* Prentice Hall, New York, 1988

Galbiati, L.J., *Machine Vision and Digital Image Processing*, Prentice Hall, Englewood Cliffs, 1990

Gonzalez, R.C. and Woods, R.E.: *Digital Image Processing*. Addison-Wesley, Massachusetts. 1992.

Low, A., *Introductory Computer vision and image processing*, McGraw-Hill, London, 1991

Niblack, W., *An Introduction to Digital Image Processing*, Prentice-Hall Int., Englewood Cliffs, New Jersey, 1986

Russ, J.C., *The image processing handbook*, CRC Press, Boca Raton, 1992.

Schalkoff, R.J., *Digital Image Processing and Computer Vision*, Wiley & Sons, New York, 1989

Vernon D., *Machine Vision: Automated Visual Inspection and Robot Vision*, Prentice Hall, New York, 1991

Developments in measurement science - especially in the field of computer vision - occur with increasing rapidity. Most developments are reported in periodicals. The following list is of particular interest.

Periodicals

IEEE transactions on pattern analysis and machine intelligence, IEEE Computer Society, New York.

IEEE transactions on systems, man and cybernetics, IEEE Systems, Man, and Cybernetics Society, New York.

Pattern Recognition. The journal of the Pattern Recognition Society, Pergamon Press, Oxford.

Pattern Recognition Letters. An official publication of the IAPR, North-Holland, Amsterdam

CVGIP: Image Understanding, Academic Press, San Diego.

CVGIP: Graphical Models and Image Processing, Academic Press, San Diego.

Machine vision and applications. An international journal, Springer, New York.

Image and vision computing. An international journal, Butterworth Scientific, Guildford.

2
IMAGE FORMATION

Meaningful interpretations of an image can only be given if there is a thorough understanding of the physical process that maps the 3-D scene to a 2-D image and the process that transforms this image into a digital representation. The first topic (image formation) is dealt with in this chapter.

The purpose of this chapter is to arrive at a mathematical model describing the relevant parts of the image formation process. Visible light, together with non-transparent objects in the scene and a pinhole camera, will serve as an example throughout this chapter. The last section discusses some image conditioning techniques.

2.1   INTERACTION BETWEEN LIGHT AND MATTER

2.1.1   Radiometry and photometry

Radiometry is the measurement of the flow and transfer of radiant (electromagnetic) energy emitted from a surface or incident upon a surface Haralick and Shapiro [1993], Longhurst [1970], and RCA [1974]. Table 2.1 presents the standard physical quantities that are needed to define the radiometric quantities.

The solid angle ω defined by a cone is equal to the area intercepted by the cone on a unit radius sphere centred at the top of the cone. The total solid angle about a point in space is $4\pi \ sr$. If a surface of area A is at a distance d from a point and the surface normal makes an angle of θ with respect to the cone axis, and, if $d^2 >> A$, then the solid angle is: $\omega \approx A\cos\theta/d^2$.

One of the characteristic quantities of an electromagnetic wave is its wavelength λ. The electromagnetic spectrum for wavelengths ranges from $10^{-10} \mu m$ to $10^5 \ km$. The spectrum is given in table 2.2. Usually, the electromagnetic energy is composed of a large number of harmonic waves, each having its own wavelength and its (randomly fluctuating) magnitude and phase. In this *incoherent* situation most optical devices behave linearly with respect to the electromagnetic energy, i.e. when dealing with incoherent light, it suffices to consider the situation with fixed

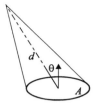

Figure 2.1 The solid angle defined by a cone

wavelength. The response to incident radiant energy with a broad spectrum can be found by decomposing the energy with respect to its wavelengths and by calculating the response to each component. Summing these individual responses yields the radiant energy of the output of the device. In the *coherent* situation (in which the magnitude and phase is deterministic) the superposition principle does not hold. This situation may occur when conditioned light sources such as lasers are used. Throughout this chapter we will assume incoherent light. We will consider monochromatic light (composed of waves with energy concentrated around a single wavelength). However, because of the superposition principle, the result can easily be extended to the polychromatic case (light with various wavelengths).

Radiometric quantities have meaning throughout the entire electromagnetic spectrum. In contrast, *photometric* quantities are meaningful only in the visible portion ($0.38\mu m < \lambda < 0.77\mu m$). The purpose of photometry is to obtain quantities which are useful for *human vision*. This book concerns (automatic) measurement systems. Within this context, human visual perception is less relevant. Nevertheless, many optical devices are specified with photometric quantities. Therefore we will consider both radiometric and photometric quantities.

Human *color* vision is still not fully understood. In most models the retina of the eye consists of a mosaic of three different receptor elements. Each element has its own spectral responsivity. It is instructive (but not necessary) to think of a red-, a green-, and a blue-sensitive receptor type. The three responsitivities appear to overlap considerably. Nevertheless, the degree of freedom in the spectral domain remains three.

In computer vision, multiple spectral bands, eventually outside the visible portion, can be very useful. The degree of freedom is not necessarily restricted to three. In some applications two bands will suffice. In other applications more than three bands are desirable. Often, color systems equipped with three bands (e.g. RGB: red, green and blue) are cheap because of the huge consumer market for color cameras and related equipment.

Table 2.1 Basic physical quantities

Quantity	Symbol	SI unit	Symbol
Length	various	meter	m
Area	A	square meter	m^2
Solid angle	ω	steradian	sr
Wavelength	λ	meter	m
Time	t	second	s

Table 2.2 The electromagnetic spectrum

EM-portion	Wavelength [μm]	EM-portion (visible part)	Wavelength [nm]
long electrical oscillations	$\begin{cases} 10^{14} \\ 10^{11} \end{cases}$	middle infrared	$\begin{cases} 6000 \\ 1500 \end{cases}$
radio waves	$\begin{cases} 10^{11} \\ 10^{6} \end{cases}$	near infrared	$\begin{cases} 1500 \\ 770 \end{cases}$
microwaves	$\begin{cases} 10^{6} \\ 10^{3} \end{cases}$	red	$\begin{cases} 770 \\ 622 \end{cases}$
infrared	$\begin{cases} 10^{3} \\ 1 \end{cases}$	orange	$\begin{cases} 622 \\ 597 \end{cases}$
visible	$\begin{cases} 1 \\ 10^{-1} \end{cases}$	yellow	$\begin{cases} 597 \\ 577 \end{cases}$
ultraviolet	$\begin{cases} 10^{-1} \\ 10^{-2} \end{cases}$	green	$\begin{cases} 577 \\ 492 \end{cases}$
X-rays	$\begin{cases} 10^{-2} \\ 10^{-6} \end{cases}$	blue	$\begin{cases} 492 \\ 455 \end{cases}$
gamma-rays	$\begin{cases} 10^{-4} \\ 10^{-8} \end{cases}$	violet	$\begin{cases} 455 \\ 390 \end{cases}$
cosmic rays	$\begin{cases} 10^{-8} \\ 10^{-10} \end{cases}$	near ultraviolet	$\begin{cases} 390 \\ 300 \end{cases}$

Referring to the superposition property of incoherent light (stated above), the modeling of multi-spectral imagery can be kept quite simple. First, we model the image formation for each wavelength resulting in a spectral image. Next, for each receptor, we multiply the spectral image with the spectral responsivity of the receptor. Finally, we integrate the thus obtained spectral energy distribution. In conclusion, it suffices to restrict ourselves to discuss single band imagery. Models of multi-spectral imagery turn out to be a repetition of a model of single band imagery.

The most important radiometric and photometric quantities, definitions, symbols and units are summarised in table 2.3. The radiometric quantities and the corresponding photometric quantities have equal symbols, except in the case in which this would be confusing. In those (exceptional) cases radiometric and photometric quantities become a sub-index "e" and "v", respectively.

The radiant energy Q is the basic quantity. It is the amount of energy that is radiated by some source at a given time t. The rate at which this amount is increasing is called the radiant flux Φ. If the source is conceived as a point from which the energy is radiated, then the radiant flux can be regarded as the flow of energy through a sphere surrounding that point.

Table 2.3a Radiometric quantities

Quantity	Symbol	Description	Unit
Radiant energy	Q or Q_e		J
Radiant flux	Φ or Φ_e	$\Phi = dQ/dt$	$Js^{-1} = W$
Radiant exitance	M	Radiant flux from a surface: $M = \partial\Phi/\partial A$	Wm^{-2}
Irradiance	E	Radiant flux at a surface: $E = \partial\Phi/\partial A$	Wm^{-2}
Radiant intensity	I	$I = \partial\Phi/\partial\omega$	Wsr^{-1}
Spectral radiant flux	$\Phi(\lambda)$	Radiant flux per unit wavelength: $\Phi(\lambda) = \partial\Phi/\partial\lambda$	$W\mu m^{-1}$
Spectral radiant exitance	$M(\lambda)$	Radiant exitance per unit wavelength	$W(m)^{-2}(\mu m)^{-1}$
Spectral irradiance	$E(\lambda)$	Irradiance per unit wavelength	$W(m)^{-2}(\mu m)^{-1}$
Spectral radiant intensity	$I(\lambda)$	Radiant intensity per unit wavelength	$W(sr)^{-1}(\mu m)^{-1}$

Table 2.3b Photometric quantities

Quantity	Symbol	Description	Unit
Luminous energy	Q or Q_v	$Q_v = \int K(\lambda)Q_e(\lambda)d\lambda$	$lm\ s$
Luminous flux	Φ or Φ_v	$\Phi = dQ/dt$	lm
Luminous exitance	M	Luminous flux from a surface: $M = \partial\Phi/\partial A$	$lm(m)^{-2}$
Illuminance	E	Luminous flux at a surface: $E = \partial\Phi/\partial A$	$lx = lm(m)^{-2}$
Luminous intensity	I	$I = \partial\Phi/\partial\omega$	$cd = lm(sr)^{-1}$

If the source is not a point, but a surface instead, the radiant exitance M is defined as the radiant flux per unit area of that surface. The flux of that surface is:

$$\Phi = \iint_{surface} M dA$$

If we have a surface that is irradiated by some external source, then the flux intercepted per unit area is called the irradiance E.

The radiant intensity I refers to a point source. It is the radiant flux emitted per unit steradian. Suppose that the radiant intensity of a point source is given by $I(\varphi,\psi)$, see figure 2.2. Then, $d\omega$ is a small solid angle through which a radiant flux $I(\varphi,\psi)d\omega$ flows. The full radiant flux of the point source is:

$$\Phi = \int_{total\ solid\ angle} I(\varphi,\psi)d\omega$$

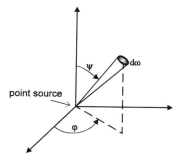

Figure 2.2 Geometry of point source

Spectral radiometric quantities give the densities of the quantities with respect to the wavelength. For instance, the radiant flux can be obtained by:

$$\Phi = \int_\lambda \Phi(\lambda) d\lambda$$

Similar expressions hold for other quantities.

As shown in table 2.3, almost all radiometric quantities have their counterparts in photometry. Spectral photometric quantities are not tabulated, but are defined similar to their radiometric counterparts. The spectral radiant flux $\Phi_e(\lambda)$ of an electromagnetic wave can be converted to the photometric quantity *spectral luminous flux* $\Phi_V(\lambda)$ [*lm*] with the so-called *spectral luminous efficacy*:

$$K(\lambda) = \Phi_V(\lambda)/\Phi_e(\lambda) \tag{2.1}$$

The unit of $K(\lambda)$ is [*lm/W*]. A graphical representation of $K(\lambda)$ is given in figure 2.3.

The basic photometric standard is the candela (*cd*). One candela is the luminous intensity of $1/60\ cm^2$ of a black body radiator operating at the temperature of the solidification of platinum. The unit of luminous flux is the lumen (*lm*). One lumen is the amount of luminous flux emanating from a point source with an intensity of one candela through one steradian. The unit of illumination is lux (*lx*). One lux is one lumen per square meter.

Example 2.1 Illumination of a table-top

A light bulb has a luminous flux of 3140 *lm*. The construction of the lamp is such that it emits light isotropically (i.e. independent of orientation). A plane (e.g. a table-top) is located at a distance *h* of 1 meter from the bulb. What is the illuminance at the surface of the plane ?

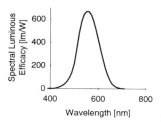

Figure 2.3 Spectral luminous efficacy against wavelength

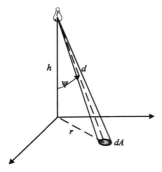

Figure 2.4 Geometry in example 2.1

The situation is as shown in figure 2.4. Since the lamp is isotropic, the luminous intensity is: $I = \Phi/4\pi = 3140/4\pi = 250$ cd. The normal of a small area dA on the plane at a distance d from the lamp makes an angle ψ with respect to the direction of the lamp. Hence, the solid angle corresponding to this area is: $d\omega = dA \cos\psi/d^2$. The luminous flux through this angle is: $Id\omega$. Therefore, the illuminance on the surface at a distance r from the centre of the plane is: $E = d\Phi/dA = Id\omega/dA = I\cos\psi/d^2$. Expressing $\cos\psi$ and d in h and r we finally obtain: $E = Ih/(h^2+r^2)^{3/2} = I\cos^3\psi/h^2$ lx. For instance, if $h=1$ m, at the centre of the table ($r = 0$) we have $E = 250$ lx.

Example 2.2 Extended light source

A small surface patch with area dA_2 is illuminated by a surface radiator dA_1 having a radiant intensity of $dI(\theta_1) = L\cos\theta_1 dA_1$ W/sr, see figure 2.5. θ_1 is the angle between the normal of the surface radiator and the direction of the patch. The distance between the two surfaces is d. What is the irradiance of the illuminated surface?

The solid angle at the radiator with respect to the illuminated patch is: $d\omega = dA_2 \cos\theta_2/d^2$. The radiant flux flowing through this angle is $d\Phi = dI(\theta_1)d\omega = L\cos\theta_1 dA_1 dA_2 \cos\theta_2/d^2$. Hence, the irradiance at the patch equals: $dE = d\Phi/dA_2 = L\cos\theta_1 \cos\theta_2 dA_1/d^2$.

An extended light source is a radiator with finite area. The irradiance due to such a source is found as the integral of the contributions from infinitesimal radiator patches making up the area of the source.

In practice, many diffusely radiating surfaces have a radiant intensity obeying $dI(\theta_1) = L\cos\theta_1 dA_1$. These surfaces are called *Lambertian*. The quantity *radiance L* (in photometry *luminance*) of an area dA is defined as the radiant intensity per foreshortened area:

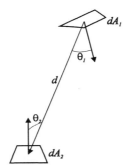

Figure 2.5 Geometry in example 2.2

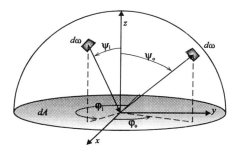

Figure 2.6 Geometry of a reflecting area

$$L = dI/dA \cos\theta. \tag{2.2}$$

The unit of radiance is $W/m^2\, sr$. The unit of luminance is the nit ($nt = cd/m^2$). From (2.2) it follows that the radiance of a Lambertian radiator is independent of the orientation θ. The total exitance of Lambertian radiator is: $M = \pi L$.

2.1.2 Reflection models

Three different processes characterise the interaction of an electromagnetic wave with an object:
- absorption (i.e. the conversion from electromagnetic energy into heat energy)
- reflection
- transmission

If there is no transmission, the object is non-transparent. In that case, reflection is the only physical process to deal with.

Suppose a reflecting area dA is irradiated by an electromagnetic wave with an incident angle of (φ_i, ψ_i) flowing through a solid angle $d\omega$, see figure 2.6. The irradiance from that direction is $dE(\varphi_i, \psi_i)$ W/m^2. In case of a perfectly mirroring surface the reflected wave will have a direction: $\psi_o = \psi_i$, $\varphi_o = \varphi_i$. Most materials are not perfectly mirroring. Instead, they reflect in all directions. This can be quantified by the introduction of the *reflectance distribution* $\rho(\varphi_i, \varphi_o, \psi_i, \psi_o)$ defined as the ratio between the radiance from an area to the irradiance incident on that area. Suppose that the radiance in a direction (φ_o, ψ_o) due to the incident wave is given by $dL(\varphi_i, \varphi_o, \psi_i, \psi_o)$. Then we have:

$$dL(\varphi_i, \varphi_o, \psi_i, \psi_o) = \rho(\varphi_i, \varphi_o, \psi_i, \psi_o) dE(\varphi_i, \psi_i) \tag{2.3}$$

The unit of $\rho(.)$ is $1/sr$. For many surfaces the dependence of $\rho(.)$ on ψ_i and ψ_o is only a dependence on $\psi_i - \psi_o$, thus we may write $\rho(\varphi_i, \varphi_o, \psi_i - \psi_o)$ instead of $\rho(\varphi_i, \varphi_o, \psi_i, \psi_o)$.

For a perfect Lambertian diffuser the reflectance distribution is: $\rho(.) = \rho_d(\varphi_i)$. In that case dL will be fully independent on the direction. In contrast, for a perfectly mirroring surface, $\rho(.)$ is zero everywhere, except for $\varphi_i = \varphi_o$ and $\psi_i = \psi_o$, hence: $\rho(.) = \text{constant} \times \delta(\varphi_i - \varphi_o)\delta(\psi_i - \psi_o)$. This situation is called *specular reflection*.

Many natural and man-made materials follow the Torrance-Sparrow model for reflectance distributions [Torrance and Sparrow 1967]. This model states that the interaction of light with material is according to two distinct processes. The first

16 IMAGE BASED MEASUREMENT SYSTEMS

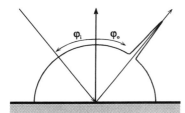

Figure 2.7 Reflectance distribution

process is scattering of photons inside the material. The part of these photons that are not absorbed leave the material surface at a random orientation giving rise to a diffuse exitance component. The second process is reflection at the material surface. If the surface is absolutely flat, this process give rise to a perfectly specular reflection. However, the roughness of the surface (measured at a micrometer scale) may lead to small deviations with respect to the specular direction. The process is called *off-specular reflection*. Glossy spots, often encountered at surface locations with high curvature, are caused by off-specular reflection.

In conclusion, the reflectance distribution often consists of two components:

$$\rho(\varphi_i, \varphi_o, \psi_i, \psi_o) = \rho_d(\varphi_i) + \rho_s(\varphi_i, \varphi_o, \psi_i - \psi_o) \qquad (2.4a)$$

Figure 2.7 gives an example of the reflectance distribution in dependence on φ_o. The differential radiant intensity of a surface equals:

$$dI(\varphi_o, \psi_o) = \{\rho_d(\varphi_i) + \rho_s(\varphi_i, \varphi_o, \psi_i - \psi_o)\} \cos\varphi_o \; dE(\varphi_i, \psi_i) \qquad (2.4b)$$

Integrating this expression with respect to φ_i and ψ_i gives the full radiant intensity $I(\varphi_o, \psi_o)$.

2.2 PROJECTION

With a few exceptions, all imaging techniques involve a mapping from the 3-dimensional space (of the scene) into a 2-dimensional space. Perspective projection is the mapping technique that is applied most widely, for instance, in consumer camera-equipment, but also in X-ray imagery.

2.2.1 World and camera co-ordinates

In measuring the position and orientation of objects with computer vision methods, we have to couple the co-ordinates of the camera system (e.g. the optical axis and the co-ordinates of the image plane) to some reference co-ordinates (generally called *world co-ordinates*) in the scene, see figure 2.8.

The co-ordinates of the camera system are denoted $\vec{x} = [x \; y \; z]^t$. Usually, the z-axis is aligned with the optical axis orthogonal to the image plane. The world co-ordinates are denoted: $\vec{X} = [X \; Y \; Z]^t$.

The two co-ordinate systems are coupled by two linear transformations: a translation and a rotation. The translation is a shift of origin (having three degrees of freedom) and can be described with a vector $\vec{t}$. If the two origins coincide, the

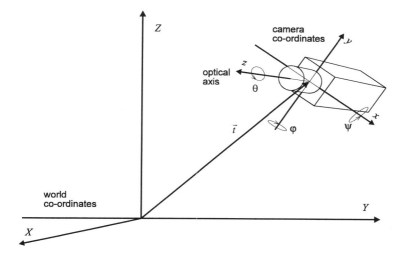

Figure 2.8 World and camera co-ordinates

remaining differences can be neutralised by rotations. Here, we have also three degrees of freedom: φ, ψ and θ. Mathematically, rotation corresponds to multiplication of the co-ordinate vector with a 3×3 orthonormal matrix **R**. Clearly, this matrix depends non-linearly on the three rotation parameters.

As a whole, the coupling between world co-ordinates and camera co-ordinates is given by the following expression:

$$\vec{x} = \mathbf{R}(\vec{X} - \vec{t}) \tag{2.5}$$

2.2.2 Pinhole camera model

Perspective projection is a simple model describing the image formation with a lens system. It is equivalent to a pinhole camera model. Such a camera consists of a non-transparent plane with a small hole (figure 2.9). Parallel to this plane, at a distance d, the image plane is located. Light emitted from the surfaces of objects in the scene passes through the hole and illuminates the image plane. If the pinhole is small

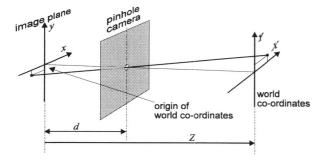

Figure 2.9 Perspective projection with a pinhole camera

enough, an infinitesimal small surface patch of an object is mapped onto a small spot of light at the image plane. The collection of all surface patches will give rise to an irradiance called "image".

In figure 2.9 the surface patch has world co-ordinates $\vec{X} = [X\ Y\ Z]^t$. In this figure the world co-ordinates are coupled to the camera without rotation or translation. The projection of $\vec{X}$ into the image plane is at $\vec{x} = [x\ y]^t$. The z-co-ordinate of the image plane is constant, and can be kept from the discussion. The relationship between world co-ordinates and image co-ordinates is:

$$x = -\frac{Xd}{Z-d} \quad y = -\frac{Yd}{Z-d} \tag{2.6}$$

From figure 2.9, one can conclude that each point in the image plane corresponds exactly to one (light emitting) surface patch in the scene (unless objects are transparent). However, from a particular image point one cannot reconstruct the original world co-ordinates of that patch. One can only deduce the ratio between position (X and Y) and the depth Z of the emitting patch.

2.2.3 Homogeneous co-ordinates

Homogeneous co-ordinates are a powerful concept to describe the transformations discussed above. Translation, rotation, scaling and perspective projection can be handled mathematically with a single matrix multiplication.

For that purpose the 3-dimensional co-ordinates are expanded with a fourth (virtual) co-ordinate having a value 1, together with a scaling factor α: $\tilde{X} = [\alpha X\ \alpha Y\ \alpha Z\ \alpha]^t$. Thus, homogeneous co-ordinates constitute a 4-dimensional vector, from which the ordinary co-ordinates can be derived by dividing the first three components by the fourth. The transformations discussed so far can be expressed as:

$$\tilde{x} = \mathbf{M}\tilde{X} \tag{2.7}$$

The 4×4 matrix $\mathbf{M}$ is consists of a concatenation of simpler matrices that perform elementary transformations like translation, rotation about a co-ordinate axis, perspective projection and scaling. The following matrices can be derived:

Translation by $\vec{t} = \begin{bmatrix} -t_x & -t_y & -t_z \end{bmatrix}^t$:

$$\mathbf{T} = \begin{bmatrix} 1 & 0 & 0 & -t_x \\ 0 & 1 & 0 & -t_y \\ 0 & 0 & 1 & -t_z \\ 0 & 0 & 0 & 1 \end{bmatrix}$$

Rotation about X

$$\mathbf{R}_x = \begin{bmatrix} 1 & 0 & 0 & 0 \\ 0 & \cos\psi & -\sin\psi & 0 \\ 0 & \sin\psi & \cos\psi & 0 \\ 0 & 0 & 0 & 1 \end{bmatrix}$$

Rotation about Y

$$\mathbf{R}_y = \begin{bmatrix} \cos\varphi & 0 & -\sin\varphi & 0 \\ 0 & 1 & 0 & 0 \\ \sin\varphi & 0 & \cos\varphi & 0 \\ 0 & 0 & 0 & 1 \end{bmatrix}$$

Rotation about Z

$$\mathbf{R}_z = \begin{bmatrix} \cos\theta & -\sin\theta & 0 & 0 \\ \sin\theta & \cos\theta & 0 & 0 \\ 0 & 0 & 1 & 0 \\ 0 & 0 & 0 & 1 \end{bmatrix}$$

Scaling by $\vec{s} = \begin{bmatrix} s_x & s_y & s_z \end{bmatrix}^t$

$$\mathbf{S} = \begin{bmatrix} s_x & 0 & 0 & 0 \\ 0 & s_y & 0 & 0 \\ 0 & 0 & s_z & 0 \\ 0 & 0 & 0 & 1 \end{bmatrix}$$

Perspective projection

$$\mathbf{P} = \begin{bmatrix} 1 & 0 & 0 & 0 \\ 0 & 1 & 0 & 0 \\ 0 & 0 & 0 & 0 \\ 0 & 0 & -1/d & 1 \end{bmatrix}$$

For instance the perspective projection of a point $\vec{x}$ yields a vector:

$$\vec{x} = \mathbf{P}\vec{X} = \begin{bmatrix} 1 & 0 & 0 & 0 \\ 0 & 1 & 0 & 0 \\ 0 & 0 & 0 & 0 \\ 0 & 0 & -1/d & 1 \end{bmatrix} \begin{bmatrix} \alpha X \\ \alpha Y \\ \alpha Z \\ \alpha \end{bmatrix} = \begin{bmatrix} \alpha X \\ \alpha Y \\ 0 \\ \alpha(1 - Z/d) \end{bmatrix}$$

Transferring this back to ordinary co-ordinates we have the co-ordinates given in equation (2.6). If we have a cascade of transformations, the corresponding operation is a cascade of matrix multiplications. For example, a translation, rotation about the z-axis, perspective projection and finally scaling of a point $\vec{x}$ is given by: $\vec{x} = \mathbf{SPR}_z\mathbf{T}\vec{x}$.

2.2.4 Point spread function

In practice, the hole in a pinhole camera is always finite. If not, there would be no radiant flux through the hole. In addition, a pinhole with a diameter on the order of the wavelength of light will scatter the light in various directions (Huygens' principle: diffraction). Therefore, in order to have a practical camera we must assume that the diameter is much larger than the wavelength.

This situation is shown in figure 2.10. A ray of light emitted from a surface at position (X_1, Y_1, Z_1) in the scene hits the image plane at position (x_1, y_1) according to the perspective projection stated in equation (2.6). Clearly, to each position (x_1, y_1) there is one (and only one) corresponding surface patch in the scene. This patch can be found by backtracking the ray from (x_1, y_1) through the pinhole until the surface of the object is hit. Consequently, from the co-ordinates (X_1, Y_1, Z_1) of that particular

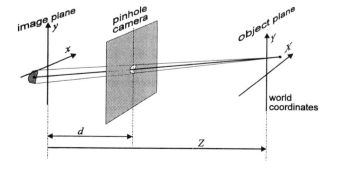

Figure 2.10 Pinhole camera with finite aperture

patch only two of them, for instance (X_1, Y_1), can be chosen freely. The other co-ordinate (Z_1) is uniquely defined by the choice of (X_1, Y_1). We can make this explicit by writing $z_1(X_1, Y_1)$ instead of Z_1. In that case $z_1(X_1, Y_1)$ is called the *depth map*.

Suppose that such a surface patch has a radiant intensity of $I(X_1, Y_1)dA$ in the direction of the pinhole. dA is the area of the patch. The area of the pinhole is called the *aperture*, and will be denoted by S. If $X_1 \ll Z_1$ and $Y_1 \ll Z_1$, the solid angle ω of the pinhole equals: $\omega \approx S/(Z_1 - d)^2$. The radiant flux passing through the pinhole is: $\Phi(X_1, Y_1) = \omega I(X_1, Y_1)dA$.

The radiant flux results in an irradiation of the image plane at a position (x_1, y_1) as given by the perspective projection. However, the spot will have a spatial extent due to the finite size of the pinhole. In fact, the shape of the pinhole is fully reflected in the shape of the spot with a scaling factor $Z_1/(Z_1 - d)$. We may take this into account by the introduction of a function $h(x - x_1, y - y_1, Z_1)$ defined in the image plane:

$$h(x - x_1, y - y_1, Z_1) = \begin{cases} 1 & \text{if } (x - x_1, y - y_1) \text{ is inside the spot} \\ 0 & \text{elsewhere} \end{cases} \quad (2.8)$$

The function describes the shape of the pinhole, projected to the image plane and shifted to the position (x_1, y_1). The crucial point to observe is that the function does not depend on the absolute values of x, y, x_1 or y_1. It depends solely on the differences $x - x_1$ and $y - y_1$ (and on Z_1). Hence, the shape of the image of the pinhole is *shift invariant*. The function $h(.)$ is called the *point spread function*. It reflects the response of the imaging device to a single point source.

The irradiance $dE(x, y)$ at the image plane is proportional to $h(x - x_1, y - y_1, Z_1)$ and $\omega I(X_1, Y_1)dA$. It is inversely proportional to the area $SZ_1/(Z_1 - d)$ of the spot. Transforming (x_1, y_1) to world co-ordinates (see equation 2.6), and absorbing all irrelevant factors into single constants, we have:

$$dE(x, y) \approx C \cdot h(x - sX_1, y - sY_1, Z_1) \cdot I(X_1, Y_1)dX_1dY_1 \quad \text{with:} \quad s = d/(d - Z_1) \quad (2.9)$$

Since, for incoherent light, radiance energy is additive, the irradiance at the image plane can be found by integrating this expression on both sides:

$$E(x, y) \approx C \iint_{X_1 Y_1} h(x - sX_1, y - sY_1, Z_1) \cdot I(X_1, Y_1)dX_1dY_1 \quad (2.10)$$

The most important fact to observe is that this image formation system is *shift invariant* (with respect to X_1 and Y_1) and *additive*. However, the point spread function $h(.)$ depends on the objects in the scene. Therefore, the image formation system is also *object dependent*. If the distance between pinhole and objects is large compared with the variations of the depth, the scaling factors do not change significantly, and the point spread function becomes almost independent on the objects. In that case equation (2.10) evolves into a *2-dimensional convolution*. The system is *linear* and shift (or space) invariant.

The effect of a point spread function (PSF) is that objects loose their sharpness in the image. Equation (2.9) shows that the image of a tiny object resembles the PSF. For instance, an image of a starry night made by a pinhole camera will show a

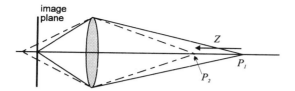

Figure 2.11 The point spread function of a thin cylindrical lens

number of spots, the shapes of which equal the PSF instead of the real shapes and sizes of the stars and planets.

The image formation process of a pinhole camera is tentative for other imaging processes, including optical (lens) systems and X-ray imaging. For instance in a lens system, at a certain depth, objects are imaged sharply (within the diffraction limits), while outside this depth of focus objects become more and more blurred. Here too, the PSF of the optical system depends on the depth of the objects. The imaging process is much like equation (2.10) except that the dependency of the PSF on Z differs (see figure 2.11). The point P_1 is in focus. The point P_2 is out-of-focus. Its point spread function will be a cylindrical function (a disk), the radius of which depends on z.

2.2.5 2-dimensional convolution

In this section, the 2-dimensional convolution integral will be discussed further. This integral not only serves as a model of many image formation systems, it also appears to be the basis of many digital image operations. For simplicity equation (2.10) will be written[1] as:

$$g(x,y) = \int\int_{\xi\,\eta} h(x-\xi, y-\eta) f(\xi,\eta) d\xi d\eta = h(x,y) * f(x,y) \qquad (2.11)$$

The symbol "∗" denotes 2-dimensional convolution. $f(x,y)$ may represent the radiant intensity of an object surface (as in equation 2.10), or it may represent an image that undergoes a convolution operation. $h(x,y)$ is the point spread function (or impulse response), sometimes called the *convolution kernel*. $g(x,y)$ is the resulting image. It is easy to show that the following properties hold:

linearity: $\qquad h(x,y)*\bigl(A_1 f_1(x,y) + A_2 f_2(x,y)\bigr) = A_1\bigl(h(x,y)* f_1(x,y)\bigr) + A_2\bigl(h(x,y)* f_2(x,y)\bigr)$

commutativity: $\qquad h(x,y)* f(x,y) = f(x,y)* h(x,y)$

separability[2]: $\qquad \bigl(h_1(x)h_2(y)\bigr)* f(x,y) = h_1(x) \bullet h_2(y) \bullet f(x,y)$

The last property states that if a convolution kernel is separable in x and y, the convolution can be performed as a cascade of two 1-dimensional convolutions. A few particular convolution integrals and related functions deserve special attention:

---

[1] It is understood that world co-ordinates are scaled so as to assert (2.11).
[2] The symbol "•" is used here to denote 1-dimensional convolution. That is $h(x) \bullet f(x) = \int h(x-\xi) f(\xi) d\xi$.

Correlation

$$\iint_{\xi\,\eta} h(x+\xi, y+\eta) f(\xi, \eta) d\xi d\eta = h(-x,-y) * f(x,y) \qquad (2.12)$$

This property states that the correlation between two functions can be written as a convolution.

The Sifting integral

$$h(x,y) * \delta(x-a, y-b) = h(a,b) \qquad (2.13)$$

The function $\delta(x,y)$ is a generalised function (called the *2-dimensional Dirac function*) defined by:

$$\delta(x,y) = 0 \quad \text{if } x \neq 0 \text{ or } y \neq 0$$
$$\iint \delta(x,y) dx dy = 1 \qquad (2.14)$$

From (2.13) it can be seen that $\delta(x,y)$ is the mathematical representation of a point source.

The line spread function (LSF)

The *line spread function* of an optical system is its response to a narrow light strip on a dark background. The LSF can be derived from the PSF by integrating the latter in the direction of the strip. A narrow strip parallel to (for instance) the x-axis can be represented by a 1-dimensional Dirac function: $\delta(y)$. Then:

$$h_{lsf}(y) = h(x,y) * \delta(y) = \int_x h(x,y) dx \qquad (2.15)$$

In general, the LSF depends on the orientation of the strip. Only if the PSF is rotationally symmetric (i.e. solely dependent on $x^2 + y^2$), the LSF does not depend on the orientation. In optical devices, the LSF is sometimes used to measure the PSF of the system experimentally.

The edge spread function (ESF)

The *edge spread function* of an optical system is its response to two halfplanes, one dark and one light. If $u(x)$ represents a 1-dimensional unit step function, then the two halfplanes aligned to (for instance) the x-axis can be represented by $u(y)$. Then, the ESF can be derived from the PSF in the following way:

$$h_{esf}(y) = h(x,y) * u(y) = \int_{\xi=-\infty}^{y} \int_x h(x,\xi) dx d\xi \qquad (2.16)$$

Here too, in general the ESF depends on the orientation of the halfplane except if the PSF is rotationally symmetric. Comparing (2.15) with (2.16) it follows that $h_{lsf}(y) = dh_{esf}(y)/dy$.

2.3 FOURIER ANALYSIS

The 2-dimensional Fourier transform is widely used, for instance to specify the resolving power of an imaging device, to design and to analyse (2-dimensional) image filters, to model the noise characteristics, and to model the image of textured surfaces.

2.3.1 Harmonic functions

We start the discussion with the introduction of 2-dimensional, complex, harmonic functions. Figure 2.12 gives an illustration. The harmonic function $g(x,y)$ is defined:

$$g(x,y) = G\exp(2\pi j(xu+yv)) \qquad G = A\exp(j\phi) \quad \text{and} \quad j=\sqrt{-1} \qquad (2.17)$$

The constants A, u, v and ϕ are real scalars. A is called the *amplitude*, ϕ is the *phase*, u and v are the *spatial frequencies* (measured in $1/mm$) in the x- and y-direction, respectively. These quantities can also be transferred to polar co-ordinates, i.e.:

$$g(x,y) = G\exp(2\pi j\rho(x\cos\theta + y\sin\theta)) \qquad \text{with:} \quad \begin{cases} \rho^2 = x^2+y^2 \\ u = \rho\cos\theta \\ v = \rho\sin\theta \end{cases} \qquad (2.18)$$

ρ is the spatial frequency of the harmonic function. The wavelength of the function equals $1/\rho$; see figure 2.12. The orientation of the function is given by θ. The real and imaginary part of an harmonic function is:

$$\begin{aligned}\Re(g(x,y)) &= A\cos(2\pi(xu+yv)+\phi) \\ \Im(g(x,y)) &= A\sin(2\pi(xu+yv)+\phi)\end{aligned} \qquad (2.19)$$

A real function can also be constructed from two complex conjugated harmonic functions: $2A\cos 2\pi(xu+yv) = G\exp(2\pi j(xu+yv)) + G^*\exp(-2\pi j(xu+yv))$.

There are at least two properties that make the harmonic functions interesting:

Property 1:
For arbitrary PSF: the response of convolution to an harmonic function is an harmonic function. The spatial frequencies remain unaltered

$$h(x,y) * G_1\exp(2\pi j(xu+yv)) = G_2\exp(2\pi j(xu+yv)) \qquad (2.20)$$

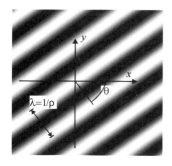

Figure 2.12 The real part of an harmonic function

24 IMAGE BASED MEASUREMENT SYSTEMS

Property 2:
Harmonic functions are orthogonal (see appendix A.3):

$$\int_{x=-\infty}^{\infty}\int_{y=-\infty}^{\infty} g_1(x,y)g_2(x,y)dxdy = \delta(u_1-u_2,v_1-v_2) \qquad (2.21)$$

Property 2 is illustrated in figure 2.13. Figures 2.13a and b show the real parts of two harmonic functions $A\cos 2\pi(xu_1+yv_1)$ and $A\cos 2\pi(xu_2+yv_2)$ differing only in the frequencies: $(u_1,v_1) \neq (u_2,v_2)$. The product $A^2 \cos 2\pi(xu_1+yv_1)\cos 2\pi(xu_2+yv_2)$ is shown in figure 2.13c. This product is rewritten as:

$$\tfrac{1}{2}A^2\{\cos 2\pi(x(u_1+u_2)+y(v_1+v_2))+\cos 2\pi(x(u_1-u_2)+y(v_1-v_2))\}$$

The product contains two terms having frequencies consisting of the sums and differences of the original frequencies. If the image in figure 2.13c is integrated, both terms vanish. Figure 2.13d shows the product of the first harmonic function with itself, i.e. $A^2\cos^2 2\pi(xu_1+yv_1)$. This product may be rewritten as:

$$\tfrac{1}{2}A^2\{\cos 4\pi(xu_1+yv_1)+1\}$$

Upon integration, the first term vanishes, but the last term retains. This is fully in accordance with equation (2.21).

Figure 2.13 The orthogonality of two harmonic functions
a-b) Two harmonic functions with different frequencies
c) The product of figure a) and b)
d) The square of the hamonic function in a)

2.3.2 The Fourier transform

Consider an image that is composed of a number of cosine functions with various amplitudes, phases and frequencies.

$$f(x,y) = \sum_n A_n \cos 2\pi(xu_n + yv_n) + \varphi_n$$

Suppose that the image $f(x,y)$ is available, but that the amplitudes, phases and frequencies are, as yet, still unknown. The question is, how to obtain these quantities.

The orthogonality of harmonic functions can be used to answer this question. Suppose we multiply the image with a harmonic function and integrate the product. The result will depend on the frequency (u,v) of this harmonic:

$$F(u,v) = \int_x \int_y \exp(-2\pi j(xu+yv)) f(x,y) dx dy$$

$$= \int_x \int_y \exp(-2\pi j(xu+yv)) \sum_n A_n \cos(2\pi(xu_n+yv_n) + \varphi_n) dx dy$$

$$= \sum_n A_n \int_x \int_y \exp(-2\pi j(xu+yv)) \cos(2\pi(xu_n+yv_n) + \varphi_n) dx dy$$

$$= \sum_n \frac{A_n}{2} \int_x \int_y \exp(-2\pi j(xu+yv)) \{\exp(2\pi j(xu_n+yv_n) + \varphi_n) + \exp(-2\pi j(xu_n+yv_n) - \varphi_n)\} dx dy$$

$$= \sum_n \frac{A_n}{2} \int_x \int_y \exp(2\pi j(x[u_n-u]+y[v_n-v]) + \varphi_n) + \exp(2\pi j(x[u_n+u]+y[v_n+v]) - \varphi_n) dx dy$$

$$= \sum_n \frac{A_n}{2} \{\exp(2\pi j \varphi_n) \delta(u_n-u, v_n-v) + \exp(-2\pi j \varphi_n) \delta(u_n+u, v_n+v)\}$$

$$= \sum_n G_n \delta(u_n-u, v_n-v) + G_n^* \delta(u_n+u, v_n+v)$$

Hence, the frequencies in $f(x,y)$ are visible in $F(u,v)$ as Dirac functions, the weights of which are proportional to the complex amplitudes (i.e. amplitude and phase) of the corresponding cosine functions. $F(u,v)$ can be interpreted as a *distribution function* of the complex amplitudes in the (u,v)-plane.

The path can also be walked on in the reverse direction. Knowing $F(u,v)$, the original image $f(x,y)$ can be reconstructed. For that purpose, multiply $F(u,v)$ with an harmonic function and repeat this for every possible frequency. Then, since $F(u,v)$ will pick out the right frequencies with the right amplitude and phase, summing all these harmonic functions together will yield $f(x,y)$:

$$\int_u \int_v \exp(2\pi j(xu+yv)) F(u,v) du dv =$$

$$= \int_u \int_v \exp(2\pi j(xu+yv)) \sum_n G_n \delta(u_n-u, v_n-v) + G_n^* \delta(u_n+u, v_n+v) du dv$$

$$= \sum_n \int_u \int_v \exp(2\pi j(xu+yv)) (G_n \delta(u_n-u, v_n-v) + G_n^* \delta(u_n+u, v_n+v)) du dv$$

$$= \sum_n G_n \exp(2\pi j(xu_n + yv_n)) + G_n^* \exp(-2\pi j(xu_n + yv_n))$$

$$= \sum_n A_n \cos(2\pi j(xu_n + yv_n) + \phi_n)$$

$$= f(x,y)$$

Apparently, the relations between $f(x,y)$ and $F(u,v)$ hold if $f(x,y)$ is composed of a countable number of harmonic functions. It is proven that they also hold for a much wider class of 2-dimensional function. In fact, for every continuous function, the following transformation is valid:

$$f(x,y) = \iint_{u\,v} F(u,v)\exp(2\pi j(xu+yv))dudv$$

$$F(u,v) = \iint_{x\,y} f(x,y)\exp(-2\pi j(xu+yv))dxdy \qquad (2.22)$$

The transform from $f(x,y)$ to $F(u,v)$ is called the *2-dimensional Fourier transform*. It is symbolically written: $F(u,v) = F\{f(x,y)\}$. The inverse transform, from $F(u,v)$ to $f(x,y)$, is written: $f(x,y) = F^{-1}\{F(u,v)\}$.

The Fourier transform of a real function is complex:

$$F(u,v) = \Re(F(u,v)) + j\Im(F(u,v)) = |F(u,v)|\exp(j\phi(u,v))$$

$$|F(u,v)| = \sqrt{\Re^2(F(u,v)) + \Im^2(F(u,v))} \qquad (2.23)$$

$$\phi(u,v) = \tan^{-1}\left(\frac{\Im(F(u,v))}{\Re(F(u,v))}\right)$$

The magnitude $|F(u,v)|$ is called the *amplitude spectrum*, the angle $\phi(u,v)$ is the *phase spectrum*. The square of the amplitude spectrum $|F(u,v)|^2$ is sometimes called the power spectrum, although in fact this term is reserved for stochastic processes only, see appendix B.4.1.

Figure 2.14 shows examples of some 2-dimensional functions and their corresponding amplitude spectrum. The dynamic range of the amplitude spectrum is large. Therefore, in displaying these spectra, usually a logarithmic scale is used, e.g. $c\log(1+|F(u,v)|)$.

In figure 2.14a the 2-dimensional *rectangle-function* is shown:

$$\text{rect}(x,y) = \text{rect}_1(x)\text{rect}_1(y) \quad \text{with:} \quad \text{rect}_1(x) = \begin{cases} 1 & \text{if } -0.5 < x < 0.5 \\ 0 & \text{elsewhere} \end{cases} \qquad (2.24)$$

Hence, the rectangle-function is separable in x and y. Its Fourier transform is a separable function made up by two *sinc-functions* (see figure 2.14b):

$$F\{\text{rect}(x,y)\} = \text{sinc}(u)\text{sinc}(v) \quad \text{with:} \quad \text{sinc}(u) = \frac{\sin(\pi u)}{\pi u} \qquad (2.25)$$

The separability is also seen in the frequency domain.

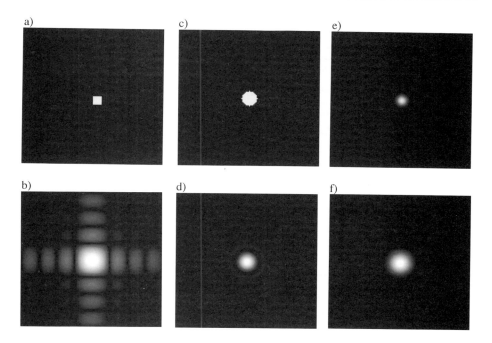

Figure 2.14 Fourier pairs
a) rectangle-function   c) cylinder-function   e) Gauss function
b) sinc-function   d) Fourier of cylinder function   f) Gauss function

The function shown in figure 2.14c is a *cylinder-function* defined by:

$$\mathrm{cyl}(x,y) = \begin{cases} 1 & \text{if } x^2 + y^2 < 1 \\ 0 & \text{elsewhere} \end{cases} \quad \text{with:} \quad F\{\mathrm{cyl}(x,y)\} = \frac{J_1\left(2\pi\sqrt{u^2+v^2}\right)}{\sqrt{u^2+v^2}} \quad (2.26)$$

It is a rotationally symmetric function. Its Fourier transform, the *first order Bessel function of the first kind* $J_1(.)$, is also rotationally symmetric. The amplitude spectrum is shown in figure 2.14d.

The *Gauss function*, or Gaussian, is also well-known in image processing techniques:

$$\mathrm{gauss}(x,y) = \frac{1}{2\pi\sigma^2}\exp\left(-\frac{x^2+y^2}{2\sigma^2}\right) \quad \text{with:} \quad F\{\mathrm{gauss}(x,y)\} = \exp\left(-\frac{\sigma^2(u^2+v^2)}{2}\right) \quad (2.27)$$

The parameter σ, called the *standard deviation*, controls the spatial extent of the Gaussian. The function is separable in x and y, and rotationally invariant. Equation (2.27) shows that the Fourier transform of a Gaussian is a Gaussian too. The Dirac function, defined in equation (2.14), is a Gaussian in the limiting case. If $\sigma \to 0$, then $\mathrm{gauss}(x,y) = \delta(x,y)$. This implies that $F\{\delta(x,y)\} = 1$.

The 2-dimensional Fourier transform has a lot of properties, the most important of which are presented in table 2.4. Properties related to stochastic processes like noise etc. will be discussed in chapter 3.

Table 2.4 Properties of the 2-dimensional Fourier transform

Property	$f(x,y)$	$F(u,v)$
linearity	$af(x,y)+bg(x,y)$	$aF(u,v)+bG(u,v)$
translation	$f(x-a,y-b)$	$F(u,v)\exp(-2\pi j(ua+vb))$
rotation	$f(x\cos\theta+y\sin\theta, x\cos\theta-y\sin\theta)$	$F(u\cos\theta+v\sin\theta, u\cos\theta-v\sin\theta)$
conjugate symmetry	$f(x,y)$ is real	$F(u,v)=F^*(-u,-v)$
rotational symmetry	$f(r)$ with $r^2=x^2+y^2$	$F(\rho)$ with $\rho^2=u^2+v^2$
scaling	$f(ax,by)$	$\dfrac{1}{|ab|}F(u/a,v/b)$
separability	$f(x,y)$	$\int_x\left(\int_y f(x,y)\exp(-2\pi jyv)dy\right)\exp(-2\pi jxu)dx$
separability	$g(x)h(y)$	$G(u)H(u)=$ $\left(\int_x g(x)\exp(-2\pi jxu)dx\right)\left(\int_y h(y)\exp(-2\pi jyv)dy\right)$
periodicity	$f(x,y)=f(x+a,y+b)\quad\forall x,y$	$\sum_n\sum_m F_{nm}\delta\left(u-\dfrac{n}{a},v-\dfrac{m}{b}\right)$
convolution	$f(x,y)*h(x,y)$	$F(u,v)H(u,v)$
multiplication	$f(x,y)h(x,y)$	$F(u,v)*H(u,v)$
correlation	$f(x,y)*h(-x,-y)$	$F(u,v)H^*(u,v)$
first derivative	$\dfrac{df(x,y)}{dx}$	$2\pi ju F(u,v)$
second derivative	$\dfrac{d^2f(x,y)}{dx^2}$	$-4\pi^2 u^2 F(u,v)$
Laplacian	$\nabla^2 f(x,y)=\dfrac{d^2f(x,y)}{dx^2}+\dfrac{d^2f(x,y)}{dy^2}$	$-4\pi^2(u^2+v^2)F(u,v)$
Parseval	$\iint f^2(x,y)dxdy$	$\iint |F(u,v)|^2 dudv$

Most of the properties also hold in the reverse direction. For instance, if $F(u,v)$ is real, then $f(x,y)$ is conjugate symmetric in the x-y domain, i.e. $f(x,y)=f^*(-x,-y)$. This also implies that if $f(x,y)$ is real and point symmetric, $f(x,y)=f(-x,-y)$, then so is $F(u,v)$.

The first property on separability states that the 2-dimensional Fourier transform can be performed as a cascade of two 1-dimensional Fourier transforms. The second property states that if $f(x,y)$ is separable in x and y, the Fourier transform is separable in u and v too.

An important property is the convolution theorem. It states that convolution in the space domain is equivalent to multiplication in the frequency domain. If $h(x,y)$ is the PSF of an optical device, then its Fourier transform $H(u,v)$ is called the *optical transfer function* (OTF), or the *modulation transfer function* (MTF). It is a measure to quantify the resolving power of an optical device. Usually the unit of frequency

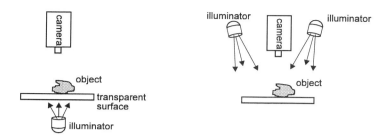

Figure 2.15 Rear and front illumination

is *lp/mm*, i.e. line pairs per millimetre (a line pair corresponds to one period of a cosine function). Most often, the OTF is normalised: $|H(0,0)|=1$. Then, the resolving power is defined as the spatial frequency at which the OTF becomes smaller than some threshold.

A particular advantage of the OTF to specify the resolving power of an optical device is that it can be cascaded. For instance, if an optical device consists of an optical lens system followed by an image intensifier with OTF $H_1(u,v)$ and $H_2(u,v)$, respectively, then the overall OTF is simply: $H_1(u,v)H_2(u,v)$.

2.4 IMAGE CONDITIONING

The sections above discuss a particular kind of image formation, namely the formation based on radiant energy, reflection at the surfaces of the objects, and perspective projection. The information about the scene is found in the contrasts (local differences in irradiance). It is well known that a lot of difficulties in the design of a computer vision system can be circumvented by carefully choosing the illumination of the objects [Novini 1986]. This section deals with a few practical considerations with respect to illumination and related topics. In the techniques discussed, the attention is focused on machine vision (i.e. the visual inspection of industrial product parts).

2.4.1 Front or rear illumination

The first choice to make is whether to illuminate the object from the rear or from the front; see figure 2.15. For non-transparent objects, illumination from the rear is advantageous if the silhouette of the object has to be extracted. (Quality control of printed circuits boards for electronics is a possible application.) Disadvantage is that the reflectance distribution of the surfaces cannot be measured. This implies a loss of information.

The purpose of front illumination is to illuminate the objects such that the reflectance distribution of the surface becomes the defining features in the image. There are many applications which require this kind of illumination: detection of flaws, scratches and other damages at the surface of material, etc. One particular branch in 3-dimensional vision, called *shape from shading* [Horn and Brooks 1989], is based on the reflection models of the surfaces. Clearly, this technique requires front illumination.

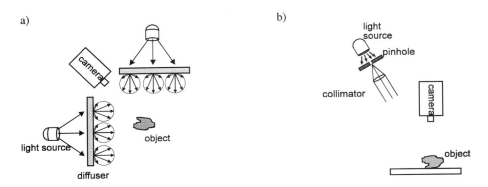

Figure 2.16 Diffuse and specular illumination

2.4.2 Specular or diffuse illumination

Another choice to make is whether to use either specular or diffuse illumination (figure 2.16). Without precaution, a combination of the two is most likely to occur. Therefore, this kind of illumination is low cost. In many applications, however, the illumination must be conditioned in order to arrive at feasible solutions. Figure 2.16a illustrates diffuse illumination. A diffuser is a transparent surface that scatters the light from the rear to the front in all directions with constant radiance (Lambertian surface). The situation is comparable with outdoor illumination during a cloudy day. In some applications, diffuse illumination is advantageous since there will be no cast shadows. In other applications, specular illumination is preferable. Specular illumination can be obtained with a collimator (figure 2.16b). The purpose of it is to create a parallel bundle of light. The result is a scene with clearly defined cast shadows.

Dark field illumination is a particular application of specular illumination (figure 2.17a). The light striking a specular or mirror type surface reflects off at the angle of the incident beam. The only light that is returned to the camera is diffusely reflected light. This technique works well for detection of defects (cracks etc.) at surfaces of materials.

Light field illumination uses the same principle as in dark field illumination (figure 2.17b). However, the camera is positioned in line with the reflected beam. This method is also used in surface defect detection. Normally, a high specular reflection component is anticipated. However, at the position of a deformation of the surface, this component is disturbed.

2.4.3 Structured light

This technique is used for applications in which depth information is required. The illumination system consists of an imager (e.g. a slide projector) that superimposes a well defined pattern of light upon the object surface. An example is shown in figure 2.18. The perspective projection of this pattern on the image plane of the camera contains information from which some 3-dimensional geometric properties

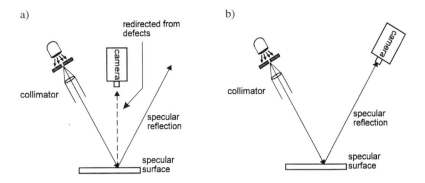

Figure 2.17 Dark field and light field illumination

of the object can be extracted. Some options for the choice of the patterns are:

- line projection
- point projection
- multiple line or point projection
- grid projection

The projection of a line from the imager defines a plane whose geometric parameters are known. Projection of this line on the surface of an object causes a stripe of light on that surface. This is equivalent to slicing the object with the plane. If the stripe at the surface of the object is projected at the image plane of the camera, then the 3-dimensional world co-ordinates of one point on the stripe can be constructed from the corresponding (2-dimensional) camera co-ordinates. The world co-ordinates of that point can be found by calculating the geometric parameters of the ray from the point at the image plane through the pinhole. The intersection of that line with the plane defined by the imager gives the 3-dimensional co-ordinates. Repeating this process with stripes of light at various positions (or equivalently, using multiple stripes) gives a depth map seen from the camera. A stripe of light may be occluded by parts of the object. This implies that parts of the depth map may be missing.

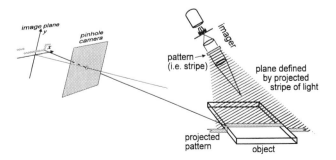

Figure 2.18 Structured light in a ranging device

The measurement of a depth map is called *range imaging*. The projection of lines is not the only way to obtain a depth map. Another possibility is point projection (flying spot). Radar devices are an example. Here, the depth information is obtained by sending a short burst of radiant energy in a particular direction, and by measuring the difference in time between transmission and return (an alternative is to use harmonic waves and to measure the shift of phase). In a radar device, the direction of the beam is modulated (mechanically or electrically) so as to scan the whole space to obtain a full depth map. This principle can also be applied in sonic devices (i.e. ultrasound radar), and in laser devices.

The use of laser devices in range imaging is not restricted to the example given above. For instance, the direction of a laser beam can be modulated with a system of rotating mirrors so as to scan the 3-dimensional space. Then, the perspective projection of the spot at the surfaces of the objects, as seen by a camera, is a clue for the 3-dimensional position of that spot.

REFERENCES

Haralick, R.M. and Shapiro, L.G., *Computer and Robot Vision, Vol. II*, Addison-Wesley, Massachusetts, 1993.

Horn, B.K.P. and Brooks, M.J., *Shape from Shading*, MIT Press, Massachusetts, 1989.

Longhurst, R.S., *Geometrical and Physical Optics*, Longman, London, 1970.

Novini, A., *Fundamentals of Machine Vision Lighting*, SPIE, Vol 728: Optics, Illumination and Image Sensing for Machine Vision, 1986.

RCA, *Electro-Optics Handbook*, Technical Series EOH-11, RCA Lancaster 1974.

Torrance, K.E. and Sparrow, E.M., *Theory for Off-Specular Reflection from Roughened Surfaces*, Journal of the Optical Society of America, Vol 56, 916-925, 1966.

3
IMAGE MODELS

Only if a simple visual task is involved, the design of a vision system is trivial. In all other cases, the arrangements of objects in the scene, together with the image formation process and the illumination, are too complex to give a direct interpretation of the image. The purpose of this chapter is to arrive at a general description of images such that the interpretation of the image can be broken down in a number of successive steps, the complexities of each are controllable. Such a description should be concise and convenient. "Concise" means that it should contain all the details that are needed for the visual task at hand. But at the same time it also means that details not relevant should be omitted.

The literature contains a large number of image models [Ahuja and Schachter 1983; Dainty and Shaw 1974]. Most models are applicable only in specific domains. For instance, Markov Random Fields to describe textures of natural objects (wood, grass, cork, clouds, etc.), or fractal models to describe a certain class of silhouettes (of mountains, moon craters, twigs of a tree, etc.).

Although these kind of models are very valuable within their domain, we will not discuss them. Instead, we will focus the attention to a model which is widely applicable, and which is adopted in most computer vision literature. The basic idea behind this model is that the surfaces of the objects in the scene can be partitioned. This partitioning is such that each part of the surface is seen in the image as an homogeneous area in the image plane. As an example, consider a nut (figure 3.1). This object consists of six sidewalls, a top plane, a bottom plane and a cylindrical hole. We may assume that each area is illuminated from a different direction, implying that, most likely, the radiant intensity of each plane or wall will differ from another. At the same time, the intensity within each plane will be nearly constant, linear, or perhaps quadratic. Of course, in this model, the appearance of shadow casts, causing areas with low intensities, also have to be envisaged.

The implication of the partitioning of object surfaces is that the image plane can be partitioned accordingly. Each area in the image plane that corresponds to a part of an object surface is called a *region* or *segment*. A convenient way to represent a partitioning of the image plane is with *labels*. Suppose that the image plane consists of K regions. These regions may be named symbolically: $\Omega = \{\omega_1, \omega_2, \cdots, \omega_K\}$. Then, the partitioning of the image plane x,y can be represented by

a)    b)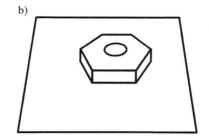

Figure 3.1. A symbolic image of a nut   a) Region oriented   b) Edge map

a function $r(x,y)$ that maps each position of the image plane to a region[1]: $r(.)$: $\mathbb{R}^2 \to \Omega$. The function $r(x,y)$ is called a *symbolic image*, or simply a *map*. In daily life, an example of a symbolic image is a road map. This map corresponds to images made by satellites.

The remaining part of this chapter deals with models for the irradiance within a certain region. As we shall see, there are various ways to define the homogeneity which must exist within such a region. The simplest model states that the irradiance within a region is constant except for fluctuations due to additional white noise. This model, together with some extensions, are discussed in section 3.1. The next section deals with the second order statistics, which is needed minimally to describe correlated fluctuations, like in textured regions.

An alternative way to characterise homogeneity is to describe the positions for which homogeneity fails; i.e. the boundary between two regions, see figure 3.1b. This description is called an *edge map* in contrast with the *region map* shown in figure 3.1a. Edges and other discontinuities are discussed in section 3.3. Finally, section 3.4 gives an example that shows how these models may be used to quantify the performance of a segmentation algorithm.

3.1   FIRST ORDER STATISTICS

The common factor in the models discussed next is that the irradiance within each region is assumed to consist of two components: one is deterministic, the other is stochastic. The deterministic component is due to a regular irradiation of the object surface and to a reflectance distribution which is smooth all over the surface. The stochastic component represents the deviations between the irradiance according to the deterministic component and the measured irradiance.

Suppose the measured irradiance within the region is given by $f(x,y)$ and the deterministic component is $g(x,y)$. Modeling $g(x,y)$ such that it accurately reflects the underlying radiometric process, is very difficult. Therefore, it is common practise to adopt a general parametric form of $g(x,y)$ that can be adapted to the situation at hand by adjusting its parameters. The commonly used general forms for

---

[1] Instead of using a set $\Omega = \{\omega_1, \omega_2, \cdots, \omega_K\}$, one may wish to use a simple numbering system: $1, 2, \cdots, K$. In software realisations this is convenient, but in mathematical discussions the former is preferred.

the deterministic part are constant (flat model), linear (sloped model), quadratic and cubic [Haralick and Shapiro 1992]. In the flat model, the irradiance is assumed constant everywhere in the region:

$$g(x,y) = c_k \quad \text{for } (x,y) \in \text{Region } \omega_k \tag{3.1}$$

In the sloped model, each ideal region has an irradiance that is a sloped plane (i.e. linearly varying with x and y). Similarly, in quadratic and cubic models, a region has irradiance that are bivariate quadratic and cubic, respectively. For instance, a quadratic model gives:

$$g(x,y) = c_k + a_k x + b_k y + d_k x^2 + e_k y^2 + f_k xy \quad \text{for } (x,y) \in \text{Region } \omega_k \tag{3.2}$$

In these expressions, the constants a_k, b_k, etc. are parameters that depend on the region considered. Clearly, if the order of $g(x,y)$ is increased, the approximation of $f(x,y)$ can be made more accurate. However, the complexity of the model increases rapidly with the order of the $g(x,y)$. Therefore, a low order is desirable.

The deviations between $f(x,y)$ and $g(x,y)$ are taken into account by a stochastic component: $\underline{n}(x,y)$. The physical background of these deviations may be manifold:
- thermal noise in the image sensor,
- quantum noise caused by the discrete nature of radiance (photons) and electrical charge (electrons),
- fluctuations in the reflectance distribution of the surface,
- the general form of $g(x,y)$ does not fully fit $f(x,y)$.

For some of these physical aspects it may be argued that the deviations are *multiplicative*, e.g. $f(x,y) = g(x,y)\underline{n}(x,y)$ or $f(x,y) = g(x,y)(1+\underline{n}(x,y))$. For instance, small fluctuations in the reflectance distribution that are not covered by $g(x,y)$ may take the latter form. Thermal noise is an example of a physical process having an *additive* nature:

$$f(x,y) = g(x,y) + \underline{n}(x,y) \tag{3.3}$$

Here, $\underline{n}(x,y)$ is a stationary zero-mean Gaussian stochastic process (see appendix B.4) with variance σ_n^2. The probability density of $\underline{n}(x,y)$ equals[2]:

$$p_{\underline{n}}(n) = \frac{1}{\sigma_n \sqrt{2\pi}} \exp\left(\frac{-n^2}{2\sigma_n^2}\right) \tag{3.4}$$

The noise model of (3.3) and (3.4) is frequently used as an approximation, even if it is known that the real stochastic process is not additive and the probability density of $\underline{n}(x,y)$ deviates somewhat from being Gaussian. The reason for using an approximation is that its mathematical analysis is much easier than (for example) a multiplicative model. Of course, such an approximation is only allowed if the errors induced by it are acceptable.

---

[2]Formally, the notation of the first order probability density is $p_{\underline{n}(x,y)}(n)$. In case of stationarity, and if confusion is unlikely, it is abbreviated to $p_{\underline{n}}(n)$.

Table 3.1 First order statistical parameters

Feature	Definition	Symbolically
expectation	$\int_n n p_{\underline{n}}(n) dn$	μ_n, $E\{\underline{n}(x,y)\}$ or $\bar{n}$
variance	$\int_n (n-\mu_n)^2 p_{\underline{n}}(n) dn = \overline{(n-\bar{n})^2} = \overline{n^2} - \bar{n}^2$	σ_n^2 or $\text{Var}\{\underline{n}(x,y)\}$
standard deviation	$\sqrt{\int_n (n-\mu_n)^2 p_{\underline{n}}(n) dn}$	σ_n
skewness	$\dfrac{1}{\sigma_n^3} \int_n (n-\mu_n)^3 p_{\underline{n}}(n) dn = \overline{(n-\bar{n})^3}/\sigma_n^3$	
kurtosis	$\dfrac{1}{\sigma_n^4} \int_n (n-\mu_n)^4 p_{\underline{n}}(n) dn - 3 = \overline{(n-\bar{n})^4}/\sigma_n^4 - 3$	

The first order statistics of the stochastic process $\underline{n}(x,y)$ is fully defined by its probability density $p_{\underline{n}}(n)$. From this density, features can be derived that offer a concise description of the probability density. Some of these features are given in table 3.1. The standard deviation is a measure of the width of the density. It is the square root of the variance. The skewness and the kurtosis are two parameters that are normalised with respect to the width. If $p_{\underline{n}}(n)$ is symmetrical, the skewness is zero. Otherwise, the skewness is negative (the tail on the left hand side is larger than the tail on the right hand side), or positive (vice versa). The kurtosis is a measure of peakedness. It has been defined such that Gaussian densities have zero kurtosis.

Example 3.1 Image model of a blood cell

Figure 3.2a shows the image of a blood cell. This image can roughly be partitioned into three regions: background, cytoplasm and nucleus. A symbolic image is given in figure 3.2b. For each region, the model stated in equation (3.1) and (3.3) has been applied. Given the images in figure 3.2a and 3.2b, one way to estimate the unknown parameters c_k $k=1,2,3$ is to calculate the average irradiance in each region. This yields the parameters c_k as shown in table 3.2. Once these parameters are known, the stochastic component follows from: $\underline{n}(x,y) = f(x,y) - c_k$. This is depicted in figure 3.2c. The probability densities of $f(x,y)$ within each region can be estimated using a technique called *histogramming* (see Chapter 5). The results are shown in figure 3.2d. Finally, from the densities, the parameters mentioned in table 3.1 can be calculated (table 3.2).

In table 3.2, the expectations of the stochastic processes are zero. This is implied in the model, since an hypothetical non-zero expectation would immediately be absorbed in the deterministic component. From the three densities, the one corresponding to "background" is most symmetrical. This follows from its skewness. The density is a little bit peaked compared with a Gaussian density. The other two densities, "cytoplasm" and "nucleus", are a little skewed. The reason for this is that irradiance is a non-negative quantity. Hence, on the left hand side the tails of the densities must fall down to zero to insure this. As a whole, the Gaussian stationary model is quite acceptable (perhaps with a few exceptions near the boundaries between the regions).

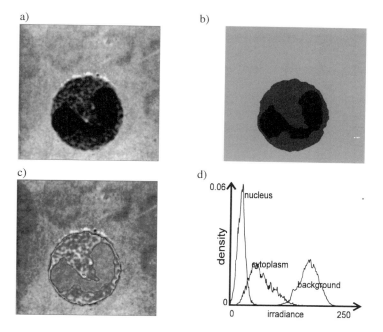

Figure 3.2 Modeling images of blood cells
a) Original
c) Stochastic component
b) Deterministic component
d) Probability densities

3.2 SECOND ORDER STATISTICS

The first order probability density $p_{\underline{n}}(n)$ defines the statistical behavior of the irradiance at one position, regardless of the behavior of the irradiance at another position. In general, however, irradiances at different positions are statistically related. In this section, we consider pairs of positions in the image plane, and discuss the statistical relations between the irradiances at these positions. This is called *the second order statistics*.

The discussion can be simplified if *stationarity* is assumed. Let (x,y) and $(x+a, y+b)$ be two points in the image plane. Then, a stochastic process $\underline{n}(x,y)$ is called stationary with respect to its second order statistics, if its statistical behavior of a pair $\underline{n}(x,y)$ and $\underline{n}(x+a, y+b)$ only depends on the separation (a,b), and not on the

Table 3.2 Estimated statistical parameters of the irradiance in the regions shown in figure 3.2

Region	c_k	Expectation	Variance	Standard deviation	Skewness	Kurtosis
background	154	0	375	19	-0.15	0.44
cytoplasm	61	0	519	23	0.63	0.01
nucleus	22	0	57	7.6	0.30	0.45

38 IMAGE BASED MEASUREMENT SYSTEMS

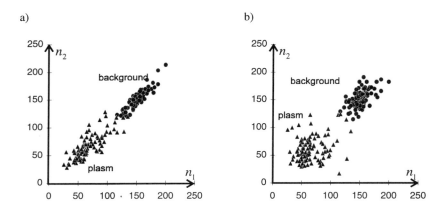

Figure 3.3 Scatter diagrams of the irradiances from two different regions in figure 3.2a
a) Separation between n_1 and n_2 small
b) Separation between n_1 and n_2 large

absolute positions. In other words, the joint probability density of $\underline{n}(x,y)$ and $\underline{n}(x+a,y+b)$ depends only on (a,b). Therefore, we may write briefly:

$$p_{\underline{n}(x,y),\underline{n}(x+a,y+b)}(n_1,n_2) = p_{\underline{n},\underline{n}}(n_1,n_2;a,b) \tag{3.5}$$

The notation on the right hand side is convenient if the process is stationary. If not, the notation on the left hand side is needed.

If the separation distance between the two points becomes wide (i.e. a or b are sufficiently large), it is likely that the statistical behavior of $\underline{n}(x,y)$ and $\underline{n}(x+a,y+b)$ is independent from each other: knowledge of the irradiance at (x,y) does not supply any knowledge about the irradiance at $(x+a,y+b)$. In that case, the joint probability density is the product of the two marginal probability densities:

$$p_{\underline{n},\underline{n}}(n_1,n_2;a,b) = p_{\underline{n}}(n_1)p_{\underline{n}}(n_2) \tag{3.6a}$$

On the other hand, if the two points are near neighbors (i.e. a and b are sufficiently small), the irradiance at (x,y) almost equals the irradiance at $(x+a,y+b)$. Hence: $\underline{n}(x,y) \approx \underline{n}(x+a,y+b)$. If this is the case, then:

$$p_{\underline{n},\underline{n}}(n_1,n_2;a,b) \approx p_{\underline{n}}(n_1)\delta(n_1-n_2) \tag{3.6b}$$

Example 3.1 Image model of a blood cell (continued)

With fixed separation (a,b), the samples from a second order joint probability density $p_{\underline{n},\underline{n}}(n_1,n_2;a,b)$ may be visualised by a so-called *scatter diagram*. This is a 2-dimensional plot in which the two axes correspond to n_1 and n_2, respectively. A number of pairs of positions are selected at random, but such that for each pair the separation between them is given by (a,b). Then, for each pair of positions, the corresponding irradiances n_1 and n_2 are determined and marked in the plot. This is illustrated for the regions "background" and "cytoplasm" in figure 3.3. In figure 3.3a the separation is $a = 0.01X$, $b = 0.01Y$ for both regions. (X and Y are the sizes of the image in horizontal and vertical direction, respectively.) In figure 3.3b this is:

Table 3.3 Second order statistical parameters

Feature	Definition	Symbolically
autocorrelation function	$\iint\limits_{n_1 n_2} n_1 n_2 p_{\underline{n},\underline{n}}(n_1,n_2;a,b) dn_1 dn_2$	$R_{nn}(a,b) = \mathrm{E}\{\underline{n}(x,y)\underline{n}(x+a,y+b)\}$
autocovariance function	$\iint\limits_{n_1 n_2} (n_1-\mu_n)(n_2-\mu_n) p_{\underline{n},\underline{n}}(n_1,n_2;a,b) dn_1 dn_2$	$C_{nn}(a,b) = R_{nn}(a,b) - \mu_n^2$
normalised autocovariance (correlation coefficient)	$C_{nn}(a,b) / C_{nn}(0,0)$	$r_{nn}(a,b)$
first order (or marginal) probability density	$\int\limits_{n_2} p_{\underline{n},\underline{n}}(n_1,n_2;a,b) dn_2$	$p_{\underline{n}}(n_1)$
variance	$C_{nn}(0,0)$	σ_n^2

$a = 0.05X$, $b = 0.05Y$. Note that the samples in figure 3.3a are more statistically dependent (that is, more lined up at the diagonal) than in figure 3.3b (as expected). Furthermore, it can be seen that the dependency of the irradiances in the region "cytoplasm" deviates from the ones in the region "background". Each region has its own specific second order statistics. Therefore, in this example, the stochastic component does not need to be regarded as a noisy disturbance of the measurement data. It carries some information about the positions of the regions. This information is additional to the information provided by the deterministic component.

The joint probability density $p_{\underline{n},\underline{n}}(n_1,n_2;a,b)$ fully defines the second order statistics. As in the first order case, a number of features can be derived from this density. Some of them are shown in table 3.3. Both the autocorrelation and the autocovariance try to express the rate of dependency between $\underline{n}(x,y)$ and $\underline{n}(x+a,y+b)$. The autocorrelation function incorporates the dependency due to the expectation, whereas the autocovariance function only considers the stochastic part. Note that if $\underline{n}(x,y)$ and $\underline{n}(x+a,y+b)$ are fully independent $R_{nn}(a,b) = \mu_n^2$ and $C_{nn}(a,b) = 0$. On the other hand, if $\underline{n}(x,y)$ and $\underline{n}(x+a,y+b)$ are fully dependent $R_{nn}(a,b) = \mu_n^2 + \sigma_n^2$ and $C_{nn}(a,b) = \sigma_n^2$. The reverse is not necessarily true. If $\underline{n}(x,y)$ and $\underline{n}(x+a,y+b)$ are *uncorrelated* (i.e. $C_{nn}(a,b) = 0$), then this does not imply that $\underline{n}(x,y)$ and $\underline{n}(x+a,y+b)$ are fully independent.

Another useful feature is the correlation coefficient, defined as the normalised autocovariance function. The following inequality holds: $-1 \leq r_{nn}(a,b) \leq 1$. If $\underline{n}(x,y)$ and $\underline{n}(x+a,y+b)$ are linearly connected (i.e. a relationship of the following form exists: $\underline{n}(x,y) = \alpha + \beta \underline{n}(x+a,y+b)$), then either $r_{nn}(a,b) = 1$ or $r_{nn}(a,b) = -1$. Note that both autocorrelation and autocovariance are symmetrical functions. For example, $R_{nn}(a,b) = R_{nn}(-a,-b)$

Example 3.1 Image model of a blood cell (continued)

Table 3.4 gives the statistical features of the samples in figure 3.3. The results are as expected. Elongated clusters in the scatter diagram have relative large correlation and covariance. More concentric clusters have smaller correlation and covariance. This is most pronounced in the correlation coefficients. However, this feature has the disadvantage that it lacks an external reference (it has no unit). This is in contrast with the autocovariance and the autocorrelation which are provided with a

Table 3.4 Second order features of the irradiances in the images given in figure 3.2a. The numerical quantities are derived from the scatter diagrams in figure 3.3

	$R_{nn}(a,b)$	$C_{nn}(a,b)$
background $a=0.01X, b=0.01Y$	315	315
background $a=0.05X, b=0.05Y$	221	221
cytoplasm $a=0.01X, b=0.01Y$	425	425
cytoplasm $a=0.05X, b=0.05Y$	120	120

unit (for instance: lx^2). In this example, the autocorrelation and the autocovariance are equal. The reason is that the deterministic component of the samples in figure 3.3 are kept apart in the region parameters c_k. The samples in figure 3.3 appear to be clustered in subareas that resemble ellipses. Therefore, $\underline{n}(x,y)$ and $\underline{n}(x+a,y+b)$ more or less correspond to two random variables with a *bivariate* Gaussian probability density (see appendix B.2). In case of stationarity this density takes the form:

$$p_{\underline{n},\underline{n}}(n_1,n_2;a,b) =$$

$$\frac{1}{2\pi\sqrt{C_{nn}^2(0,0) - C_{nn}^2(a,b)}} \exp\left(-\frac{C_{nn}(0,0)n_1^2 - 2C_{nn}(0,0)n_1n_2 + C_{nn}(0,0)n_2^2}{2\left(C_{nn}^2(0,0) - C_{nn}^2(a,b)\right)}\right) \quad (3.7)$$

Although this Gaussian density is approximate only, it is used very often, because of its mathematical convenience.

3.2.1 Stochastic processes and 2-dimensional filtering

In this section we consider a linear optical device (or more general a linear filter) with PSF $h(x,y)$, and discuss its behavior in the context of stationary stochastic processes. The 1-dimensional analogy of this section is considered in appendix B.4. Let the input of the filter be given by the process $\underline{n}(x,y)$ with autocorrelation function $R_{nn}(a,b)$. The output of the filter is denoted by $\underline{m}(x,y)$. The expectation of $\underline{m}(x,y)$ simply follows from:

$$E\{\underline{m}(x,y)\} = E\{\underline{n}(x,y) * h(x,y)\} = E\{\underline{n}(x,y)\} * h(x,y) \quad (3.8)$$

Thus, if $\underline{n}(x,y)$ is stationary with respect to its expectation $E\{\underline{n}(x,y)\} = \mu_n$, the expectation of $\underline{m}(x,y)$ is constant and is given by $\mu_m = \mu_n \iint h(x,y)dxdy$.

To determine the autocorrelation function of $\underline{m}(x,y)$, we shall first determine the correlation between $\underline{m}(x,y)$ and $\underline{n}(x+a,y+b)$. The quantity is called the *cross correlation* and is denoted: $R_{mn}(a,b)$ by:

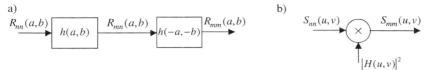

Figure 3.4 2-dimensional filtering of stochastic processes
a) Calculations of the autocorrelation function
b) Calculation of the power spectrum

$$\begin{aligned} R_{mn}(a,b) &= E\{\underline{m}(x,y)\underline{n}(x+a,y+b)\} \\ &= E\{(\underline{n}(x,y)*h(x,y))\underline{n}(x+a,y+b)\} \\ &= R_{nn}(a,b)*h(a,b) \end{aligned} \quad (3.9)$$

Similarly we can show that for the autocorrelation function $R_{mm}(a,b)$ of $\underline{m}(x,y)$ the following equality holds:

$$R_{mm}(a,b) = E\{\underline{m}(x,y)\underline{m}(x+a,y+b)\} = R_{nn}(a,b)*h(a,b)*h(-a,-b) \quad (3.10)$$

The calculations of the correlation functions of (3.9) and (3.10) are depicted in figure 3.4a. We can give it the following interpretation. Applying $R_{nn}(a,b)$ to a linear filter with PSF $h(a,b)$, we obtain as output $R_{mn}(a,b)$. With $R_{mn}(a,b)$ as input of a filter with PSF $h(-a,-b)$, the output is $R_{mm}(a,b)$.

We conclude this section with two remarks:

- The variance of $\underline{m}(x,y)$ can be found simply by the following relation: $\text{Var}\{\underline{m}(x,y)\} = C_{mm}(0,0) = R_{mm}(0,0) - \mu_m^2$. Hence, in order to find this first order feature of $\underline{m}(x,y)$, the second order feature $R_{nn}(a,b)$ has to be known.
- If $\underline{n}(x,y)$ has a second order Gaussian probability density like in equation (3.7), so has $\underline{m}(x,y)$.

3.2.2 Autocorrelation and power spectrum

The power spectrum[3] $S_{nn}(u,v)$ of a stationary stochastic process $\underline{n}(x,y)$ is defined as the Fourier transform of its autocorrelation function $R_{nn}(a,b)$:

$$S_{nn}(u,v) = F\{R_{nn}(a,b)\} \quad (3.11)$$

The Fourier transform takes place with respect to the variables (a,b).

The significance of the power spectrum becomes clear especially when the behavior of linear filters is studied. Let $h(x,y)$ be the PSF of a linear filter. The OTF of this filter is given by $H(u,v)$. The input of the filter is a stationary stochastic process $\underline{n}(x,y)$ with power spectrum $S_{nn}(u,v)$. The output is a stochastic process $\underline{m}(x,y)$ with autocorrelation function $R_{mm}(a,b)$ and power spectrum (see figure 3.4b):

$$S_{mm}(u,v) = |H(u,v)|^2 S_{nn}(u,v). \quad (3.12)$$

---

[3]In this section the word "spectrum" refers to spatial frequencies. This is not to be confused with the word "spectrum" in the context of electromagnetic waves in which it refers to temporal frequencies or wavelengths.

This follows immediately from the definition of the power spectrum, the convolution theorem (table 2.4), and the result from the previous section (equation 3.10).

The autocorrelation function $R_{mm}(a,b)$ can be found by application of the inverse Fourier transform:

$$R_{mm}(a,b) = F^{-1}\{S_{mm}(u,v)\} \tag{3.13}$$

Suppose that $\underline{m}(x,y)$ has zero expectation, then its variance is:

$$\sigma_m^2 = R_{mm}(0,0) = \iint_{u\ v} |H(u,v)|^2 S_{nn}(u,v)dudv \tag{3.14}$$

The power spectrum can be interpreted as a density function. It gives the distribution of the signal energy over the spatial frequencies (u,v). To see this, assume that the filter is such that it acts as an ideal band-pass system:

$$H(u,v) = \begin{cases} 1 & \text{if } u_0 < u < u_0 + \Delta u \text{ and } v_0 < v < v_0 + \Delta v \\ 0 & \text{elsewhere} \end{cases}$$

Then, if Δu and Δv are sufficiently small, $\sigma_m^2 = S_{nn}(u_0,v_0)\Delta u\Delta v$. In other words, $S_{nn}(u,v)$ represents the variance of $\underline{n}(x,y)$ per unit bandwidth. Therefore, the power spectrum is also called the *spectral power density*. Note that since the power spectrum measures variance per bandwidth (its physical unit may be lx^2m^2), this function can never be negative.

A model frequently used in image processing is that of Gaussian *white noise*. This is a stationary stochastic process with Gaussian probability densities, and with constant power spectrum: $S_{nn}(u,v) = V$. Physically, white noise does not exist since it would require an infinitely broad bandwidth. Also, it would imply a variance that tends to infinite. However, in practise, a stochastic process is considered white if its power spectrum is constant within the spatial resolution of the imaging device. Suppose the OTF of this imaging device is given by $H(u,v)$, and the input consists of noise, then this noise is considered white if the variance of the output equals:

$$\sigma_m^2 = V \iint |H(u,v)|^2 dudv \tag{3.15}$$

The variance can also be obtained by application of Parseval's theorem (table 2.4):

$$\sigma_m^2 = V \iint h^2(x,y)dxdy \tag{3.16}$$

Example 3.1 Image model of a blood cell (continued)

As indicated above, the autocorrelation function $R_{nn}(a,b)$ is a measure of the dependency between two points separated by a distance (a,b). Figure 3.5a shows a rectangle subarea of the region "background" of the image in figure 3.2a. From this image the autocorrelation is estimated and shown as a 3-dimensional plot in figure 3.5b. The autocorrelation can also be visualised as an intensity image with co-ordinates (a,b). This is illustrated in figure 3.5c. Fourier transformation of the autocorrelation function yields an estimate of the power spectrum. The estimate is shown in figure 3.5d as an intensity image with co-ordinates (u,v).

In the given example, a stationary Gaussian noise model appears to be quite adequate. This is demonstrated in figure 3.5e and 3.5f which show two artificial stochastic processes. The first one (figure 3.5e) is obtained with a Gaussian pseudo-random generator. The samples in the image are uncorrelated. Hence, the stochastic process may be considered as white. The second image (figure 3.5f) is a filtered version of the white noise. The OTF of the filter is: $\sqrt{\hat{S}_{nn}(u,v)}$ in which $\hat{S}_{nn}(u,v)$ is the estimate of the power spectrum shown in figure 3.5d. The result is that the power spectrum of figure 3.5f approximates the one of figure 3.5a. Note the similarity between the two processes.

3.3 DISCONTINUITIES

Instead of modeling the irradiance of one part of the surface of an object one could also model the *changes* of irradiances between two neighboring parts. As an example we consider the image of a cube (figure 3.6a). The figure also shows a plot of the irradiance of the image taken along the horizontal line indicated in the figure (i.e. a horizontal cross section). In this example, the boundary between neighboring faces of the cube appear as step-like transitions in the irradiances.

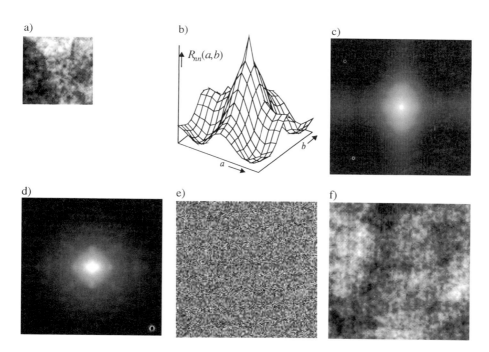

Figure 3.5 Stationary autocorrelation function
a) Subregion from figure 3.2a. For visibility the contrast and spatial resolution has been increased
b) Plot of autocorrelation function estimated from the image in figure 3.5a
c) Autocorrelation function shown as intensity image
d) Power spectrum estimated from 3.5c. The log-power is shown as an intensity image
e) A realisation of a stationary Gaussian white noise process
f) Version of 3.5e filtered such that the power spectrum matches the one shown in figure 3.5d

These transitions of the irradiance are frequently modeled by the edge spread function (ESF; see section 2.2.5). Suppose that it is known that the perspective projection of an object boundary yields a step-like transition of irradiances in the image plane at a position (x_1, y_1) with orientation φ. Then, the model of the image in the surroundings of this position is:

$$f(x,y) = C_0 + C_1 h_{esf}(x - x_1, y - y_1, \varphi) + \underline{n}(x,y) \tag{3.17}$$

C_0 and C_1 are two constants that are defined by the reflectance and illumination of the two faces of the object. The function $h_{esf}(.)$ is the ESF measured at an angle φ. As before, the term $\underline{n}(x,y)$ models the stochastic component. In figure 3.6a the circled area denotes a subregion for which the edge model approximately holds.

There are certain limitations in the edge model. First of all, the model only holds in the vicinity of the point (x_1, y_1), because the further away from this point, the more uncertain we are concerning the exact course of the boundary. Secondly, the model does not cover junctions or branches in the boundary. Thirdly, if the surfaces of the objects are glossy, the irradiance at the boundary between two faces of the object is likely to show a peak caused by the specular reflection component.

In some applications, the objects in the scene are so thin that their widths fall down the resolution of the imaging device. The image of these objects are governed by the line spread function (LSF; see section 2.2.5). Figure 3.6b shows the image of a line drawing together with a cross section. If in this image one would try to measure the widths of the lines, one would actually measure the width of LSF of the imaging device with which the image is obtained. The model to describe this phenomenon is similar to equation (3.17):

$$f(x,y) = C_0 + C_1 h_{lsf}(x - x_1, y - y_1, \varphi) + \underline{n}(x,y) \tag{3.18}$$

The terms are virtually equal to the one in equation (3.17). The ESF is replaced

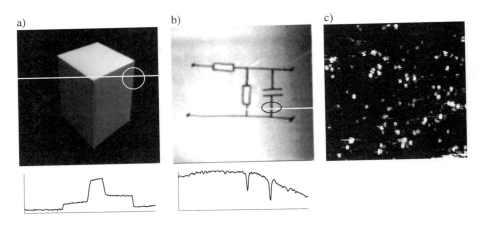

Figure 3.6 Modeling discontinuities
a) Image of a cube containing step edges
b) Image of a line drawing
c) Radar image containing spots caused by birds (Reproduced by permission of TNO-FEL, The Hague).

with the LSF $h_{lsf}(.)$. Some of the limitation of the edge model also apply to the line model stated in (3.18). In figure 3.6b the circled area denote a subregion for which the line model approximately holds.

Some domains in which the line model applies are the medical domain (e.g. images of small blood vessels), remote sensing (e.g. satellite images of roads and rivers), and the industrial domain (quality control of electrical wiring in printed circuit boards, detection of flaws and scratches at the surface of materials, etc.)

In other applications, the object of interest is so small that in <u>all</u> dimensions the size falls down the resolution of the imaging device. For instance, a picture from a starry night may contain stars the appearance of which is primarily governed by the imaging device. The stars appear as *spots*, fully defined by the PSF, and not by the shape of the stars themselves. This phenomenon also may happen in radar applications where the target (e.g. aeroplanes) are much smaller than the width of the scanning beam. The image model for such a spot is:

$$f(x,y) = C_0 + C_1 h(x - x_1, y - y_1) + \underline{n}(x,y) \qquad (3.19)$$

or in case that many spots exist:

$$f(x,y) = C_0 + \sum_{i=1} C_i h(x - x_i, y - y_i) + \underline{n}(x,y) \qquad (3.20)$$

The function $h(x,y)$ is the PSF of the imaging device. Figure 3.6c gives an example. It is a radar image in which the bright spots are caused by the echo of birds.

Example 3.2 Finding scratches at material surfaces

We conclude this chapter with an example which demonstrates the use of image models. In this example, we consider an image of a single *line-like* object, for instance a scratch on the surface of some material. We assume that this scratch is imaged such that it is vertically or almost vertically oriented in the image plane. Furthermore, we assume that the image is corrupted by additive Gaussian white noise. The task of a computer vision system is to detect the scratch; i.e. to check to see whether there is such a scratch, and if so, to localise it.

The strategy is to find all the points in the image plane that makes up the scratch. This is done for each point individually. In a later stage, the points found may be grouped together, but this stage (i.e. edge linking) will be discussed only in Chapter 7. To find the individual points, the system first applies a low-pass filter[4] to reduce the influence of noise. In the resulting image, a position is marked as a *line element* if the irradiance at that position is larger than a reference (called a *threshold*). The thresholding-procedure is based on the premise that the image of the scratch has a positive contrast with respect to its surroundings. This is the case if dark field illumination is applied (see section 2.4.2). An illustration of the computer vision system is given in figure 3.7. The following problems will be addressed:
- What will be the quality of the detection ?
- What is an appropriate choice of the low-pass filter ?

These questions can be answered only if a mathematical model is available. So, the starting point of the discussion will be the development of a model.

If a scratch exists, we can describe it as a locus of points in the image plane with an (almost) vertical orientation. This implies that the locus can be described

---
[4] Chapter 5 deals with the problem how to implement this filter digitally.

with the x co-ordinate given as a function of the y co-ordinate: $x = l(y)$. Furthermore, if the orientation does not change rapidly, we may assume that in every small neighborhood the scratch is on a straight line segment. If (x_1, y_1) is a point of the line, the line is approximated by: $x - x_1 = (y - y_1)\sin\alpha$. Here, the quantity α is the angle between the line and the y-axis. The angle is assumed to be very small (the line is almost vertical). Hence, we do not introduce a large error if, in our analysis, we ignore this angle. Then, locally the line is given by: $x = x_1$.

We assume that the PSF $h(x, y)$ of the noise-reduction filter is much more influential than the PSF of the imaging device. In that case, the PSF of the imaging device can be ignored, and the line can be described with:

$$f(x,y) = C_0 + C_1 \delta(x - x_1) + \underline{n}(x, y) \tag{3.21}$$

This image passes the noise-reduction filter. The output of this filter is:

$$s(x,y) = C_0 + C_1 \int_y h(x - x_1, y) dy + \underline{n}(x,y) * h(x,y) \tag{3.22}$$

This completes the mathematical model of the line.

Detection quality

The threshold operation determines whether a position is marked as a line element, or not. Therefore the quality of the detection particularly depends on this operation. In the noiseless case, the irradiance measured exactly on the line ($x = x_1$) will be:

$$C_0 + C_1 \int_y h(0, y) dy$$

In the absence of a line, the irradiance will be simply C_0. Hence, in the noiseless case, the detection would be perfect as long as the threshold is chosen just above

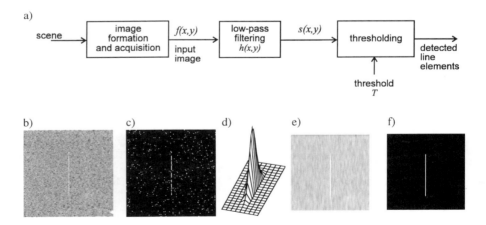

Figure 3.7 Finding vertically oriented scratches at material surfaces
a) Functional structure of the computer vision system at hand
b) Input image
c) Map of detected line elements obtained by directly thresholding the input image
d) PSF of the low-pass filtering
e) Filtered image
f) Map of detected line elements obtained by thresholding the filtered image

C_0. However, the presence of noise will degrade the performance of the detection. Suppose that the standard deviation of the noise in $s(x,y)$ is σ_s. In order to have a high detection quality this deviation must be small compared with the magnitude of the line. Therefore, a good measure of the detection quality is the so-called *signal-to-noise ratio* (*SNR*)[5]:

$$SNR = \frac{C_1}{\sigma_s} \int_y h(0,y) dy \tag{3.23}$$

The choice of the low-pass filter

As an example, our first choice of the low-pass filter will be a rotationally symmetric Gaussian with width σ_p (see eq. 2.27). In that case the magnitude of the line becomes:

$$C_1 \int_y h(0,y) dy = \int_y \frac{C_1}{2\pi\sigma_p^2} \exp\left(-\frac{y^2}{2\sigma_p^2}\right) dy = \frac{C_1}{\sigma_p\sqrt{2\pi}}$$

The variance σ_s^2 of the noise can be obtained from equation (3.15) or (3.16):

$$\sigma_s^2 = V \iint |h(x,y)|^2 dx dy = V \iint \frac{1}{4\pi^2\sigma_p^4} \exp\left(-\frac{x^2+y^2}{\sigma_p^2}\right) dx dy = \frac{V}{4\pi\sigma_p^2}$$

Therefore, the *SNR* is:

$$SNR = \frac{C_1\sqrt{2}}{\sqrt{V}} \tag{3.24}$$

Hence, the detection quality is proportional with C_1 and inversely proportional with the square root of V, the variance per unit bandwidth of the white noise. Most surprisingly, the quality does not depend on the width σ_p of the noise-reduction filter. The reason is that the filter is rotationally symmetric. It reduces the noise at the same rate as it attenuates the signal.

If a filter was chosen which enhances the signal in a preferred direction (i.e. in the vertical direction) the detection quality can be improved. To show this, suppose that the filter $h(x,y)$ is built from the cascade of two 1-dimensional filters: $h(x,y) = h_1(x)h_2(y)$. As an example, we assume that both $h_1(x)$ and $h_2(y)$ are 1-dimensional Gaussians with width σ_1 and σ_2, respectively:

$$h_1(x) = \frac{1}{\sigma_1\sqrt{2\pi}} \exp\left(-\frac{x^2}{2\sigma_1^2}\right) \qquad h_2(y) = \frac{1}{\sigma_2\sqrt{2\pi}} \exp\left(-\frac{y^2}{2\sigma_2^2}\right)$$

The magnitude of the line becomes:

$$C_1 \int_y h(0,y) dy = C_1 \int_y h_1(0)h_2(y) dy = \frac{C_1}{\sigma_1\sqrt{2\pi}}$$

The variance σ_s^2 of the noise is:

---

[5]Sometimes signal-to-noise ratio is defined as a ratio of powers, i.e. $SNR = \left(\frac{C_1}{\sigma_s} \int_y h(0,y) dy\right)^2$.

$$\sigma_s^2 = V\iint |h_1(x)|^2 |h_2(y)|^2 \, dxdy$$

$$= V\int_x \frac{1}{\sigma_1^2 2\pi} \exp\left(-\frac{x^2}{\sigma_1^2}\right) dx \int_y \frac{1}{\sigma_2^2 2\pi} \exp\left(-\frac{y^2}{\sigma_2^2}\right) dy = \frac{V}{4\pi\sigma_1\sigma_2}$$

Hence, the *SNR* becomes:

$$SNR = C_1 \sqrt{\frac{2\sigma_2}{V\sigma_1}} \tag{3.24}$$

Again, the detection quality is proportional with C_1 and inversely proportional with the square root of V. However, this time it is also proportional with the square root of the ratio between σ_2 and σ_1. Hence, it is favorable to choose a PSF that is elongated as much as possible: it must take the form of a vertically oriented line segment. Of course, the length of the line segment is limited, because if the PSF chosen is too lengthy, the model stated in equation (3.21) no longer holds.

The procedure is illustrated in figure 3.7. Figure 3.7b shows the image of a vertical line contaminated by white noise. Simply thresholding this image yields a map with detected line elements shown in figure 3.7b. Low-pass filtering with the elongated, vertically oriented PSF shown in figure 3.7c yields the image in figure 3.7d. Finally, thresholding gives a map with detected elements (figure 3.7e). The result is much better than in figure 3.7b.

The conclusion is as follows: in finding line segments with a certain orientation and a certain length, it is advantageous to filter the image with a PSF that fully matches the shape of such a line segment. The filter we have found to be suitable is a so-called *matched filter*.

REFERENCES

Ahuja, N. and Schachter, B.J., *Pattern Models*, J. Wiley & Sons, New York, 1983.

Haralick, R.M. and Shapiro, L.G., *Computer and Robot Vision, Vol. I*, Addison-Wesley, Massachusetts, 1992.

Dainty, J.C. and Shaw, R.: *Image Science*, Academic Press, London, 1974.

4
IMAGE ACQUISITION

The images considered so far are defined as densities of radiant energy upon 2-dimensional planes; irradiances or illuminances. As such an image is a physical quantity defined everywhere in the image plane, and represented by a real (non-negative) 2-dimensional function. This *continuous* representation is to be contrasted with a *discrete* and *digital* representation of a computer vision system in which an image is represented by a finite number of bits; see [Park and Schowengerdt 1982; Huck et al 1985; Pratt 1991].

The transformation of an image from the physical continuous domain (say irradiances) to a discrete domain is the topic of this chapter. Loosely speaking, such a transformation consists of four functional elements:

1. *Sensoring*: transforming irradiance to an electrical quantity (e.g. charge or voltage).
2. *Scanning:* transforming the 2-dimensional space domain to the 1-dimensional time domain.
3. *Spatial sampling*: transforming the continuous space domain to a discrete space domain.
4. *Analog-to-digital conversion*: transforming the electrical quantity to a digital representation (e.g. integer numbers).

These four functional elements need not to be housed in four separate organs. For instance, a 2-dimensional CCD combines the first three functional elements within one device. In a vidicon-camera tube the image plane is scanned in a sequence of horizontal lines; in so doing irradiance is converted into electrical charge. Hence, a vidicon-camera accomplishes the first two and a half functional elements.

At present, the image acquisition systems most widely used are based on CCDs. We will discuss some of the properties of this device in section 4.1. The next section concerns discretisation in the spatial domain. This section includes an error analysis. Furthermore, it discusses the effect of discretisation in the frequency domain. Finally, section 4.3 concludes this chapter with a discussion about discretisation in the radiometric domain, i.e. amplitude quantisation.

4.1 THE CCD-CAMERA

A *charge-coupled device* (CCD) consists of an array of *charge-storage elements*, a *transport mechanism* and an *output device*. The charge-storage elements are either arranged linearly (the line-scan CCD; see figure 4.1a), or arranged in a matrix structure (the area CCD; see figure 4.1b). Each element can transform the radiant energy incident on its surface into an electrical charge. Underneath the elements, an electronic transport mechanism can move the charge of each element to the output terminal of the device.

The operational structure of the line-scan camera is as follows. Suppose that at a certain moment the electrical charges in all storage elements are zero. Then, from that moment the radiant flux incident on the surface of an element starts charging that element at a rate which is proportional to the incident flux. At the end of the integration period, the charge in each element is pushed into the analog shift register. This occurs simultaneously for all storage elements. Then, during each cycle of the clock, the charges shift simultaneously one position to the right. In doing so, the charge most to the right is pushed into the output circuitry, which converts this charge into a voltage appearing at the output terminal. As a result, in sequential clock cycles, the output voltage is proportional to the radiant energy of the consecutive elements.

Line-scan CCDs need a linear motion of either the device or the scene, in order to obtain a 2-dimensional scan of the image plane. In quality control applications, products are often transported by means of a conveyor. If the motion of the conveyor is constant, a 2-dimensional image of the products can be obtained provided that the array of the CCD is aligned orthogonally to the direction of the conveyor. The same technique is used in satellite-imagery. Image scanners, used in optical character readers (OCRs), also rely on line-scan CCDs. The image is acquired by feeding the paper through the scanner. Alternatively, in some hand-held scanners, the CCD is moved across the paper. The advantage of line-scan CCDs is the resolution. A linear array can easily contain 4000 elements, yielding an image of (say) 4000×4000 elements. An area CCD contains typically about 600×600 elements. A drawback of the line-scan CCD is that the acquisition of a single image takes more time compared with that of an area CCD. Therefore, in video applications (motion pictures) area CCDs are used more frequently.

The area CCD is equipped with a more sophisticated electronic transport mechanism. This mechanism allows the acquisition of a full image without a mechanical scan. The charge-storage elements are connected to the parallel inputs

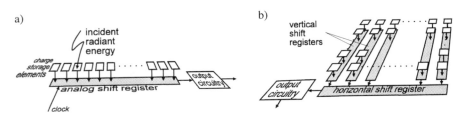

Figure 4.1 Charge coupled devices
a) Line-scan CCD
b) Area CCD

of a bank of shift registers (the so-called *frame storage*). The outputs of these shift registers are connected to another shift register aligned orthogonally to the parallel bank. An image acquisition cycle consists of two stages:

1. An integration period. In this period the radiant flux at each element is converted to electrical charge. At the end of the period the charges are loaded into the frame storage.

2. A charge transport period. The charges in the frame storage are shifted in vertical direction, thereby loading the horizontal shift register with the data from one row. Subsequently, the horizontal shift register is read out at a high rate. This process repeats until all rows in the frame storage are read.

Normally these two stages are pipelined. At the beginning of the transport period, the integration period of the next image already starts.

Most area CCDs have a provision to separate the odd rows of the storage elements from the even ones. With this provision it is possible to load the odd rows (the so-called *odd field*) solely, while the even rows (*even field*) are still integrating the radiant flux (figure 4.2a). After the read out of all odd rows, the even rows can be loaded and read out (figure 4.2b), while the odd rows are integrating. This way of scanning an image plane is called *interlacing*. With this arrangement, the CCD complies with video standards like the CCIR TV standard. Its main advantage is to reduce the awareness of flickering of video images.

CCDs for color image acquisition are also widely available. One way to obtain color CCDs is to equip the device with three horizontal shift registers (figure 4.3a) together with a color stripe filter having a pattern as shown in figure 4.3b. Each color filter has its own spectral transmittance which, together with the spectral properties of the storage elements, results in a spectral response specific for the output of the corresponding horizontal shift register. By forming linear combinations of the three outputs (corresponding to green, yellow and cyan) other spectral responses (e.g. red, green and blue) can be obtained. In the same way, compatibility to a color standard (e.g. PAL TV or SECAM TV) is achieved.

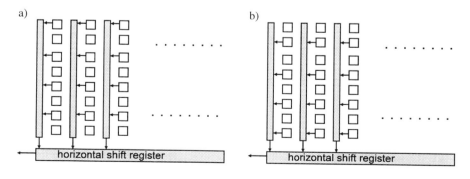

Figure 4.2 Interlacing
a) The odd field is loaded in the vertical shift registers
b) The even field is loaded in the vertical shift registers

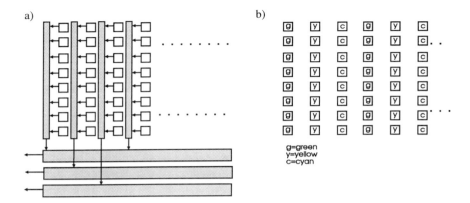

Figure 4.3 Color CCDs
a) Layout for a color CCD
b) Color stripe filters on the chip

CCDs are prone to many types of disturbing factors[1]:
- *Dynamic range.* The maximum number of electrons that can be held by a storage element is limited to (typically) 4×10^5. Therefore, above a certain illumination level the storage elements are saturated.
- *Blooming.* When part of the image plane is illuminated above the saturation level, the area of saturation increases due to diffusion of the excess-charge.
- *Smear.* When the charges are shifted vertically the integration of radiant flux continues. Therefore a bright spot makes a vertical stripe on the image. This effect is called smear.
- *Dark current noise.* Each storage element may have a random number of electrons due to thermal fluctuations (recombinations and generations of electron-hole pairs). Typically the standard deviation of these fluctuations is about 50 electrons.
- *Fixed-pattern noise.* Impurities in the crystals of the CCD may produce electron-hole pairs. The standard deviation of the number of produced electrons varies from storage element to storage element. But, typically, this deviation is about 700.
- *Reset noise.* When resetting the charge in a capacitor there will be a fluctuating number of electrons that remains due to thermal activities. The standard deviation of this number is inversely with the square root of the capacitance. Resetting occurs at each clock cycle at the input of the output-circuitry. Typically the standard deviation is about 100 electrons.
- *Thermal noise in the output amplifier.* Even an optimised amplifier design cannot prevent a thermal noise contribution. This noise level lies at around 100 electrons.
- *Quantum noise.* The number of electrons in a charge packet obeys the Poisson

---

[1] The figures given below are based on the Philips NXA1021. This is a color CCD with the following features [Philips 1987]: element size: $10 \mu m \times 15.6 \mu m$, image area: $6.0 mm \times 4.5 mm$, number of elements 604x576, interlaced, color stripe filter.

distribution. This implies that the standard deviation due to the discrete nature of the charge is proportional to the square root of the mean. Consequently, without illumination this standard deviation is almost zero, but at saturation level, the standard deviation becomes about 700 electrons.

4.2 SPATIAL SAMPLING AND RECONSTRUCTION

In a large number of image acquisition systems spatial sampling is defined on an orthogonal, equidistant grid, i.e. $(n\Delta x, m\Delta y)$ with $n = 1, \cdots, N$ and $m = 1, \cdots, M$. The sampling periods in the x- and y-direction are denoted by Δx and Δy, respectively. For the sake of brevity we will assume that these periods are equal, so that we can define: $\Delta = \Delta x = \Delta y$. Each point $(n\Delta, m\Delta)$ of the grid is called a *pixel* (picture element).

Physically, there are two restrictions to the sampling process. First of all, each sample needs a finite amount of radiant energy. Secondly, the number of samples per image is limited. The first restriction implies that integration with respect to time and area is needed to convert the irradiance $f(x,y)$ of an image to radiant energy per pixel. The second restriction implies that the useful area of the image plane is bounded.

4.2.1 Impulse modulation

If, as yet, these restrictions are discarded, the sampled image $f_{n,m}$ can be defined as a 2-dimensional sequence of samples taken from the irradiance:

$$f_{n,m} = f(n\Delta, m\Delta) \quad \begin{cases} n = \cdots, -1, 0, 1, 2, \cdots \\ m = \cdots, -1, 0, 1, 2, \cdots \end{cases} \tag{4.1}$$

In order to analyse the sampling process mathematically, it is advantageous to describe this process as an *impulse modulation* process. In that case, the sampled version of $f(x,y)$ is symbolically written $f^s(x,y)$, and defined as[2]:

$$\begin{aligned} f^s(x,y) &= f(x,y)\Delta^2 \sum_{n=-\infty}^{\infty} \sum_{m=-\infty}^{\infty} \delta(x - n\Delta, y - m\Delta) \\ &= \Delta^2 \sum_n \sum_m f_{n,m} \delta(x - n\Delta, y - m\Delta) \end{aligned} \tag{4.2}$$

Here, the sampling process is regarded as a multiplication of the image with a 2-dimensional *Dirac comb-function*. Such a comb-function consists of an infinite number of 2-dimensional Dirac functions arranged at the grid points of the orthogonal grid (figure 4.4a). Equation 4.2 shows that the samples are found as the weights of the Dirac functions.

Expansion of the comb-function into a 2-dimensional Fourier series reveals that the Fourier transform of the comb-function is again a comb-function (figure 4.4b), i.e.:

---

[2] The factor Δ^2 is added to conserve the correspondence between integrals and summation.

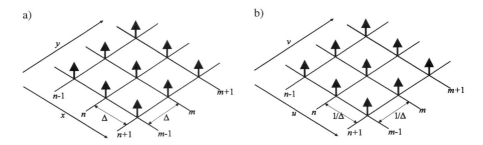

Figure 4.4 The Fourier transform of a comb-function is a comb-function
a) Graphical representation of a comb-function b) Its Fourier transform

$$F\left\{\Delta^2 \sum_n \sum_m \delta(x - n\Delta, y - m\Delta)\right\} = \sum_n \sum_m \delta(u - n/\Delta, v - m/\Delta) \qquad (4.3)$$

Suppose that the Fourier transforms of $f(x,y)$ and $f^s(x,y)$ are denoted by $F(u,v)$ and $F^s(u,v)$, respectively. Then, using the multiplication/convolution theorem stated in table 2.4, $F^s(u,v)$ is found as:

$$\begin{aligned} F^s(u,v) &= F(u,v) * \sum_n \sum_m \delta(u - n/\Delta, v - m/\Delta) \\ &= \iint_{\xi\,\eta} F(u - \xi, v - \eta) \sum_n \sum_m \delta(\xi - n/\Delta, \eta - m/\Delta) d\xi d\eta \\ &= \sum_n \sum_m F(u - n/\Delta, v - m/\Delta) \end{aligned} \qquad (4.4)$$

The Fourier transform of the sampled image consists of the Fourier transform of the original ($n=0$, $m=0$) and an infinite number of aliases ($n \neq 0$ and/or $m \neq 0$) shifted along the frequency-axis by n/Δ and m/Δ. This phenomenon is called *aliasing*. Figures 4.5a shows a map of the frequency domain in which the Fourier transform $F(u,v)$ is symbolised by a grey shaded circle. The Fourier transform $F^s(u,v)$ of the sampled image is shown in figure 4.5b.

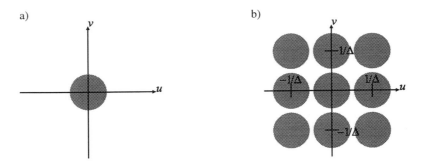

Figure 4.5 Aliasing
a) Symbolic representation of F(u,v)
b) The original spectrum F(u,v) and eight of its aliases

Figure 4.6 gives an example. Figure 4.6a is a reproduction of a (part of) an engraving. Figure 4.6b is a representation of the magnitude of its Fourier transform. Note that some hatched areas of the original are found in the frequency domain as bright spots. Figure 4.6c, d and e are graphical representations of the original with sampling rates varying from fine to coarse. Figures 4.6f, g and h give the corresponding Fourier transforms. It can be seen that with decreasing sampling rate the aliases become more and more dense.

4.2.2 Reconstruction

In many application fields of digital image processing (e.g. image coding, image enhancement, interactive image analysis) reconstruction of the image from a

Figure 4.6 Sampling
a) Original
c) Sampled
f) Amplitude spectrum of c
b) Amplitude spectrum of original
d) Sampled
g) Amplitude spectrum of d
e) Sampled
h) Amplitude spectrum of e

sampled version is needed. The aim of the reconstruction is to create a luminous intensity image that is a close replica of the original, and that can be presented to human observers. Such an image may be obtained electronically with (for instance) a CRT (cathode ray tube). It may also be obtained chemically with (photo)graphical techniques.

A reconstruction can be realised in various ways. An example from daily life is a newspaper picture. Here, the samples are mapped to black dots, the radius of each is made proportional to the irradiance of the corresponding pixel. If CRT-monitors, or other electronic display technology, are involved, the reconstruction is usually based on interpolation. Suppose that the reconstructed image is denoted by $f^r(x,y)$, then:

$$f^r(x,y) = \Delta^2 \sum_n \sum_m f_{n,m} r(x-n\Delta, y-m\Delta) \tag{4.5}$$

A 1-dimensional example of this process is given in figure 4.7. The reconstruction is the sum of a number of interpolation functions $r(x-n\Delta)$ shifted by $n\Delta$. The interpolation functions are weighted by the factors f_n. In figure 4.7 the interpolation function is a triangle function leading to linear interpolation between consecutive samples. The 2-dimensional generalisation is to apply a linear interpolation in the horizontal direction followed by linear interpolation in the vertical direction. This method is called *bilinear interpolation*. This is the case if in equation (4.5) a pyramid function is substituted for $r(x,y)$. Optically such a pyramid function is difficult to realise. In CRT monitors the interpolation function is an electron beam that scans the screen of the tube. The shape of this beam is Gaussian-like. In LCD-displays the interpolation function is a 2-dimensional rectangle function.

Equation (4.5) can be written as:

$$f^r(x,y) = \Delta^2 \sum_n \sum_m f_{n,m} \delta(x-n\Delta, y-m\Delta) * r(x,y) \tag{4.6}$$
$$= f^s(x,y) * r(x,y)$$

i.e. the convolution of the interpolation function and the impulse modulated image. Hence, interpolation can also be regarded as a convolution or filter operation. Therefore, the operation is also called the *reconstruction filter*. The PSF of this filter equals the interpolation function $r(x,y)$. The OTF of the filter is the Fourier transform $R(u,v)$.

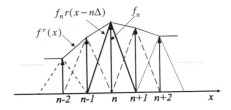

Figure 4.7 Reconstruction in the 1-dimensional case

In the frequency domain the reconstructed image can be described as:

$$F^r(u,v) = F^s(u,v)R(u,v) = R(u,v)\sum_n \sum_m F(u-n/\Delta, v-m/\Delta)$$

which can also be written as:

$$F^r(u,v) = F(u,v) + [R(u,v)-1]F(u,v) + R(u,v)\sum_{n,m\neq 0,0} F(u-n/\Delta, v-m/\Delta) \qquad (4.7)$$

The first term on the right hand corresponds to the original image. Therefore, if the objective is to restore the original image, the two other terms give rise to deviations, called *reconstruction errors*. These errors are suppressed if the following two constraints are fulfilled:

- The reconstruction filter should pass the original spectrum:

$$R(u,v) = 1 \quad \text{if} \quad F(u,v) \neq 0.$$

- The reconstruction filter should not pass the aliases:

$$R(u,v) = 0 \quad \text{if} \quad F(u-n/\Delta, v-m/\Delta) \neq 0; \quad (n,m) \neq (0,0)$$

These constraints can be fully fulfilled if the image complies with the so-called *Nyquist criterion*:

$$F(u,v) = 0 \quad \text{if} \quad \begin{cases} |u| > u_c & u_c = 1/2\Delta \\ |v| > v_c & v_c = 1/2\Delta \end{cases} \qquad (4.8)$$

This is the case in the example shown in figure 4.5b. Here, the reconstruction filter can be chosen so as to suppress both error terms in equation (4.7): the original spectrum can be fully recovered.

However, if the sampling rate is chosen too coarse, the aliases show overlap (see figure 4.8a). In this circumstance the original spectrum cannot be recovered. The constraints given above are conflicting. Choosing a reconstruction filter which suppresses all aliases will give rise to a large *resolution error* (the second term in equation 4.7). Increasing the cut-off frequencies of the filter will decrease this error, but at the same time the influence of the third term (the *interference* or *aliasing error*), will increase. One has to choose a trade-off between resolution error and aliasing error. A stylistic graphical representation of this is shown in figure 4.8.

Figures 4.9a, b and c show the reconstructed images of figure 4.9c, d and f, respectively. All images are reconstructed with an (almost) ideal low-pass filter with cut-off frequency equal to the Nyquist frequency (i.e. half the sampling frequency). In figure 4.9a the reconstruction is without severe loss of information. In figure 4.9b the hatched areas are reconstructed erroneous. Note the change in frequency due to the aliasing error. In figure 4.9c this error becomes even more severe.

Figure 4.10 shows the effect of choosing a reconstruction filter in which the cut-off frequency is not properly matched to the sampling frequency. The figure shows two reconstruction from figure 4.6d. In figure 4.10a the cut-off frequency is half the Nyquist frequency. Here, the resolution error is dominating. In figure 4.10b the cut-off frequency is twice the Nyquist frequency. This time, the aliasing error is dominant.

58 IMAGE BASED MEASUREMENT SYSTEMS

4.2.3 Area sampling and presampling filtering

In impulse modulation, the physical fact that each sample requires a finite amount of energy is ignored. In real image acquisition systems integration with respect to time and area is required. As an example we will consider this phenomenon in the case of a CCD. Suppose that the area of each storage element of the CCD can be described with a rectangle having a size: $X \times Y \, [m^2]$. If the irradiance at the target of the CCD is denoted $f(x,y) \, [W/m^2]$, the radiant flux incident on a storage element with center position $x = n\Delta, y = m\Delta$ is:

$$f_{n,m}^a = \int_{\xi=n\Delta-X/2}^{n\Delta+X/2} \int_{\eta=m\Delta-Y/2}^{m\Delta-Y/2} f(\xi,\eta) d\xi d\eta \quad [W]$$

This can be written more general by introduction of a so-called *aperture function* $a(x,y)$. For CCDs this is a rectangle function with size $X \times Y \, [m^2]$. For other image acquisition systems this function has other shapes (for instance, in a vidicon camera tube, a photoconductive target is scanned with an electron beam having a Gaussian-like density). With the introduction of such an aperture function the radian flux can be written as:

$$f_{n,m}^a = \int_{\xi=-\infty}^{\infty} \int_{\eta=-\infty}^{\infty} f(\xi,\eta) a(n\Delta-\xi, m\Delta-\eta) d\xi d\eta = f(x,y) * a(x,y)\big|_{x=n\Delta, y=m\Delta} \quad (4.9)$$

Regarded as an impulse modulation, the area sampled image becomes:

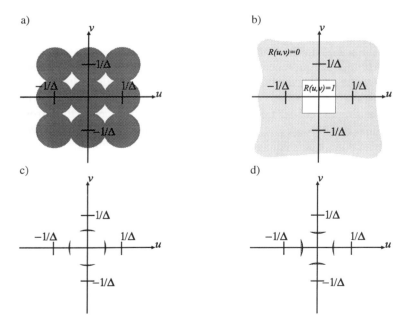

Figure 4.8 Resolution error versus aliasing error
a) Overlap of aliases due to undersampling b) Reconstruction filter
c) Resolution error   d) Aliasing error

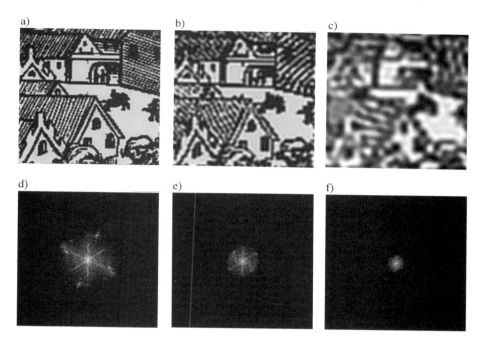

Figure 4.9 Reconstruction
a) Reconstruction of 4.6a from 4.6c
b) Reconstruction from 4.6d
c) Reconstruction from 4.6e
d) Amplitude spectrum of a
e) Amplitude spectrum of b
f) Amplitude spectrum of c

$$f^a(x,y) = \Delta^2 \sum_n \sum_m f^a_{n,m}\delta(x-n\Delta, y-m\Delta)$$
$$= f(x,y)*a(x,y)\Delta^2 \sum_n \sum_m \delta(x-n\Delta, y-m\Delta) \quad (4.10)$$

In order to arrive at radiant energy (or number of photons) the flux has to be integrated with respect to time. If the scene does not change too rapidly, the radiant energy is proportional to the exposure time and the radiant flux.

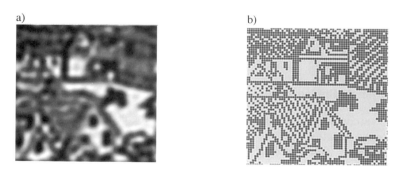

Figure 4.10 Reconstruction error versus resolution error
a) Reconstruction of 4.6a from 4.6d. The resolution error is dominant
b) Reconstruction of 4.6a from 4.6d. The aliasing error is dominant

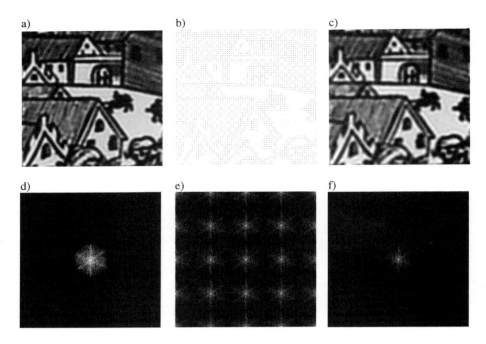

Figure 4.11 Application of presampling filterering
a) Figure 4.6a low-pass filtered
b) Sampled version of a
c) Reconstruction of 4.6a from c
d-f) Amplitude spectra of a-c

Area sampling can be regarded as a convolution operation of the irradiance and the aperture function, followed by impulse modulation. In general, a convolution that takes place before sampling is called *presampling filtering*. If it is properly chosen, a presampling filter is advantageous. It may help to decrease the aliasing error. This is illustrated in figure 4.11. Figure 4.11a is the result $f(x,y)*a(x,y)$ of a presampling filter. The cut-off frequency of this filter matches the Nyquist frequency. Figure 4.11d is the Fourier transform. Figure 4.11b represents a sampled version of the figure 4.11a. The sampling period is equal to the one in figure 4.6d. The spectrum, shown in figure 4.11e, does not show any overlap of aliases. This is due to the presampling filter. Therefore, the reconstruction (figure 4.11c) does not contain an aliasing error.

4.2.4 2-dimensional sampled stochastic processes

In this section we consider a 2-dimensional, stationary stochastic process $\underline{f}(x,y)$ with autocorrelation function $R_{ff}(x,y)$ and power spectrum $S_{ff}(u,v)$. Suppose that this process is sampled with sampling period Δ in both the horizontal and vertical direction. For the sake of simplicity we will assume impulse modulation. This is without loss of generality because if area sampling is involved we can take this into account by first convolving $\underline{f}(x,y)$ with the aperture function. The problem addressed in this section is how to obtain the statistical parameters of the sampled process.

Impulse modulation of $f(x,y)$ yields a sequence of samples $\underline{f}_{nm} = \underline{f}(n\Delta, m\Delta)$. The correlation between two samples at positions $(n\Delta, m\Delta)$ and $(i\Delta, j\Delta)$ is:

$$E\{\underline{f}_{n,m} \underline{f}_{n+i,m+j}\} = R_{ff}(i\Delta, j\Delta) \tag{4.11}$$

If for the sake of convenience we assume that $\underline{f}(x,y)$ has zero expectation, then the variance of the samples is:

$$\text{Var}\{\underline{f}_{n,m}\} = R_{ff}(0,0) \tag{4.12}$$

Equation (4.11) gives the correlations related to the sequence of numbers $\underline{f}_{n,m}$. Since stationarity is assumed, these correlations $R_{ff}(i\Delta, j\Delta)$ form again a sequence (indexed by i and j). In mathematical analysis it is convenient[3] to attach an impulse modulated function to these correlations. This function can be defined similar to the impulse modulated image in equation (4.2):

$$R_{ff}^s(x,y) = \Delta^2 \sum_i \sum_j R_{ff}(i\Delta, j\Delta)\delta(x - i\Delta, y - j\Delta) \tag{4.13}$$

The *power spectrum* of the sequence $\underline{f}_{n,m}$ is defined as the Fourier transform of $R_{ff}^s(x,y)$:

$$S_{ff}^s(u,v) = F\{R_{ff}^s(x,y)\} = \sum_i \sum_j S_{ff}(u - i/\Delta, v - j/\Delta) \tag{4.14}$$

The last equation follows from the same argument as in (4.4).

Since the autocorrelation function $R_{ff}^s(x,y)$ consists of a weighted comb function, the power spectrum can be expanded in a Fourier series:

$$S_{ff}^s(u,v) = \Delta^2 \sum_i \sum_j R_{ff}^s(i\Delta, j\Delta)\exp(-2\pi j(iu\Delta + jv\Delta)) \tag{4.15}$$

As such, this power spectrum is periodic along both the u- and the v-frequency-axis. The periodicity is $1/\Delta$. Inverse Fourier transformation of this periodic function yields:

$$R_{ff}^s(i\Delta, j\Delta) = \int_{u=-u_c}^{u_c} \int_{v=-v_c}^{v_c} S_{ff}^s(u,v)\exp(2\pi j(iu\Delta + jv\Delta))dudv \quad \text{where} \quad u_c = v_c = 1/2\Delta \tag{4.16}$$

$$\text{Var}\{\underline{f}_{n,m}\} = R_{ff}^s(0,0) = \int_{u=-u_c}^{u_c} \int_{v=-v_c}^{v_c} S_{ff}^s(u,v)dudv \tag{4.17}$$

---

[3] This convenience is outstanding if the propagation of noise (and other stochastic processes) through digital filters is to be studied. Chapter 5 deals with 2-dimensional digital filters.

As an example, consider a sampling period chosen so large that: $R_{ff}(i\Delta, j\Delta) = 0$ for all (i,j) except $(i,j)=(0,0)$. In that case, according to (4.15) and (4.17):

$$S_{ff}^s(u,v) = \Delta^2 R_{ff}^s(0,0) = \Delta^2 \text{Var}\{\underline{f}(x,y)\} \tag{4.18}$$

That is, undersampling of a stochastic process tends to yield a flat power spectrum. A sequence of numbers for which equation (4.18) holds, is called a *white noise sequence*. In contrast with its continuous counterpart, a sequence of white noise can be realised physically.

If Δ is chosen small enough, then $S_{ff}(u,v) = 0$ for $|u| > u_c$ and $|v| > v_c$ (Nyquist criterion). Consequently $S_{ff}^s(u,v) = S_{ff}(u,v)$ for $|u| < u_c$ and $|v| < v_c$. In this case, the aliasing error is reduced to zero.

4.2.5 Finiteness of the image plane

Another physical restriction in the acquisition process is that images can only be scanned at a finite part of the image plane. This implies that the sequence of samples $f_{n,m}$ is only defined for indices $n = 1, 2, \cdots, N$ and $m = 1, 2, \cdots, M$. Mathematically, this can be taken into account with the introduction of a *window function*. This is a function $w(x,y)$ that is unity if (x,y) is inside the area scanned, and zero elsewhere. Often the scanned area is a rectangle, so that $w(x,y)$ becomes a rectangle function, the size of which is $N\Delta \times M\Delta$ [m^2]:

$$w(x,y) = \text{rect}(x/N\Delta, y/M\Delta) \tag{4.19}$$

The finiteness of the image plane is modeled by multiplying the irradiance at the image plane with the window function. All samples not encompassed by the window function are turned to zero, thereby kept from further considerations:

$$f^w(x,y) = f(x,y)w(x,y) \quad \text{with Fourier transform:} \quad F^w(u,v) = F(u,v) * W(u,v) \tag{4.20}$$

The consequence of the finiteness of the image plane is *smearing* (also called *leakage*) of Fourier components in the frequency domain. As an example, suppose that the irradiance consists of a discrete number of harmonic functions:

$$f(x,y) = \sum G_n \exp 2\pi j(xu_n + yv_n) \qquad F(u,v) = \sum G_n \delta(u_n - u, v_n - v)$$

The area being scanned is assumed to be a rectangle, so that:

$$f^w(x,y) = f(x,y)\text{rect}(x/N\Delta, y/M\Delta)$$

$$F^w(u,v) = F(u,v) * NM\Delta^2 \text{sinc}(uN\Delta, vM\Delta) = NM\Delta^2 \sum G_n \text{sinc}((u_n - u)N\Delta, (v_n - v)M\Delta)$$

It can be seen that the original Dirac functions in the frequency domain are replaced with sinc functions, the width of which is inversely proportional with the size of the area scanned. In the limiting case, the area becomes infinite, and the sinc functions become Dirac functions. But in all practical situations, a single frequency component is spread out to a function with finite size. Due to this smearing it becomes difficult to discriminate two components with almost the same frequencies: the resolving power in the frequency domain is limited.

4.2.6 Discrete Fourier transform

The smearing in the frequency domain enables the possibility to sample the Fourier transform of the windowed image without loss of information. That is, if the Fourier transform $F^w(u,v)$ is sampled at frequencies:

$$u_i = i/N\Delta \quad v_j = j/M\Delta \quad i = \cdots, -1, 0, 1, 2, \cdots \quad j = \cdots, -1, 0, 1, 2, \cdots \quad (4.21)$$

the original spectrum $F^w(u,v)$ can be reconstructed from this sampled version. This statement is the dual form of the sampling theorem expressed in equation (4.8).

The Fourier transform $F^s(u,v)$ of a sampled image $f^s(u,v)$ can be expanded into a series of harmonic functions:

$$F^s(u,v) = F\{f^s(x,y)\} = \sum_{n=-\infty}^{\infty} \sum_{m=-\infty}^{\infty} f_{n,m} \exp(-2\pi j(un\Delta + vm\Delta))\Delta^2 \quad (4.22)$$

Therefore, the Fourier transform $F^{ws}(u,v)$ of a windowed and sampled image $f^{ws}(x,y)$ (i.e. the finite sequence $f_{n,m}$ with $n = 0, 1, \cdots, N-1$ and $m = 0, 1, \cdots, M-1$)[4] equals:

$$F^{ws}(u,v) = \sum_{n=0}^{N-1} \sum_{m=0}^{M-1} f_{n,m} \exp(-2\pi j(un\Delta + vm\Delta))\Delta^2 \quad (4.23)$$

Note that $F^{ws}(u,v)$ is periodic in u and v with period $1/\Delta$.

A sampled version of $F^{ws}(u,v)$ can be obtained by combination of (4.21) and (4.23):

$$F^{ws}(u_i, v_j) = \sum_{n=0}^{N-1} \sum_{m=0}^{M-1} f_{n,m} \exp(-2\pi j(u_i n\Delta + v_j m\Delta))\Delta^2 \quad (4.24)$$

However, since $F^{ws}(u,v)$ is periodic with period $1/\Delta$, it suffices to evaluate $F^{ws}(u_i, v_j)$ at a finite number of samples, for instance: $u_i = i/N\Delta$, $v_j = j/M\Delta$, $i = 0, 1, \cdots, N-1$, $j = 0, 1, \cdots, M-1$. The resulting finite sequence is called the *discrete Fourier transform* (DFT).

Stated more explicitly, the discrete Fourier transform of the finite sequence $f_{n,m}$ is defined as $F_{i,j} \equiv F^{ws}(i/N\Delta, j/M\Delta)$:

$$F_{i,j} = \Delta^2 \sum_{n=0}^{N-1} \sum_{m=0}^{M-1} f_{n,m} \exp\left[-2\pi j\left(\frac{ni}{N} + \frac{mj}{M}\right)\right] \quad (4.25a)$$

The reverse path is known as the *inverse discrete Fourier transform* (IDFT):

$$f_{n,m} = \frac{1}{NM\Delta^2} \sum_{i=0}^{N-1} \sum_{j=0}^{M-1} F_{i,j} \exp\left[2\pi j\left(\frac{ni}{N} + \frac{mj}{M}\right)\right] \quad (4.25b)$$

---

[4] In fact, this method of enumeration implies a translation of the origin to one of the corners of the windowing function.

In conclusion, the DFT is a sampled version of the Fourier transform of a finite sequence. The DFT is reversible (i.e. an original sequence can be reconstructed from its DFT). The number of samples in the DFT chosen is as small as possible, yet guaranteeing reversibility.

The merit of the DFT is that it can be accomplished numerically. Therefore, the DFT paves the way to digital implementations of image operations in the frequency domain. However, in the design of such implementations, one should bear in mind that the DFT refers to images defined on a finite image plane, i.e. to windowed images. The discontinuities at the boundary of the window function leads to extra frequency components (smearing). Moreover, since the DFT is a *sampled* version, it refers to a *periodically repeated* version of this windowed image. In fact, the DFT applies to the following image:

$$\sum_i \sum_j f^{ws}(x+iN\Delta, y+jM\Delta) \qquad (4.26)$$

This follows from the duality of the sampling theorem stated in equation (4.4). Equation (4.26) is fully in line with equation (4.25b). This last expression implies that $f_{n+kN,m+lM} = f_{n,m}$ for all integer values k and l.

The popularity of the DFT is due to the existence of an efficient algorithm to implement equation (4.25). A blunt, direct implementation of (4.25) would require $N^2 M^2$ operations (complex additions and multiplication), which is on the order of 10^{11} operations for an image with moderate size. Exploiting the separability of the harmonic functions reduces this number to $NM^2 + N^2 M$, which is still on the order of 10^9 operations. A further reduction is obtained by using the so-called FFT (*fast Fourier transform*). This is an algorithm that exploits the symmetries existing in harmonic functions. The number of required operation reduces to $(^2\log N + {}^2\log M)NM$ (provided that N and M are powers of 2). This number is on the order of 10^7.

An example of a realisation of the DFT given in pseudo-C is given in listing 4.1.

4.3 AMPLITUDE DISCRETISATION

Usually, the last step in the image acquisition process is the conversion of electrical quantities to integer numbers. This task is accomplished by a device called *ADC* (analog-digital converter). A functional description of such a device is as follows [Widrow 1961]. The input consists of the (analog) samples $f_{n,m}$ of an image (electrically encoded in a time-discrete electrical signal; see section 4.1). Each sample is compared with a number of decision levels d_k with $k = 1, 2, \cdots, K-1$. The result of the comparison is the index k for which the following relation holds:

$$d_k < f_{n,m} \leq d_{k+1}$$

The output of the device consists of a digital (integer) representation of that particular index, i.e. $k_{n,m}$. Note that if $f_{n,m}$ is such that $f_{n,m} < d_0$ or $f_{n,m} \geq d_{K-1}$, the output is assigned the index 0 or $K-1$, respectively. Sometimes, the output of the ADC is called the *grey level* $k_{n,m}$ of *pixel* (n,m). A grey level is without a physical unit. However, sometimes a unit [*DN*] (digital number) is used.

Most often, the decision levels are chosen *equidistant*, that is: $d_k = (k-0.5)\Delta d$. The difference between two consecutive levels is Δd. In that case, quantisation of the irradiance can be regarded as a numerical round off process. In case the quantity Δd is known, the relation between grey levels and irradiances (e.g. $1[DN] = \Delta d[W/m^2]$) can be used to reconstruct the original irradiances from the grey levels. As an example, figure 4.12b shows a reconstruction of a quantised image. The original image is shown in figure 4.12a. The number of quantisation levels equals $K = 8$. Therefore, in figure 4.12b, three bits are required to represent a single sample. The graph in figure 4.12d is an estimate of probability density of the irradiances of figure 4.12a. Likewise, the graph in figure 4.12e is an estimate of the probability of grey levels in figure 4.12b. It can be seen that Δd is chosen such that only six grey levels are occupied.

Similar to numerical round off processes, quantisation introduces a *quantisation error*. This error is the difference between the original irradiances and the reconstructed. Figure 4.12c shows the quantisation errors of figure 4.12b. The probability density of the error (given in figure 4.12f) tends to be *uniform* with

Listing 4.1 2-dimensional DFT

```
void dft(f,N,M)            /* replaces the input image with its DFT            */
struct cmplx **f;          /* Pointer to an image, declared as a 2-dimensional
                              array of complex numbers. Initially, the input image is
                              stored in the real part, the imaginary part is zero.   */
int N,M;                   /* number of rows and columns, respectively         */
{
        int i;
        struct cmplx **tmp;                 /* temporary array used to store the
                                               transposed of **f                    */
        initialise(tmp,N,M);                /* allocate memory space for **tmp     */
        for (i=0;i<N;i++) fft(f[i],M);      /* fft on the rows of **f              */
        transpose(f,tmp,N,M);               /* transposition: tmp[i][j]:=f[j][i]   */
        for (i=0;i<M;i++) fft(tmp[i],N);    /* fft on the rows of **tmp */
        free(tmp);                          /* free memory allocated to **tmp      */
}


static void fft(f, n)  /* FFT on a 1-dimensional array *f with n elements       */
struct cmplx *f;
int n;
{
        struct cmplx t, u, w, dummy;
        float pi = -3.1415926;              /* Changing this constant to 3.1415926
                                               implements the inverse FFT          */
        int nv, i, j, k, l, le, le1, ip, ln;
        ln=2log(n);
        nv = n/2;
        for (i = 0, j = 0; i < n - 1; i++, j += k)
        {
                if (i < j)
                {
                        t = f[i];
                        f[i]=f[j];
                        f[j]=t;
                }
                for (k = nv; k <= j; j -= k, k /=2);
        }
        for (l = 1; l <= ln; l++)
        {
                for (le = 1, i = 1; i <= l; le *= 2, i++);
                le1 = le / 2;
                u=complex(1.0,0.0);
                w=complex(cos(pi/le1),sin(pi/le1));
                for (j = 0; j < le1; j++)
                {
                        for (i = j; i < n; i += le)
                        {
                                ip = i + le1;
                                t = f[ip]*u;
                                f[ip] = f[i] - t;
                                f[i] = f[i] + t;
                        }
                        u = u*w;
                }
        }
}
```

66 IMAGE BASED MEASUREMENT SYSTEMS

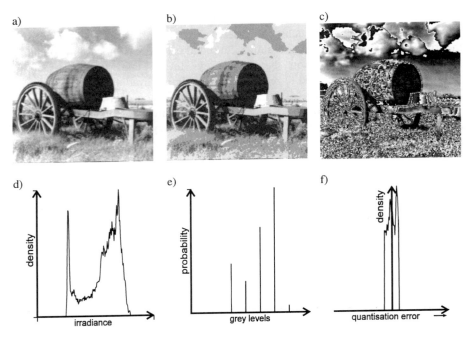

Figure 4.12 Amplitude discretisation
a) Original imageb) Quantised to eight levelsc) Quantisation error
d) Probability density of ae) Probabilities of grey levelsf) Probability density of c

width Δd. Therefore, the expectation of the quantisation error is zero, and the standard deviation is $\Delta d\sqrt{3}/6$.

In a well designed acquisition system the quantisation error is on the same order of, or less than, the noise due to other random sources. Suppose that the standard deviation of the other noise sources is by denoted σ_n. The quantisation error is negligible if:

$$\frac{\Delta d\sqrt{3}}{6} << \sigma_n \qquad (4.26)$$

On the other hand, if the full range of the irradiances in the image equals f_{range}, in order to prevent underflow and overflow of the ADC the number of decision levels must be at least:

$$K\Delta d = f_{range} \qquad (4.27)$$

From these two equations the required K and Δd can be derived.

As an example, consider the CCD discussed in section 4.1. Here, the dynamic range of a storage element is typically 4×10^5 electrons. The main contributions to the noise come from the fixed pattern noise (700 electrons) and the quantum noise (at most 700 electrons). Since these noise sources are independent the total standard deviation of the noise is around 1000 electrons. Hence, if the quantisation error is allowed to be on the same order as σ_n, it follows that:

$$\Delta d = \frac{6\sigma_n}{\sqrt{3}} \approx 3500 \text{ electrons}$$

The ADC must be equipped with about $K \approx 4 \times 10^5 / 3500 \approx 120$ number of decision levels. This implies that each pixel must be represented with at least 7 bits.

In practice, most image acquisition systems have 8 bits AD-converters, i.e. 256 grey levels. Seen in the light of the discussion above, this number of levels suffices. Correspondingly, the storage capability of an acquired image should involve at least 8 bits per pixel. However, in order to prevent a propagation of quantisation errors in further image processing, more bits are required. Most image processing systems designed in the early 1980s are provided with 12 bits per pixel storage capabilities. The equipment designed more recently uses memories with 16 bits per pixel. In workstations, grey levels are often represented in *long integer* (32 bits), *floating point* (32 bits) or *double* (64 bits) types.

REFERENCES

Huck, F.O., Fales, C.L., Halyo, N., Samms, R.W. and Stacy, K., *Image Gathering and Processing: Information and Fidelity*, Journal of the Optical Society of America, Vol. 2, No. 10, 1644-1666, October 1985.

Park, S.K. and Schowengerdt, R.A., *Image Sampling, Reconstruction, and the Effect of Sample-Scene Phasing*, Applied Optics, Vol. 21, No. 17, 3142-3151, September 1982.

Philips ELCOM, *Solid-State Image Sensors & Peripheral Integrated Circuits*, Philips Export B.V., Eindhoven, 1987.

Pratt, W.K., *Digital Image Processing*, Second Edition. J Wiley & Sons, New York 1991.

Widrow, B., *Statistical Analysis of Amplitude-quantized Sampled-Data Systems*, IEEE Tr. Appl. Industry, Vol. 81, January 1961.

5
IMAGE OPERATIONS

This chapter discusses operations defined on digital images. Here, the word *operation* refers to a rule which uniquely transforms one image into another. The images considered are defined on an $N \times M$ lattice (n,m) with $n = 0, 1, \cdots, N-1$ and $m = 0, 1, \cdots, M-1$. The domain of the amplitudes of the pixels will be referred to as the grey scale G. This domain may be: a finite set of integers $G = \{0, 1, \cdots, K-1\}$, the set of real, non-negative numbers, or the set of all real numbers. In Fourier transforms, it even includes the set of complex numbers. An integer image with $K=8$ is called a *byte image*. An integer image with $K=1$ is a *bitmap*.

Most operations considered in this chapter are useful in computer vision applications. Section 5.1 discusses pixel-to-pixel operations. An important sub-class of operations are linear operations (e.g. convolutions, differential operations, orthogonal transforms). These operations are considered in Section 5.2. Correlation techniques are introduced in Section 5.3. The last section concerns morphological operations.

General reading for the material discussed in this chapter are Gonzalez and Woods [1992], Haralick and Shapiro [1992], Jain [1989], Pratt [1991], and Rosenfeld and Kak [1982].

5.1 PIXEL-TO-PIXEL OPERATIONS

Pixel-to-pixel operations (also called: *point operations*) are image operations in which the amplitudes (or grey levels) of individual pixels are modified according to some rule (see figure 5.1). Most generally, the operation is denoted mathematically by:

$$g_{n,m} = T_{n,m}(f_{n,m}) \qquad (5.1)$$

The indices of the operator $T_{n,m}(\cdot)$ indicate that the rule depends on the position of the pixel. The operator is *space* or *shift variant*. An example in which a space variant operator is applied is in shading correction. Most image acquisition systems show a drop of sensitivity of the image sensors towards the edges of the image area. This so-called *shading* may be corrected by the inverse operation, which is a space variant point operation, i.e. $g_{n,m} = a_{n,m} f_{n,m} + b_{n,m}$.

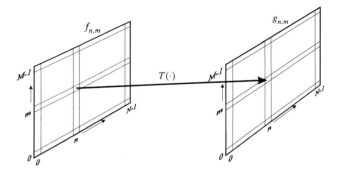

Figure 5.1 Pixel-to-pixel operations

If the rule does not depend on the position, the operation is *space invariant*. In that case we can write:

$$g_{n,m} = T(f_{n,m}) \tag{5.2}$$

This section merely discusses space invariant operations.

5.1.1 Monadic operations

Monadic operations[1] are pixel-to-pixel operations with a single operand image. This is in contrast with, for example, dyadic operations which have two operand images. Table 5.1 presents some basic operations.

Threshold operations are used in computer vision applications. A threshold operation can be applied to detect all pixels in the image that correspond to an object (or a part of an object) in the scene. However, this is only successful if the difference between the radiances of the object and its background is large enough. Usually, this condition can only be met if the illumination of the scene is designed carefully (see section 2.4). Figure 5.2c and d show the results of thresholding the images in figure 5.2a and b, respectively. In figure 5.2b, the contrasts between object and background is sufficiently high to obtain a low detection error rate. In figure 5.2a, this condition is not met.

Floor, ceiling and *clip operations* are useful to limit the dynamic range of the grey levels in an image. The *offset* and *scale* operations are similar to the brightness and contrast control of a CRT-monitor. Together with the *square, square root* and the *absolute value operations* these operations are primitive operations with which, in combination with other operations, more complicated functions can be designed.

Look-up tables

As with other operations, monadic operations can be implemented in various ways. One popular implementation is the so-called *table-look-up* operation. This

---

[1] In some image processing software library packages, monadic pixel-to-pixel operations are referred to as *unary operations*. However, in mathematical analysis, this term is reserved for orthogonal transforms.

Table 5.1 Some monadic operations

Name	Mathematical description	Look-up table	Parameters
thresholding	$g_{n,m} = \begin{cases} 1 & \text{if } f_{n,m} > t \\ 0 & \text{elsewhere} \end{cases}$		t: threshold
multi-thresholding	$g_{n,m} = l_i \quad \text{if } t_{i-1} < f_{n,m} \le t_i$		t_i: i-th threshold l_i: i-th level ($t_i < t_{i+1}$)
floor operation	$g_{n,m} = \begin{cases} f_{n,m} & \text{if } f_{n,m} > c_f \\ c_f & \text{elsewhere} \end{cases}$		c_f: floor
ceiling operation	$g_{n,m} = \begin{cases} c_c & \text{if } f_{n,m} > c_c \\ f_{n,m} & \text{elsewhere} \end{cases}$		c_c: ceiling
clip operation	$g_{n,m} = \begin{cases} c_c & \text{if } f_{n,m} > c_c \\ c_f & \text{if } f_{n,m} < c_f \\ f_{n,m} & \text{elsewhere} \end{cases}$		c_c: ceiling c_f: floor
offset (brightness) operation	$g_{n,m} = f_{n,m} + \textit{offset}$		offset
scale (contrast) operation	$g_{n,m} = \textit{scale} \times f_{n,m}$		scale
square	$g_{n,m} = f_{n,m}^2$		
square root	$g_{n,m} = \sqrt{f_{n,m}}$		
absolute value	$g_{n,m} = |f_{n,m}|$		

implementation can be applied if the grey scale consists of a finite number of integers. In that case, the operation $T(.)$ can be defined with a table. The grey level of an input pixel serves as an index in that table. The table-entry pointed to by this input grey level gives the output grey level. Listing 5.1 is a realisation of this implementation written in pseudo-C. Here, the images are considered to be byte-images with grey levels from 0 up to 255.

The implementation can also be realised in hardware. Often, the video output circuitry of a computer vision system is provided with (hardware) look-up tables

Listing 5.1 Monadic operation implemented with a look-up table

```
unsigned char table[256];       /* look-up table                                 */
unsigned char **f,**g;          /* pointer to an image f and g, declared as
                                   2-dimensional arrays of unsigned chars.       */
int N,M;                        /* number of rows and columns, respectively      */
int i,j;                        /* counters                                      */
...
        for (i=0;i<256;i++)
                table[i]=(unsigned char) sqrt((double)i);
                                /* pre-fill look-up table, in this case a square root
                                   operation                                     */
...
        for (i=0,i<N;i++)
           for (j=0;j<M;j++) g[i][j] = table[f[i][j]];      /*      table look-up */
...
```

a)  b)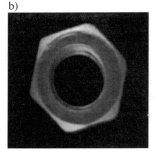

c)  d)

Figure 5.2 Thresholded images
a) Original b) Original
c) Thresholded d) Thresholded

(so-called LUTs). With these LUTs the contrasts and (pseudo) colors in the displayed images can be manipulated easily.

The advantages of using a table-look-up operation above other implementations (e.g. direct implementations) are its flexibility and its speed. It only needs a re-fill of the table to change the monadic operation. Table-look-up is a fast implementation since it only needs one indirect addressing per pixel.

Gamma correction

In consumer video equipment, it is custom to use a non-linear relationship between (il)luminance and video signal. In a wide range of luminances, the human eye can detect relative differences between luminance rather than absolute differences: the scale that is used in the human visual system is logarithmic. It is advantageous to expand (or stretch) the grey scale at low luminances, and to compress the scale at high luminances. This conversion of scale assures that quantisation errors and other additive noise sources have their influences spread out equally along the grey scale. Usually, the conversion from luminance L to grey levels g is according to the following relationship:

$$g = (constant \times L)^\gamma \qquad (5.3)$$

The exponent γ is denoted the *gamma* of the acquisition system. In most video cameras, the value of gamma is about 0.5. The non-linearity of the cameras are compensated by the grey scale transform performed by the display system. In a

CRT-monitor, the relationship between grey scale and luminance is approximately:

$$L = (constant \times g)^{\gamma} \tag{5.4}$$

The value of gamma is typically about 2, implying that equation (5.4) is the inverse of equation (5.3).

In contrast with broadcast video, in computer vision, non-linearity at the conversion from (ir)radiances to grey levels is not advantageous. Many CCD cameras offer the possibility to switch off the gamma-conversion. If this is not the case, then a square operation applied to the input image will give a *gamma-correction*. In interactive image analysis, the gamma-conversion of the display system is inconvenient, because it compresses details in areas with low grey levels. This can be corrected by loading the output LUTs with a square root function.

Histogram manipulations

The *histogram* of a digital image is a table or graph showing the percentage of pixels having certain grey levels. As such, it can be regarded as an estimate of the probabilities of these grey levels. Suppose that $f_{n,m}$ is an integer image. The number of pixels in $f_{n,m}$ with grey level f is denoted n_f. Then, the histogram of $f_{n,m}$ is defined as:

$$h_f(f) = \frac{n_f}{NM} \quad \text{for} \quad g = 0, 1, \cdots, K-1 \tag{5.5}$$

Listing 5.2 gives an implementation of this procedure, assuming that the grey levels can be represented by bytes. If the grey scale of the image consists of real numbers, the histogram is obtained by defining a number of intervals (bins) on the real line, and by counting the pixels with grey levels within each bin.

The histogram gives a quick impression of the occupation of grey levels. For instance, the histogram of an acquired image shows us whether the adjustments of the camera (e.g. gain, offset, diaphragm) matches the illumination of the scene. Figure 5.3 gives the histograms of the images shown in figure 5.2. The bimodal histogram in figure 5.3b indicates that the image can be thresholded successfully.

Obviously, if a monadic operation $T(.)$ is applied to an image $f_{n,m}$ its histogram $h_f(f)$ undergoes a change. If the operation $T(.)$ is monotonically increasing, then the histogram of the resulting image can be calculated easily. For that purpose we define the *cumulative histogram*:

Listing 5.2 Histogram determination

```
void gethis(f,his,N,M)  /* get histogram of image f returned in double array his  */
unsigned char **f;      /* pointer to an image f, declared as a 2-dimensional array of
                           unsigned chars.                                         */
double his[256];        /* double array his                                        */
int N,M;                /* number of rows and columns, respectively                */
{
        int i,j;        /* counters                                                */

        for (i=0;i<256;i++) his[i]=0;            /* initiate "his" to zero         */
        for (i=0;i<N;i++)
        for (j=0;j<M;j++) his[(int) f[i][j]]++;  /* count pixels                   */
        for (i=0;i<256;i++) his[i]/=(N*M);       /* normalise histogram            */
}
```

$$H_f(f) = \sum_{i=0}^{f} h_f(i) \tag{5.6}$$

This cumulative histogram is the percentage of pixels having grey levels less than or equal to f.

Let $g_{n,m}$ be the image resulting from the monadic operation. The cumulative histogram of this image is $H_g(g)$. Since $T(.)$ is monotonically increasing, the percentage of pixels in $g_{n,m}$ having grey levels less than or equal to g equals $H_f(f)$ with $g = T(f)$ or $f = T^{-1}(g)$. Therefore:

$$H_g(g) = H_f\big(T^{-1}(g)\big) \quad \text{and} \quad H_g(T(f)) = H_f(f) \tag{5.7}$$

The histogram of $g_{n,m}$ follows from:

$$h_g(g) = \begin{cases} H_g(0) & \text{if } g = 0 \\ H_g(g) - H_g(g-1) & \text{elsewhere} \end{cases} \tag{5.8}$$

Equations 5.6 and 5.7 can be used to construct a monadic operation such that the resulting image has an histogram in a prescribed form. Suppose that it is desired that the resulting image $g_{n,m}$ has a flat histogram:

$$h_g(g) = \frac{1}{K} \quad \text{for} \quad g \in G$$

Then, the cumulative histogram of $g_{n,m}$ must be linearly increasing:

$$H_g(g) = \frac{g}{K-1}$$

Substitution of (5.7) gives:

$$H_g(T(f)) = \frac{T(f)}{K-1} = H_f(f)$$

Hence:

$$T(f) = (K-1)H_g(f) \tag{5.9}$$

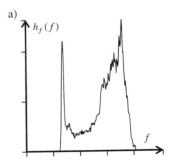

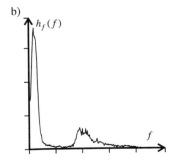

Figure 5.3 Histogram determination
a) Histogram of figure 5.2a
b) Histogram of figure 5.2b

Listing 5.3 Histogram egalisation

```
void his_egal(f,g,N,M)          /* histogram egalisation                        */
unsigned char **f,**g;          /* pointers to the input and output images f and g,
                                   declared as 2-dimensional array of unsigned chars.*/
int N,M;                        /* number of rows and columns, respectively     */
{
        int i,j;                /* counters                                     */
        double his[256];        /* double array his                             */
        int lut[256];           /* table to perform monadic operation           */

        gethis(f,his,N,M);                              /* calculate histogram          */
        for (i=1;i<256;i++) his[i] += his[i-1];         /* calculate cumulative histogram*/

        for (i=0;i<256;i++) lut[i] = (int) (255*his[i]);

        for (i=0,i<N;i++)
        for (j=0;j<M;j++) g[i][j]=lut[f[i][j]];         /* monadic operation            */
}
```

An implementation of this procedure (called *histogram egalisation*) is given in listing 5.3. Figure 5.4 gives an illustration. Note that, due to the discrete nature of the grey scale, a fully flat histogram cannot be obtained.

Another monadic operation is *normalisation*. Here, the contrast and brightness operations are applied such that one or more parameters of the histogram become normalised. An example is the normalisation with respect to the minimum and maximum grey level in the image (f_{min} and f_{max}, respectively). Suppose that it is desired to change the grey scale such that the new minimum and maximum grey level becomes 0 and 255, respectively. This is accomplished with the following contrast and brightness operation:

$$g_{n,m} = 255\left(\frac{f_{n,m} - f_{min}}{f_{max} - f_{min}}\right) \tag{5.10}$$

Often, this operation is used to convert an integer or real image into a byte image. Since most video display systems can handle only byte images, the operation is useful in display functions.

a)

b)

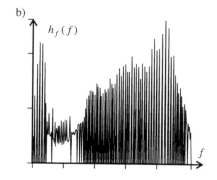

Figure 5.4 Histogram egalisation
a) Image with egalised histogram   b) Histogram of a)

5.1.2 Dyadic operations

Dyadic operations are pixel-to-pixel operations with two operand images:

$$h_{n,m} = T(f_{n,m}, g_{n,m})$$

The dyadic operations $T(.,.)$ include the arithmetic operations, like:

- addition
- multiplication
- division
- weighted addition (i.e. $\alpha f_{n,m} + \beta g_{n,m}$).

Other operations are the maximum- and minimum operations, for instance:

$$\max(f_{n,m}, g_{n,m}) = \begin{cases} f_{n,m} & \text{if } f_{n,m} > g_{n,m} \\ g_{n,m} & \text{elsewhere} \end{cases}$$

Logical operations are dyadic operations defined on bitmaps, for instance logical-and, logical-or, etc.)

Example 5.1 Color transforms

Color acquisition systems are equipped with three spectral bands: red, green and blue. A color image is captured in three band images denoted by $r_{n,m}$, $g_{n,m}$ and $b_{n,m}$, respectively. These three bands are obtained with three image sensors each one having its own spectral responsitivity (see Chapter 4). Due to the structure of the human eye and due to the additivity of radiant energy (Chapter 2), all colors can be regarded as linear combinations of three so-called *primary colors*. In 1931, the CIE (Commission International de l'Éclarage) defined three standard primaries, called **R**, **G** and **B**. These primaries have a monochromatic spectral energy distribution at wavelengths 700 nm, 546.1 nm and 435.8 nm, respectively. The color of a pixel can be represented by a linear combination of these primaries:

$$R_{n,m}\mathbf{R} + G_{n,m}\mathbf{G} + B_{n,m}\mathbf{B} \qquad (5.12)$$

The coefficients $R_{n,m}$, $G_{n,m}$ and $B_{n,m}$ are called the *tristimulus values*.

In order to establish a match between the colors seen by a camera and the corresponding colors produced by a CRT, an additive pixel-to-pixel operation suffices, i.e.

$$\begin{bmatrix} R_{n,m} \\ G_{n,m} \\ B_{n,m} \end{bmatrix} = \begin{bmatrix} m_{11} & m_{12} & m_{13} \\ m_{21} & m_{22} & m_{23} \\ m_{31} & m_{32} & m_{33} \end{bmatrix} \begin{bmatrix} r_{n,m} \\ g_{n,m} \\ b_{n,m} \end{bmatrix} \qquad (5.11)$$

where the 3×3-matrix **M** depends on the spectral energy distributions of the three phosphors of the color CRT[2].

The RGB model of the CIE has one defect: not all colors can be matched by simple linear combinations of non-negative tristimulus values. Therefore, the CIE also defined another set of primaries called **X**, **Y** and **Z** (the so-called XYZ-model)

---

[2]In contrast with CRTs, printing devices have a *subtractive* nature. A linear combination of primaries is subtracted from white to obtain a desired color. These so-called *subtractive primaries* are the complements of red, green and blue, i.e. cyan, magenta and yellow.

which does not have this defect. The RGB representation of a color can be converted into an XYZ representation by linearly transforming the tristimulus values of the RGB model.

A color may also be specified by its *luminance* and *chromaticities*. In the XYZ model, the tristimulus values $X_{n,m}$, $Y_{n,m}$ and $Z_{n,m}$ are converted to luminance by:

$$L_{n,m} = X_{n,m} + Y_{n,m} + Z_{n,m} \tag{5.13a}$$

and chromaticities:

$$\begin{aligned} x_{n,m} &= X_{n,m}/L_{n,m} \\ y_{n,m} &= Y_{n,m}/L_{n,m} \end{aligned} \tag{5.13b}$$

In psycho physics, the discussion of color perception usually involves three quantities: *hue, saturation* and *brightness*. Hue refers to the dominant wavelength of a color (as in a rainbow). Saturation refers to how far a color is from grey or white. Brightness is the perceived luminous intensity of a surface. There are several standards that are built on these three concepts: e.g. the HSV model, the HLS model and the HSI model. As an example we consider the HSV model.

In the HSV model, S refers to saturation, defined in the interval $[0,1]$:

$$S_{n,m} = \frac{\max(r_{n,m}, g_{n,m}, b_{n,m}) - \min(r_{n,m}, g_{n,m}, b_{n,m})}{\max(r_{n,m}, g_{n,m}, b_{n,m})} \tag{5.14a}$$

V (from *value*) refers to brightness. It is expressed as:

$$V_{n,m} = \max(r_{n,m}, g_{n,m}, b_{n,m}) \tag{5.14b}$$

H (hue) refers to the wavelength of the main color. Suppose that H is expressed in degrees. Then, going through the spectrum: red=0°, green=120°, blue=240°. 360° represents red again. If for some pixel (n,m) green is larger than all other tristimulus values, then H can be calculated according to:

$$H_{n,m} = 60°(2 + b_{n,m} - r_{n,m})/(g_{n,m} - \min(r_{n,m}, b_{n,m})) \tag{5.14c}$$

If red or blue is larger than all other tristimulus values, then similar expressions hold.

Figure 5.5 shows the RGB representation of the color image in figure I. The corresponding HSV representation is given in figure 5.6.

5.2 LINEAR OPERATIONS

In contrast with pixel-to-pixel operations, an output pixel from a linear operation depends on all pixel of the input image (or on a subset of these input pixels). The operation is called linear as soon as the *superposition principle* holds. To stipulate this more precisely, let **f** denote the set of all pixels in an image $f_{n,m}$:

$$\mathbf{f} = \{f_{n,m} | n = 0,\cdots,N-1 \quad m = 0,\cdots,M-1\} \tag{5.15}$$

Note that **f** can be regarded as a vector comprising *NM* elements. An operation $O(\cdot)$ is a rule that uniquely maps **f** into a new image: $\mathbf{g} = O(\mathbf{f})$. The operation is called *linear* when:

78   IMAGE BASED MEASUREMENT SYSTEMS

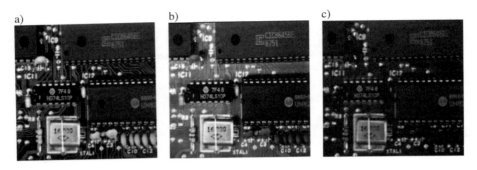

Figure 5.5 RGB representation: a) red, b) green, and c) blue

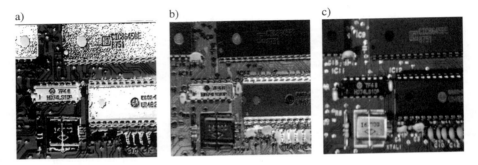

Figure 5.6 HSV representation: a) hue, b) saturation, and c) value

$$O(\alpha \mathbf{f}_1 + \beta \mathbf{f}_2) = \alpha O(\mathbf{f}_1) + \beta O(\mathbf{f}_2) \tag{5.16}$$

If f is regarded as a vector, every linear operation can be described as a matrix-vector multiplication:

$$\mathbf{g} = \mathbf{O}\mathbf{f} \tag{5.17}$$

with **O** an $NM \times NM$ matrix.

Despite the generality of (5.17), it is not used often in practice since the matrix **O** is very large ($250,000 \times 250,000$ elements if $N=M=500$). However, most linear operations involve matrices that are very sparse and/or symmetrical. Exploiting these properties reduces the complexity of the operations. Two sub-classes of linear operations, important in image analysis and computer vision, are the convolution and the unitary (orthogonal) transform. We discuss both items in subsequent sections.

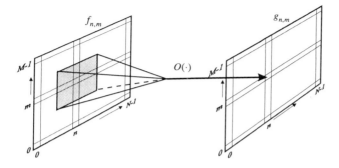

Figure 5.7 Neighborhood operations

5.2.1 Convolution

A *local neighborhood* of a pixel (n,m) is the collection of pixels in a subarea of the image enclosing the pixel (n,m). Usually, the subarea is chosen rectangular with (n,m) in the center of the area. Suppose that the size of the neighborhood is $(2K+1)\times(2L+1)$, then the local neighborhood $\mathbf{f}_{n,m}$ of the pixel (n,m) in the image $\mathbf{f}$ is:

$$\mathbf{f}_{n,m} = \begin{bmatrix} f_{n-K,m-L} & f_{n-K,m-L+1} & \cdots & f_{n-K,m+L} \\ f_{n-K+1,m-L} & & & \vdots \\ \vdots & & & \vdots \\ f_{n+K,m+L} & \cdots & \cdots & f_{n+K,m+L} \end{bmatrix} \quad (5.18)$$

In (5.18) $\mathbf{f}_{n,m}$ is arranged as a matrix. Nevertheless, $\mathbf{f}_{n,m}$ can also be regarded as a vector comprising $(2K+1)\times(2L+1)$ elements (for instance, by stacking the columns of the matrix).

A *local neighborhood operation* is an operation in which each output pixel (n,m) is assigned a grey level depending on the corresponding local neighborhood $\mathbf{f}_{n,m}$ of the input image (figure 5.7).

$$g_{n,m} = O(\mathbf{f}_{n,m}) \quad (5.19)$$

The operation is called *space invariant* (or *shift invariant*) if $O(\cdot)$ does not depend on the pixel position (n,m). If, in addition, the operation is also linear, then it is a *discrete convolution*. In that case, the operation is a weighted sum of the grey levels of the neighborhood. Suppose $\mathbf{h}$ is a $(2K+1)\times(2L+1)$ vector comprising the weight factors $h_{k,l}$:

$$\mathbf{h} = \begin{bmatrix} h_{-K,-L} & h_{-K,-L+1} & \cdots & h_{-K,L} \\ h_{-K+1,-L} & & & \vdots \\ \vdots & & & \vdots \\ h_{K,-L} & \cdots & \cdots & h_{K,L} \end{bmatrix} \quad (5.20)$$

then the linear operation can be written as the innerproduct between $\mathbf{h}$ and $\mathbf{f}_{n,m}$ ($\mathbf{h}'$ is the transposed vector $\mathbf{h}$):

$$g_{n,m} = \mathbf{h}^t \mathbf{f}_{n,m} = \sum_{k=-K}^{K}\sum_{l=-L}^{L} h_{k,l} f_{n-k,m-l} \equiv h_{n,m} * f_{n,m} \qquad (5.21)$$

The weight factors $h_{k,l}$ are the discrete analog of the PSF of a continuous convolution. Therefore, when arranged in matrix form (as in 5.20), the weight factors are called the *PSF-matrix* or the *convolution kernel*. Sometimes the term *convolution mask* is used. The operation itself is called 2-*dimensional convolution*, also called *linear image filtering*.

The convolution sum in equation (5.21) is valid when $n \geq K$, $n < N - K$, $m \geq L$ and $m < M - L$. An ambiguity occurs near the edges of the image since here the definition of the local neighborhood according to (5.18) is not valid. Three methods to solve this ambiguity are known:

- Discarding the pixels near the edge of the output image.
  Here, the image $\mathbf{f}$ is regarded as not being defined for $n < 0$, $n \geq N$, $m < 0$ and $m \geq M$. The convolution is only performed for pixels $g_{n,m}$ for which $n \geq K$, $n < N - K$, $m \geq L$ and $m < M - L$.
- Padding the input image with zeroes.
  Outside the definition area of $\mathbf{f}$, i.e. $n < 0$, $n \geq N$, $m < 0$ or $m \geq M$, the image $\mathbf{f}$ is padded with zeroes.
- Circulant convolution.
  The image is periodically repeated, thereby defining the image outside its original definition area. This method is implicitly used when the 2-dimensional convolution is implemented with the DFT (see sections 4.2.6 and 5.2.2.)

Listing 5.4 is an implementation of the 2-dimensional convolution. The number of calculations (multiplication and addition) is about $NM(2K+1)(2L+1)$. If the convolution is *separable*, i.e. $h_{n,m} = r_n c_m$, the number of calculations can be reduced to about $NM(2K+2L+2)$.

Transfer function

If in equation (5.21) the digital images $f_{n,m}$ and $g_{n,m}$ are regarded as impulse modulated functions (section 4.2.1), the convolution sum can be described in the frequency domain. In accordance with equation (4.22), let the Fourier transforms of the images and kernel be defined as:

$$F(u,v) = \sum_{n=0}^{N-1}\sum_{m=0}^{M-1} f_{n,m} \exp(-2\pi j(un\Delta + vm\Delta))\Delta^2 \qquad (5.22a)$$

$$G(u,v) = \sum_{n=0}^{N-1}\sum_{m=0}^{M-1} g_{n,m} \exp(-2\pi j(un\Delta + vm\Delta))\Delta^2 \qquad (5.22b)$$

$$H(u,v) = \sum_{k=-K}^{K}\sum_{l=-L}^{L} h_{k,l} \exp(-2\pi j(uk\Delta + vl\Delta)) \qquad (5.22c)$$

then:

$$G(u,v) = H(u,v) F(u,v) \qquad (5.23)$$

Listing 5.4 2-dimensional convolution

```
void convolve(f,g,h,N,M,K,L)    /* 2-dimensional convolution   Note: pixels near
                                   the edge of the image are discarded          */
double **f,**g,**h;             /* pointers to the images f and g, respectively,
                                   and the kernel h. These are declared as 2-dimensional
                                   arrays of doubles                             */
int N,M;                        /* number of rows and columns of f and g        */
int K,L;                        /* size of neighbourhood                         */
{
        int n,m;                /* indices in image                              */
        int k,l;                /* indices in kernel                             */
        int dk,dl;              /* number of rows and columns of kernel          */
        double sum;

        dk=2*K+1;
        dl=2*L+1;

        for (n=K; n<N-K; n++)
        for (m=L; m<M-L; m++)
        {
                sum=0.0;
                for (k=0; k<dk; k++)
                for (l=0;l<dl; l++)
                {       sum += h[k][l] * f[n-k-K][m-l-L];
                }
                g[n][m]=sum;
        }
}
```

Hence, $H(u,v)$ is the transfer function of the convolution operation. It is similar to the optical transfer function in optical devices. Note that the transfer function is periodic with periods $1/\Delta$ in both the u- and v-direction.

The convolved image can be reconstructed from its Fourier transform by the inverse transformation:

$$g_{n,m} = \int_{u=-1/2\Delta}^{1/2\Delta} \int_{v=-1/2\Delta}^{1/2\Delta} G(u,v)\exp(2\pi j(un\Delta + vm\Delta))dudv \qquad (5.24)$$

Low-pass filtering

If the transfer function $H(u,v)$ is such that high-frequency components (i.e. $u \approx 1/2\Delta$ and/or $v \approx 1/2\Delta$ are suppressed, while at the same time low-frequency components are passed unchanged, the operator is called a *low-pass filter*. Examples of low-pass filters are the *averaging operator* and the *Gaussian filter*.

The *averaging operator* has a PSF-matrix in which all elements are constant. For instance, in a 3×3 neighborhood ($K=1$ and $L=1$), the averaging operator is:

$$\mathbf{h} = \frac{1}{9}\begin{bmatrix} 1 & 1 & 1 \\ 1 & 1 & 1 \\ 1 & 1 & 1 \end{bmatrix}$$

In a $(2K+1)\times(2L+1)$ neighborhood, the averaging operator is defined by the matrix:

$$\mathbf{h} = \frac{1}{(2K+1)(2L+1)}\begin{bmatrix} 1 & 1 & \cdots & 1 \\ 1 & \ddots & & 1 \\ \vdots & & & \vdots \\ 1 & 1 & \cdots & 1 \end{bmatrix} \qquad (5.25a)$$

This PSF-matrix is separable in row- and column-direction. This implies that it can be implemented efficiently with only $2+2K+2L$ addition and one multiplication per output pixel.

The transfer function of the averaging operator is given by:

$$H(u,v) = \frac{1}{(2K+1)(2L+1)}\left(1+2\sum_{k=1}^{K}\cos(2\pi uk\Delta)+2\sum_{l=1}^{L}\cos(2\pi vl\Delta)\right)$$

$$= \frac{\sin((2K+1)\pi u\Delta)}{(2K+1)\sin(\pi u\Delta)}\frac{\sin((2L+1)\pi v\Delta)}{(2L+1)\sin(\pi v\Delta)} \qquad (5.25b)$$

For low-frequency components, i.e. $|u|\ll 1/\Delta$ and $|v|\ll 1/\Delta$, the expression can be replaced with:

$$H(u,v) \approx \text{sinc}((2K+1)u\Delta)\text{sinc}((2L+1)v\Delta) \qquad (5.25c)$$

In this approximation, the averaging operator is seen as a continuous convolution with a rectangle function (see section 2.3.2) scaled with factors $(2K+1)\Delta$ and $(2L+1)\Delta$, respectively. The 3-dB cut-off frequency (i.e. the frequency for which $|H(u,0)|^2 = 0.5$) of this filter is about $u \approx 0.4/(2K+1)\Delta$.

Examples of an image filtered by the averaging operator are given in figure 5.8. The magnitudes of the transfer functions are also shown. Note that the averaging operator does not fully suppress the high-frequency components. The high-frequency ripple in the transfer function can be seen as aliasing components due to the (relative) large side-lobs of the sinc-function.

Another low-pass filter is the *Gaussian filter*. Here, the elements in the PSF-matrix are samples from a 2-dimensional Gauss function (equation 2.27):

$$h_{n,m} = \frac{1}{2\pi\sigma_x^2\sigma_y^2}\exp\left(-\left(\frac{n^2}{2\sigma_x^2}+\frac{m^2}{2\sigma_y^2}\right)\Delta^2\right) \qquad (5.26a)$$

The Gauss function is spatially unlimited. Therefore, it must be truncated in order to obtain a finite size of the neighborhood. Usually, if the standard deviations of the Gaussian are chosen according to: $K \geq 3\sigma_x/\Delta$ and $L \geq 3\sigma_y/\Delta$, the truncation error is negligible. In that case, the transfer function of the (discrete) Gaussian is (compare with equation 2.27):

$$H(u,v) \approx \sum_i\sum_j \exp\left(-\frac{1}{2}\left(\sigma_x^2\left(u-\frac{i}{\Delta}\right)^2+\sigma_y^2\left(v-\frac{j}{\Delta}\right)^2\right)\right) \qquad (5.26b)$$

The transfer function consists of a part that corresponds to the continuous Gaussian filter and an infinite number of aliases. These aliases are negligible within the frequency domain of interest, i.e. $|u| \leq 1/2\Delta$ and $|v| \leq 1/2\Delta$ if $\sigma_x \geq 0.7\Delta$ and $\sigma_y \geq 0.7\Delta$:

$$H(u,v) \approx \exp\left(-\frac{1}{2}\left(\sigma_x^2 u^2+\sigma_y^2 v^2\right)\right) \quad \text{if} \quad \begin{cases}|u| \leq 1/2\Delta \\ |v| \leq 1/2\Delta\end{cases} \quad \begin{array}{l}\sigma_x \geq 0.7\Delta \\ \sigma_y \geq 0.7\Delta\end{array} \qquad (5.26c)$$

Examples of Gaussian filtering are given in figure 5.9. Compared with the averaging operator the transfer function of the Gaussian is much more smooth. High-frequency components are almost fully suppressed. Furthermore, if $\sigma_x = \sigma_y$,

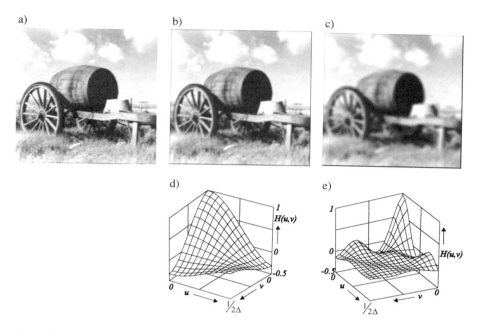

Figure 5.8 Averaging operation
a) Original
b) 3×3 operation
c) 7×7 operation
d) H(u,v) of 3×3 operation
e) H(u,v) of 7×7 operation

the Gaussian is rotationally invariant (isotropic). Altogether, this makes the Gaussian filter preferable above the averaging operator, especially in image analysis applications.

Gaussian convolution can be implemented in various ways. The 2-dimensional Gaussian function is separable. This implies that the Gaussian filter can be implemented as the cascade of a row- and column-operation. For instance, the Gaussian PSF with $\sigma_x^2 = \sigma_y^2 = 0.7\Delta^2$ is approximated by:

$$\mathbf{h} = \frac{1}{16}\begin{bmatrix} 1 & 2 & 1 \\ 2 & 4 & 2 \\ 1 & 2 & 1 \end{bmatrix} \quad (5.27)$$

This can be seen as a cascade of two filters with PSFs $\mathbf{h}_r = [1 \ 2 \ 1]$ and $\mathbf{h}_c = [1 \ 2 \ 1]'$, i.e. $\mathbf{h} = \mathbf{h}_c \mathbf{h}_r$. A Gaussian filter with $\sigma_x^2 = \sigma_y^2 = P0.7\Delta^2$ (with P some integer number) can be approximated as a cascade of P convolutions with PSF-matrices equal to the $\mathbf{h}$ given above. The approximation becomes closer when P increases. This follows readily from the *central limit theorem* stating that the cascade of a large number of identical convolutions is equivalent to one convolution with a PSF that approximates a Gaussian function.

Some applications require Gaussian filters with standard deviations that are very large. Consequently, the neighbourhood sizes are so large, that from a computational point of view it becomes advantageous to use frequency domain processing, i.e. application of the fast Fourier transform (see section 4.2.6 and 5.2.2).

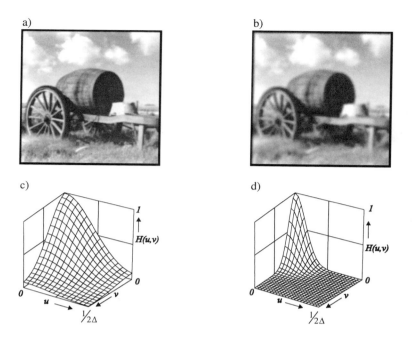

Figure 5.9 Gaussian filtering (the original is shown in figure 5.8a)
a) $\sigma_x = 0.8\Delta$, $\sigma_y = 0.8\Delta$ b) $\sigma_x = 2\Delta$, $\sigma_y = 2\Delta$
c) H(u,v) corresponding to a) d) H(u,v) corresponding to b)

High-pass filtering

If the transfer function of a convolution is such that low-frequency components are suppressed while high-frequency components are unattenuated, the filter is *high-pass*. Since low-pass filtering is complementary to high-pass filtering the latter can be achieved by subtracting of a low-pass filtered image from its original:

$$H_{high-pass}(u,v) = 1 - H_{low-pass}(u,v) \tag{5.28a}$$

In terms of PSF-matrices, equation (5.28a) is equivalent to:

$$\mathbf{h}_{high-pass} = \mathbf{1} - \mathbf{h}_{low-pass} \quad \text{with } \mathbf{1} \text{ the unity operator, i.e.: } \mathbf{1} = \begin{bmatrix} 0 & \cdots & \cdots & \cdots & 0 \\ \vdots & 0 & 0 & 0 & \vdots \\ \vdots & 0 & 1 & 0 & \vdots \\ \vdots & 0 & 0 & 0 & \vdots \\ 0 & \cdots & \cdots & \cdots & 0 \end{bmatrix} \tag{5.28b}$$

For instance, the counterpart of the Gaussian low-pass filter in (5.27) is:

$$\mathbf{h} = \frac{1}{16}\begin{bmatrix} -1 & -2 & -1 \\ -2 & 12 & -2 \\ -1 & -2 & -1 \end{bmatrix} \tag{5.29}$$

Figure 5.10a shows the image in figure 5.8a filtered by this PSF. Another example

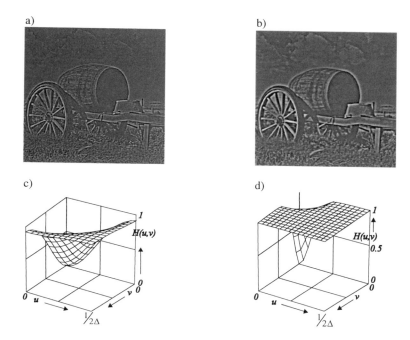

Figure 5.10 High-pass filtering using a Gaussian filter
a) $\sigma_x = 0.8\Delta$, $\sigma_y = 0.8\Delta$
b) $\sigma_x = 2\Delta$, $\sigma_y = 2\Delta$
c) $H(u,v)$ corresponding to a
d) $H(u,v)$ corresponding to b

is given in figure 5.10b. In both operators $H_{low-pass}(u,v)$ is a Gaussian transfer function. Both images in figure 5.10 show the details of the image. However, in figure 5.10a the cut-off frequency is higher than in figure 5.10b. This implies that the details revealed in figure 5.10a are smaller than those in figure 5.10b. The transfer functions of the filters are given in figure 5.10c and d.

High-emphasis filtering

High-emphasis filters enhance the high-frequency components while leaving the low-frequency components unchanged. These kind of filters are used to compensate (to some extent) the image blurring due to the OTF of the acquisition system. If the design of such filter is based on a quantitative blurring- and noise model, then the filter operation is called *image restoration*.

Since quantitative modeling of the blur- and noise characteristics is often laborious, sometimes one confines oneself to high-emphasis filters that are not based on such model. One method to obtain high-emphasis filters is by linearly combining a low-pass filtered image with its original:

$$H_{high-emphasis}(u,v) = \alpha + (1-\alpha) H_{low-pass}(u,v) \tag{5.30a}$$

In terms of PSF-matrices the filter is defined by:

$$\mathbf{h}_{high-emphasis} = \alpha \mathbf{1} + (1-\alpha) \mathbf{h}_{low-pass} \tag{5.30b}$$

The parameter α controls the amplification of the high-frequency components.

a)
b)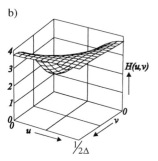

Figure 5.11 High-emphasis filtering using a Gaussian filter
a) $\sigma_x = 0.8\Delta$, $\sigma_y = 0.8\Delta$, $\alpha = 4$
b) $H(u,v)$ corresponding to a

Figure 5.11 gives an example. Here, the high-frequency components are blown up four times ($\alpha = 4$). The white bands along the edges of the objects indicate that this high-emphasis filter overcompensates the OTF of the acquisition system from which the original image stems.

Propagation of noise in discrete convolutions

The propagation of stationary noise in continuous convolutions has been discussed earlier in section 3.2. When dealing with discrete convolutions the results from that section must be adapted. Table 5.2 summarises the relationships between stochastic processes and statistical parameters in both the continuous and the discrete case. The figures in table 5.2 show the notation of the signals and parameters.

5.2.2 Orthogonal transforms

In section 2.3.2 the Fourier transform of a (continuous) 2-dimensional function was introduced. This transform appeared to be very useful in many occasions: specification of the resolving power of an optical device (section 2.3.2), description of the second order statistics of stationary random processes (section 3.2), analysis of issues concerning image sampling (section 4.2), analysis of digital image filters (section 5.2.1). In addition, a discrete version of the Fourier transform (the DFT) was defined in section 4.2.6, enabling a digital implementation of the transform, which in turn enables the implementation of image operations in the frequency domain. In this section we will discuss a class of (discrete) linear operations called *orthogonal transforms*. Appendix A.3 and A.4 summarises the mathematical background of this section.

Suppose the image $f_{n,m}$ is regarded as an NM-dimensional vector $\mathbf{f}$, defined as:

$$\mathbf{f} = [f_0 \quad f_1 \quad \cdots \quad f_{NM-1}]^t \qquad \text{with elements:} \qquad f_{n+Nm} = f_{n,m} \qquad (5.31)$$

The *inner product* of two images $\mathbf{f}$ and $\mathbf{g}$ is:

$$\mathbf{f}^t \mathbf{g} = \sum_{i=0}^{NM-1} f_i g_i \qquad (5.32)$$

IMAGE OPERATIONS   87

Table 5.2 Statistical parameters of linearly filtered, stationary stochastic processes

Continuous convolution	Discrete convolution

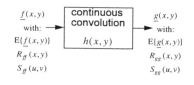
$f(x,y)$ with:
$E\{f(x,y)\}$
$R_{ff}(x,y)$
$S_{ff}(u,v)$

$g(x,y)$ with:
$E\{g(x,y)\}$
$R_{gg}(x,y)$
$S_{gg}(u,v)$

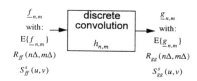
$\underline{f}_{n,m}$ with:
$E\{\underline{f}_{n,m}\}$
$R_{ff}(n\Delta,m\Delta)$
$S_{ff}^s(u,v)$

$\underline{g}_{n,m}$ with:
$E\{\underline{g}_{n,m}\}$
$R_{gg}(n\Delta,m\Delta)$
$S_{gg}^s(u,v)$

Filtered stochastic process:

$$\underline{g}(x,y) = \iint_{\xi\eta} \underline{f}(x-\xi, y-\eta)h(\xi,\eta)d\xi d\eta$$

$$= \underline{f}(x,y) * h(x,y)$$

$$\underline{g}_{n,m} = \sum_{k=-K}^{K}\sum_{l=-L}^{L} \underline{f}_{n-k,m-l} h_{k,l}$$

$$= \underline{f}_{n,m} * h_{n,m}$$

Expectation:

$$E\{\underline{g}(x,y)\} = E\{\underline{f}(x,y)\} * h(x,y)$$

$$E\{\underline{g}_{n,m}\} = E\{\underline{f}_{n,m}\} * h_{n,m}$$

Autocorrelation function:

$$R_{gg}(x,y) = R_{ff}(x,y) * h(x,y) * h(-x,-y)$$

$$R_{gg}(n\Delta,m\Delta) = R_{ff}(n\Delta,m\Delta) * h_{n,m} * h_{-n,-m}$$

Power spectrum:

$$S_{ff}(u,v) = F\{R_{ff}(x,y)\}$$

$$S_{gg}(u,v) = S_{ff}(u,v)|H(u,v)|^2$$

$$S_{ff}^s(u,v) = F\{R_{ff}(n\Delta,m\Delta)\}$$
$$= \sum_n\sum_m R_{ff}(n\Delta,m\Delta)\exp(-2\pi j(un\Delta + vm\Delta))$$
$$S_{gg}^s(u,v) = S_{ff}^s(u,v)|H(u,v)|^2$$

Variance:

$\text{Var}\{\underline{g}(x,y)\} = R_{gg}(0,0) - E^2\{\underline{g}(x,y)\}$
where:

$$R_{gg}(0,0) = \iint_{u\,v} S_{ff}(u,v)|H(u,v)|^2 \, du\,dv$$

$\text{Var}\{\underline{g}_{n,m}\} = R_{gg}(0,0) - E^2\{\underline{g}_{n,m}\}$
where:

$$R_{gg}(0,0) = \int_{u=\frac{-1}{2\Delta}}^{\frac{1}{2\Delta}} \int_{v=\frac{-1}{2\Delta}}^{\frac{1}{2\Delta}} S_{ff}^s(u,v)|H(u,v)|^2 \, du\,dv$$

The (Euclidean) *norm* of an image is:

$$\|\mathbf{f}\| = \sqrt{\mathbf{f}^t \mathbf{f}} = \sqrt{\sum_i f_i^2} \qquad (5.33)$$

As stated in equation (5.17), any linear operation can be performed as a matrix multiplication $\mathbf{g} = \mathbf{Of}$ with $\mathbf{O}$ an $NM \times NM$ matrix. Let $\mathbf{O}^t$ denote the transpose of $\mathbf{O}$. The following statements are equivalent:

- The operation given by $\mathbf{g} = \mathbf{Of}$ is an *orthogonal transformation*.
- $\mathbf{O}^t = \mathbf{O}^{-1}$.
- $\mathbf{f}^t \mathbf{g} = (\mathbf{Of})^t \mathbf{Og}$.

The first statement is merely a term used to denote the properties given in the second and third statements. The equivalence of the last two properties is shown in mathematical analysis. The proof will be omitted here. The second statement implies that the inverse of an orthogonal transformation always exists: $\mathbf{f} = \mathbf{O}^t (\mathbf{Of})$. The third statement implies that an orthogonal transform preserves the norm: $\|\mathbf{Of}\| = \|\mathbf{f}\|$. In fact, this preservation is fully in accordance with the theorem of Parseval (see table 2.4), and with the theorem of Pythagoras in geometry.

It is instructive to think of vectors in $\mathbb{R}^2$ or $\mathbb{R}^3$. Here, an arbitrary vector is represented by a linear combination of two or three orthogonal basis vectors. An orthogonal transformation in $\mathbb{R}^2$ or $\mathbb{R}^3$ is fully equivalent to a rotation of these basis vectors about the origin. Obviously, such a rotation changes the representation of an arbitrary vector, but it does not affect its length nor its norm.

Separability

In general, an orthogonal transform requires $N^2 M^2$ multiplications and addition. For most image sizes this number is prohibitive. However, a reduction to $NM^2 + N^2 M$ can be obtained if the transform is *separable*. This is the case if the transform can be thought of as a cascade of two 1-dimensional transforms: one acting on the rows, one acting on the columns. Suppose $\mathbf{R}$ is an orthogonal $M \times M$-matrix acting on the rows of the image $\mathbf{f}$. $\mathbf{C}$ is an orthogonal $N \times N$-matrix acting on the columns. From these two matrices a 2-dimensional transformation can be defined by regarding the image as an $N \times M$-matrix, i.e.:

$$\mathbf{F} = \begin{bmatrix} f_{0,0} & f_{0,1} & \cdots & \cdots & f_{0,M} \\ f_{1,0} & f_{1,1} & & & \vdots \\ \vdots & & \ddots & & \vdots \\ \vdots & & & \ddots & \vdots \\ f_{N-1,0} & \cdots & \cdots & \cdots & f_{N-1,M-1} \end{bmatrix} \qquad (5.34)$$

The 2-dimensional transform can be represented as:

$$\mathbf{G} = \mathbf{CFR}^t \qquad (5.35a)$$

which is equivalent to:

$$g_{n,m} = \sum_{i=0}^{N-1} c_{i,n} \sum_{j=0}^{M-1} r_{m,j} f_{i,j} \qquad (5.35b)$$

The so-called *left direct matrix product* ⊗ gives the connection between the $NM \times NM$-matrix **O** on one hand, and the $M \times M$- and $N \times N$-matrices, **R** and **C**, respectively, on the other hand.

$$\mathbf{O} = \mathbf{C} \otimes \mathbf{R} \equiv \begin{bmatrix} r_{0,0}\mathbf{C} & r_{0,1}\mathbf{C} & \cdots & r_{0,N-1}\mathbf{C} \\ r_{1,0}\mathbf{C} & r_{1,1}\mathbf{C} & \cdots & r_{1,N-1}\mathbf{C} \\ \vdots & \vdots & \ddots & \vdots \\ r_{N-1,0}\mathbf{C} & \cdots & \cdots & r_{N-1,N-1}\mathbf{C} \end{bmatrix} \qquad (5.36)$$

With this concept a separable orthogonal transformation can also be represented in matrix-vector style:

$$\mathbf{g} = (\mathbf{C} \otimes \mathbf{R})\mathbf{f} \qquad (5.37)$$

The inverse of a separable transform $\mathbf{g} = (\mathbf{C} \otimes \mathbf{R})\mathbf{f}$ is:

$$\mathbf{f} = (\mathbf{C} \otimes \mathbf{R})^{-1}\mathbf{g} = (\mathbf{C}^{-1} \otimes \mathbf{R}^{-1})\mathbf{g} = (\mathbf{C}^t \otimes \mathbf{R}^t)\mathbf{g} \qquad (5.38a)$$

which is equivalent to:

$$\mathbf{F} = \mathbf{C}^t \mathbf{G} \mathbf{R} \qquad (5.38b)$$

or:

$$f_{n,m} = \sum_{i=0}^{N-1} c_{n,i} \sum_{j=0}^{M-1} r_{j,m} g_{i,j} \qquad (5.38c)$$

In section 4.2.6, it has already been mentioned that an important improvement of the computational efficiency of the Fourier transform can be obtained (i.e. the fast Fourier transform). This holds true for all transforms that show certain appropriate symmetries in the columns of the matrices **R** and **C**. These so-called *fast transforms* have a computational complexity on the order of $(^2\log N + {}^2\log M)NM$.

Fourier transform

Strictly speaking, the definitions of inner product and orthogonality do not apply to this transform. The reason for this is that the Fourier transformation-matrix is complex. We have to adapt the definition of the inner product given in (5.32). For this purpose, define the *adjoint* of a complex matrix (or vector) as its *transposed* and *complex conjugated*. For example, the adjoint $\mathbf{f}^*$ of a vector $\mathbf{f}$ is defined as: $\mathbf{f}^* = [\bar{f}_0 \cdots \bar{f}_{NM-1}]$. Then, the inner product of two complex vectors becomes:

$$\mathbf{f}^* \mathbf{g} = \sum_{i=0}^{NM-1} \bar{f}_i g_i \qquad (5.39)$$

Transformation matrices **O** for which the property:

$$\mathbf{O}^* = \mathbf{O}^{-1} \qquad (5.40)$$

holds are called *unitary* transformation matrices. Note that any real, orthogonal matrix is unitary. Therefore, the class of unitary transforms encompasses the class of orthogonal transforms.

The discrete Fourier transform (DFT) is defined in equation (5.35b) with:

$$r_{i,n} = \frac{1}{\sqrt{N}} \exp\left(\frac{-2\pi j i n}{N}\right) \quad \text{and} \quad c_{j,m} = \frac{1}{\sqrt{M}} \exp\left(\frac{-2\pi j j m}{M}\right) \tag{5.41a}$$

Accordingly, the 2-dimensional Fourier transformation matrix is:

$$\mathbf{O} = \mathbf{C} \otimes \mathbf{R} \tag{5.41b}$$

Using the vector representation for images, expressed in equation (5.31), we now continue by showing that the convolution theorem for continuous signals also holds for the discrete case. For that purpose, consider a circulant convolution given by:

$$g_{n,m} = \sum_{k=-K}^{K} \sum_{l=-L}^{L} h_{k,l} f_{\text{mod}_N(n-k), \text{mod}_M(m-l)} \quad n = 0, \cdots, N-1 \quad m = 0, \cdots, M-1 \tag{5.42}$$

$\text{mod}_N(.)$ is the modulo-N operator. This modulo operation has been added so as to define the output pixels near the edge of the image according to the circulant convolution mentioned in section 5.2.1. Clearly, the operation defined in (5.42) is linear. Therefore, the circulant convolution can also be represented in matrix-vector style:

$$\mathbf{g} = \mathbf{H}\mathbf{f} \tag{5.43}$$

$\mathbf{H}$ is a $NM \times NM$-matrix equipped with the elements from the PSF-matrix $\mathbf{h}$ in equation (5.20)[3]. Consider the eigenvectors $\vec{e}_i$ and corresponding eigenvalues λ_i of the matrix $\mathbf{H}$. These eigenvectors and eigenvalues are defined by:

$$\mathbf{H}\vec{e}_i = \lambda_i \vec{e}_i$$
$$\|\vec{e}_i\| = 1 \tag{5.44}$$

The eigenvectors of the operator $\mathbf{H}$ correspond to a set of orthogonal images which are invariant under the convolution operation, except for a multiplication factor λ_i. It is not difficult to show that these images must have the functional form:

$$\frac{1}{\sqrt{NM}} \exp\left(2\pi j \left(\frac{in}{N} + \frac{jm}{M}\right)\right)$$

These images are found in the columns of the matrix $\mathbf{O}^*$. The columns in $\mathbf{O}^*$ are the eigenvectors $\vec{e}_i$:

$$\mathbf{O}^* = [\vec{e}_0 \; \vdots \; \vec{e}_1 \; \vdots \; \cdots \; \vdots \; \vec{e}_{NM-1}] \tag{5.45}$$

---

[3] $\mathbf{H}$ and $\mathbf{h}$ are matrices that describe the same convolution. However, the arrangement of elements $h_{n,m}$ in $\mathbf{H}$ is quite different from the one in $\mathbf{h}$.

IMAGE OPERATIONS 91

If we define an $NM \times NM$ diagonal matrix Λ according to:

$$\Lambda = \begin{bmatrix} \lambda_0 & 0 & 0 & \cdots & 0 \\ 0 & \lambda_1 & 0 & \ddots & 0 \\ \vdots & 0 & \ddots & 0 & \vdots \\ 0 & \ddots & 0 & \lambda_{NM-2} & 0 \\ 0 & \cdots & 0 & 0 & \lambda_{NM-1} \end{bmatrix} \qquad (5.46)$$

then equation (5.44) can be written more economically:

$$\mathbf{HO}^* = \mathbf{O}^*\Lambda \qquad (5.47)$$

Combining (5.42) with (5.47) yields:

$$\mathbf{g} = \mathbf{Hf} = \mathbf{HO}^{-1}\mathbf{Of} = \mathbf{HO}^*\mathbf{Of} = \mathbf{O}^*\Lambda\mathbf{Of} \qquad (5.48)$$

Hence, the operation **H** is equivalent to a cascade of three operations:

- **Of**: The 2-dimensional DFT applied to the image **f**, resulting in a (complex) image, say, f.
- **Λf**: Since Λ is a diagonal matrix, this operation corresponds to a "space-variant"[4] pixel-to-pixel operation applied to the complex image f. The result is denoted g.
- **$\mathbf{O}^{t*}g$**: The inverse DFT applied to the filtered image g.

Seen in this light, the diagonal matrix Λ can be regarded as a transfer function. The practical significance is that discrete convolution can be implemented with the DFT. This occurs according to the following algorithm (Listing 5.5 is a pseudo-C realisation of the algorithm):

Algorithm 5.1 2-dimensional circulant convolution with DFT

Input: Input image, PSF-matrix

1. Compute the transfer function of the PSF-matrix **h**:

$$H_{i,j} = \frac{1}{\sqrt{NM}} \sum_{n=-K}^{K} \sum_{m=-L}^{L} h_{n,m} \exp\left(-2\pi j \left(\frac{in}{N} + \frac{jm}{M}\right)\right)$$

2. Compute the DFT of the input image **f**:

$$F_{i,j} = \frac{1}{\sqrt{NM}} \sum_{n=0}^{N-1} \sum_{m=0}^{M-1} f_{n,m} \exp\left(-2\pi j \left(\frac{in}{N} + \frac{jm}{M}\right)\right)$$

3. Compute the DFT of the output image:

$$G_{i,j} = H_{i,j} F_{i,j}$$

4. Compute the inverse DFT:

$$g_{n,m} = \frac{1}{\sqrt{NM}} \sum_{i=0}^{N-1} \sum_{j=0}^{M-1} G_{i,j} \exp\left(2\pi j \left(\frac{in}{N} + \frac{jm}{M}\right)\right)$$

---

[4] Perhaps the term *frequency variant* would apply better.

Listing 5.5 2-dimensional circulant convolution with DFT

```
void circ_convolve(f,g,h,N,M,K,L)    /* 2-dimensional circulant convolution    */
double **f,**g,**h;                   /* pointers to the images f and g, respectively,
                                         and the kernel h. These are declared as
                                         2-dimensional arrays of doubles         */
int N,M;                              /* number of rows and columns of f and g   */
int K,L;                              /* size of neighbourhood                   */
{
    cmplx **tmp_h;                    /* pointer to 2-dimensional buffer, needed to pad h with
                                         zeroes and to store h as type complex   */
    double **tmp_f;                   /* pointer to 2-dimensional buffer, needed to store f
                                         as type complex                         */
    int i,j;                          /* counters                                */

    initiate(tmp_h,N,M);              /* allocate memory space and assign it to tmp_h */
    initiate(tmp_f,N,M);              /* allocate memory space and assign it to tmp_f */
    for (i=0;i<N;i++)
    for (j=0;j<M;j++)
    {   tmp_f[i][j]=complex(f[i][j],0.0); /* copy f to tmp_f*/
        tmp_h[i][j]=complex(0.0,0.0);     /* set tmp_h to zero */
    }
    for (i=-K;i=<K;i++)
    for (j=-L;j=<M;j++)
    {   tmp_h[i%N][j%L]=complex(h[i][j],0.0); /* copy h to tmp_h */
    }
    dft(tmp_h,N,M);                   /* FT of h; see listing 4.1 */
    dft(tmp_f,N,M);                   /* FT of f */
    for (i=-K;i=<K;i++)               /* calculate FT of output */
    for (j=-L;j=<M;j++) tmp_f[i][j]=tmp_f[i][j]*tmp_h[i][j];
    idft(tmp_f,N,M);                  /* inverse FT */
    for (i=0;i<N;i++)
    for (j=0;j<M;j++) g[i][j] = real(tmp_f[i][j]);  /* copy tmp_f to g */
}
```

Step 1 can be omitted if the transfer function is known at forehand. For instance, a low-pass filter can easily be defined in the frequency domain. Suppose that in a certain application a (continuous) transfer function denoted $H(u,v)$ is needed. Then, a filter with this transfer function can be implemented by the following conversion:

$$H_{i,j} = H(u_i, v_j)$$

with:

$$u_i = \begin{cases} i/N\Delta & \text{if } 0 \le i \le N/2 \\ (i-N)/N\Delta & \text{if } N/2 < i < N \end{cases}$$

$$v_j = \begin{cases} j/M\Delta & \text{if } 0 \le j \le M/2 \\ (j-M)/N\Delta & \text{if } M/2 < j < M \end{cases}$$
(5.49)

Isotropic low-pass filters with cut-off frequency ρ_c can be implemented by:

$$H(u,v) = 1 \Big/ \sqrt{\left(1 + \left(\frac{u^2+v^2}{\rho_c^2}\right)^n\right)}$$
(5.50)

This transfer is a generalisation of the 1-dimensional Butterworth filter. The order n of the filter specifies the steepness of the transfer near the cut-off frequency ρ_c.

Hadamard transform

Another orthogonal transform that has been used in image analysis is the *Hadamard transform*. For image sizes with N and M a power of 2 the 1-dimensional Hadamard transformation matrix **R** and **C** can be recursively defined according to:

$$\mathbf{H}_2 = \begin{bmatrix} 1 & 1 \\ 1 & -1 \end{bmatrix} \qquad \mathbf{H}_{2i} = \begin{bmatrix} \mathbf{H}_i & \mathbf{H}_i \\ \mathbf{H}_i & -\mathbf{H}_i \end{bmatrix} \quad \text{for} \quad i = 2,4,8,16,\cdots$$
(5.51a)

Table 5.3 1-dimensional Hadamard transformation matrices

$N=2$	$N=4$	$N=8$
$\frac{1}{\sqrt{2}}\begin{bmatrix} 1 & 1 \\ 1 & -1 \end{bmatrix}$	$\frac{1}{2}\begin{bmatrix} 1 & 1 & 1 & 1 \\ 1 & -1 & 1 & -1 \\ 1 & 1 & -1 & -1 \\ 1 & -1 & -1 & 1 \end{bmatrix}$	$\frac{1}{\sqrt{8}}\begin{bmatrix} 1 & 1 & 1 & 1 & 1 & 1 & 1 & 1 \\ 1 & -1 & 1 & -1 & 1 & -1 & 1 & -1 \\ 1 & 1 & -1 & -1 & 1 & 1 & -1 & -1 \\ 1 & -1 & -1 & 1 & 1 & -1 & -1 & 1 \\ 1 & 1 & 1 & 1 & -1 & -1 & -1 & -1 \\ 1 & -1 & 1 & -1 & -1 & 1 & -1 & 1 \\ 1 & 1 & -1 & -1 & -1 & -1 & 1 & 1 \\ 1 & -1 & -1 & 1 & -1 & 1 & 1 & -1 \end{bmatrix}$

$$\mathbf{R} = \frac{1}{\sqrt{M}} \mathbf{H}_M \qquad \mathbf{C} = \frac{1}{\sqrt{N}} \mathbf{H}_N \qquad (5.51b)$$

Table 5.3 shows the 1-dimensional Hadamard transformation matrices for $N=2, 4$ and 8. Note that the Hadamard matrices are symmetric, $\mathbf{R} = \mathbf{R}^t = \mathbf{R}^{-1}$ and $\mathbf{C} = \mathbf{C}^t = \mathbf{C}^{-1}$. Consequently, the Hadamard transform and the inverse Hadamard transform are identical. For dimensions N that are not a power of 2 the Hadamard transform also exists, but in these cases its definition cannot be given recursively. The ordering in which the rows in the matrices $\mathbf{R}$ and $\mathbf{C}$ appear is arbitrary. In the examples given in table 5.3 the ordering is according to the recursive definition given above. However, sometimes a so-called *sequency ordering* is preferred. A row is ordered according to its number of sign changes along that row. This corresponds well to the natural ordering of the Fourier transform.

The 2-dimensional Hadamard transform is separable. The transformation can be performed by the operation in equation (5.35). The 2-dimensional transformation matrix follows from equation (5.36). Figure 5.12 gives the basis functions $r_{i,n} c_{j,m}$ for $N=M=4$. Figure 5.13 is an example of a Hadamard transform of the image in figure 5.2a. Listing 5.6 is C-pseudo-code of a fast realisation of the transform.

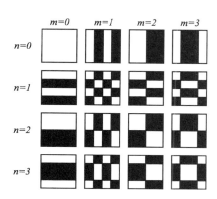

Figure 5.12
Basis functions of the 2-dimensional Hadamard transform ($N=M=4$)

Figure 5.13
Log-magnitude representation of the Hadamard transform of the image in figure 4.6a

Listing 5.6 2-dimensional Hadamard transform

```
void fht(f,N,M)        /* replaces the input image with its Hadamard transform    */
double **f;            /* Pointer to an image, declared as a 2-dimensional array of
                          doubles.
int N,M;               /* number of rows and columns, respectively
                          N and M MUST BE a power of 2                             */
{
    int i;
    double **tmp;
    initiate(tmp,N,M);                   /* tmp array needed to transpose image    */
    for (i=0;i<N;i++) fht1d(f[i],M);     /* allocate memory and initiate tmp       */
    transpose(f,tmp,N,M);                /* fast Hadamard transform on the rows    */
    for (i=0;i<M;i++) fht1d(tmp[i],N);   /* transposition: tmp[i][j]:= f[j][i]     */
    transpose(tmp,f,M,N);                /* fht on the rows of **tmp               */
    free(tmp);                           /* transposition: f[i][j]:= tmp[j][i]     */
}                                        /* free memory allocated to **tmp         */


static void fht1d(f, n)      /* fast Hadamard transform on a 1-dimensional
                                array *f with n elements                           */
double *f;
int n;
{   double dummy;
    int i, ip, j, k,l, le;
    l=1;
    while (l<n)              /* This loop does the Hadamard transform              */
    {   le=l;
        l=l*2;
        for(j=0;j<le;j++)
        for(i=j;i<n;i=i+1)
        {   ip=i+le;
            dummy=f[ip];
            f[i]=f[i]-dummy;
            f[i]=f[i]+dummy;
        }
    }
    for (i=0;i<n;i++) f[i]=f[i]/sqrt(n);
}
```

5.2.3 Differential operators

Another class of linear operators are differential operators. In mathematical analysis, these operators are applicable to continuous functions defined in $\mathbb{R}^N$. Hence, when differential operators are considered in image analysis, these concepts refer to continuous images $f(x,y)$ rather than to digital images $f_{n,m}$. Therefore, we will first consider some basic operations defined on $f(x,y)$. Later on the discrete case will be discussed.

The *gradient vector* of a 2-dimensional image is defined as the vector:

$$\vec{\nabla} f(,x,y) = \left[\frac{\partial f(x,y)}{\partial x} \quad \frac{\partial f(x,y)}{\partial y} \right]^t = \begin{bmatrix} f_x(x,y) \\ f_y(x,y) \end{bmatrix} \tag{5.52}$$

In a 2-dimensional landscape, the gradient can be regarded as a vector that for each position (x,y) points to the direction of steepest ascent. The magnitude $\|\nabla f(x,y)\|$ of the gradient is proportional to the steepness.

The *(first) directional derivative* of the function $f(x,y)$ taken in the direction α at the position (x,y) is defined as the first derivative of $f(x+t\cos\alpha, y+t\sin\alpha)$ with respect to t:

$$\dot{f}_\alpha(x,y) = \left. \frac{\partial f(x+t\cos\alpha, y+t\sin\alpha)}{\partial t} \right|_{t=0} \tag{5.53a}$$

which can be written as:

$$\dot{f}_\alpha(x,y) = f_x(x,y)\cos\alpha + f_y(x,y)\sin\alpha = \vec{n}'\vec{\nabla}f(x,y) \quad \text{where:} \quad \vec{n}' = [\cos\alpha \quad \sin\alpha] \quad (5.53b)$$

$\dot{f}_\alpha(x,y)$ can be interpreted as the first derivative of the cross section of $f(x,y)$ passing through the position (x,y) in the direction α. It is easily shown that with fixed (x,y) the direction α that maximises $\dot{f}_\alpha(x,y)$ equals the direction of the gradient. The maximum of $\dot{f}_\alpha(x,y)$ is the magnitude of the gradient. If α is orthogonal to the direction of the gradient, then $\dot{f}_\alpha(x,y) = 0$.

The *Laplacian* of a function is the sum of second derivatives:

$$\nabla^2 f(x,y) = \frac{\partial^2 f(x,y)}{\partial x^2} + \frac{\partial^2 f(x,y)}{\partial y^2} = f_{xx}(x,y) + f_{yy}(x,y) \quad (5.54)$$

The Laplacian operator is *rotationally invariant*. If the function $f(x,y)$ is rotated with respect to some center of rotation, then the Laplacian of the image does not change at that center point. The magnitude of the gradient is also rotationally invariant, but the directional derivative and the gradient vector are not.

The definition of the *second directional derivative* is similar to the one of the first derivative:

$$\ddot{f}_\alpha(x,y) = \left.\frac{\partial^2 f(x+t\cos\alpha, y+t\sin\alpha)}{\partial t^2}\right|_{t=0} \quad (5.55a)$$

Written in another form:

$$\ddot{f}_\alpha(x,y) = f_{xx}(x,y)\cos^2\alpha + 2f_{xy}(x,y)\cos\alpha\sin\alpha + f_{yy}(x,y)\sin^2\alpha \quad (5.55b)$$

As shown above, the gradient, the Laplacian and the directional derivatives can be expressed in terms of the partial derivatives $f_x(x,y), f_{xx}(x,y), f_{xy}(x,y)$, etc. The corresponding operations $\partial/\partial x$, $\partial^2/\partial x^2$, $\partial^2/\partial x \partial y$, etc. are space invariant. The derivative of a harmonic function is again a harmonic function. Consequently, the derivative operators can be analysed in the frequency domain. The transfer functions follows from table 2.4:

$$\begin{aligned}
\frac{\partial}{\partial x} &\xrightarrow{FT} H(u,v) = 2\pi j u \\
\frac{\partial^2}{\partial^2 x} &\xrightarrow{FT} H(u,v) = -4\pi^2 u^2 \\
\frac{\partial^2}{\partial x \partial y} &\xrightarrow{FT} H(u,v) = -4\pi^2 uv \\
\nabla^2 f(x,y) &\xrightarrow{FT} H(u,v) = -4\pi^2 u^2 - 4\pi^2 u^2
\end{aligned} \quad (5.56)$$

Operations based on differences

The operations based on the derivatives are not easily transformed from the continuous domain to the domain of digital images. Sometimes, the first derivative in the x-direction is approximated with the first order differences (derivatives in the y-direction are handled similar):

$$\left.\frac{\partial f(x,y)}{\partial x}\right|_{x=n\Delta, y=m\Delta} \approx \frac{f(n\Delta, m\Delta) - f((n-1)\Delta, m\Delta)}{\Delta} = \frac{f_{n,m} - f_{n-1,m}}{\Delta} \quad \text{(backward difference)}$$

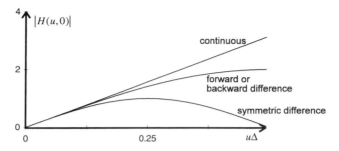

Figure 5.14 Magnitude of the transfer function of a derivative operation and two discrete approximations

$$\left.\frac{\partial f(x,y)}{\partial x}\right|_{x=n\Delta, y=m\Delta} \approx \frac{f_{n+1,m} - f_{n,m}}{\Delta} \qquad \text{(forward difference)} \qquad (5.57)$$

$$\left.\frac{\partial f(x,y)}{\partial x}\right|_{x=n\Delta, y=m\Delta} \approx \frac{f_{n+1,m} - f_{n-1,m}}{2\Delta} \qquad \text{(symmetric difference)}$$

It can be seen that these operations correspond to discrete convolutions with the following PSF-matrices and transfer functions:

$$\frac{1}{\Delta}[0 \; 1 \; -1]^t \xrightarrow{FT} H(u,v) = \frac{2j}{\Delta}\sin(\pi u\Delta)\exp(-j\pi u\Delta) \qquad \text{(backward)}$$

$$\frac{1}{\Delta}[1 \; -1 \; 0]^t \xrightarrow{FT} H(u,v) = \frac{2j}{\Delta}\sin(\pi u\Delta)\exp(j\pi u\Delta) \qquad \text{(forward)} \qquad (5.58)$$

$$\frac{1}{2\Delta}[1 \; 0 \; -1]^t \xrightarrow{FT} H(u,v) = \frac{j}{\Delta}\sin(2\pi u\Delta) \qquad \text{(symmetric)}$$

The magnitude of the transfer functions are shown in figure 5.14. For low frequencies, the transfer functions with respect to the differences correspond well to the transfer function of the first derivative. However, for high frequencies the transfer functions deviate more and more. The forward and backward differences seem to approximate the first derivative closer than the symmetric difference. An advantage of the symmetric difference is that it does not induce a shift over $\Delta/2$ in the space domain. For the backward and forward differences this shift results in a linear phase shift in the transfer function. An example of the backward difference and the symmetric difference applied to the image in figure 5.2a is shown in figure 5.15. The images in figure 5.15b and 5.15c taken together yield an approximation of the gradient.

Usually, second derivatives are approximated by a cascade of a backward and a forward difference. In the x-direction, this yields the following PSF-matrix and related transfer function:

$$1/\Delta^2 \, [-1 \; 2 \; -1]^t \xrightarrow{FT} H(u,v) = (-2 + 2\cos(2\pi u\Delta))/\Delta^2 \qquad (5.59)$$

The first two terms of a Taylor expansion series of this transfer function is: $H(u,v) = -4\pi^2 u^2 + 4/3\pi^4 u^4 + \cdots$. Again, for low frequencies the approximation is close,

Figure 5.15 First order differences
a) Backward difference, vertical direction
b) Symmetric difference, vertical direction
c) Symmetric difference, horizontal direction

see figure 5.16. Second order differences in both directions are given in figure 5.17a and b. Combination of these differences yields an approximation of the Laplacian. The PSF-matrix of the approximation is:

$$\frac{1}{\Delta^2}\begin{bmatrix} 0 & -1 & 0 \\ -1 & 4 & -1 \\ 0 & -1 & 0 \end{bmatrix} \quad \text{(discrete Laplacian)} \tag{5.60}$$

The discrete Laplacian of the image in figure 5.8a is shown in figure 5.17c.

Noise analysis

The discrete approximations of the derivative operators given above turn out to be discrete convolutions. Therefore, the influence of additive noise can be analysed with the equations given in table 5.2. In general, discrete approximations of derivative operators are very sensitive to noise.

As an example we consider figure 5.18. Figure 5.18a contains a 1-dimensional data sequence containing two step functions. The sequence is contaminated by additive noise. Sequences that are composed of a number of step functions have a power spectrum which decreases rapidly for high frequencies. In general, the noise has a much broader bandwidth. This is shown in figure 5.18b. In this figure, the

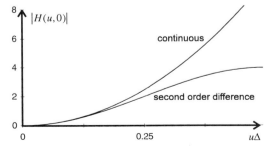

Figure 5.16 Magnitude of the transfer function of a second derivative operation and a discrete approximation

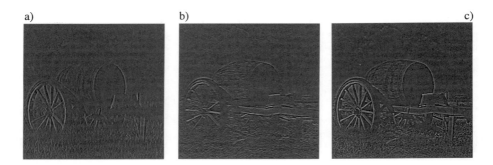

Figure 5.17 Second order differences
a) Second order difference, vertical direction
b) Second order difference, horizontal direction
c) Laplacian

SNR defined as the ratio between variances of signal and noise is seen as the ratio between the areas below the two spectra. Figure 5.18c shows the backward difference of the sequence. The power spectra of the resulting noise and signal can be deduced from the power spectra of the original data by multiplying it with the squared magnitude of the transfer function: $|H(u)|^2 = 4\sin^2(\pi u\Delta)/\Delta^2$, see figure 5.18d. Note that in this example the *SNR* has decreased dramatically.

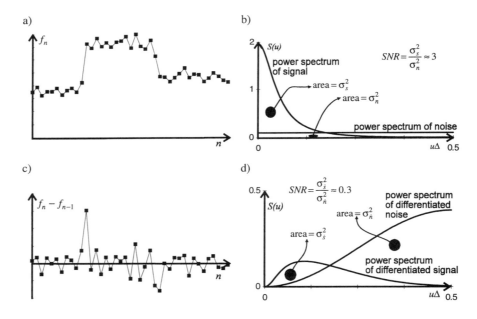

Figure 5.18 Noise sensitivity
a) 1-dimensional sequence consisting of signal and noise
b) Power spectra of signal and noise
c) Backward difference of a
d) Power spectra of differentiated signal and noise

The situation in the example is similar to the derivative operators applied to images. Here too, the Fourier transform of a noiseless image tends to be dominated by low-frequency components. Therefore, derivative operators, essentially being high-pass filters, tend to impair the *SNR*.

Operations based on derivatives of the Gaussian

Another approach to define the derivative of a digital image is to consider the grey levels $f_{n,m} = f(n\Delta, m\Delta)$ as samples from a continuous image $f(x,y)$ contaminated with noise: $f(x,y) = g(x,y) + \underline{n}(x,y)$. The problem is to find the derivatives of $g(x,y)$ given the samples $f(n\Delta, m\Delta)$. One way to find a solution is to reconstruct the original image $f(x,y)$ first, and then to differentiate the reconstruction. As an example, suppose that the first derivative in the x-direction is sought. The reconstructed image is given in equation (4.5):

$$f^r(x,y) = \sum_n \sum_m f_{n,m} h(x - n\Delta, y, -m\Delta) \tag{5.61}$$

where $h(x,y)$ is the PSF of the reconstruction filter. An estimate of the first derivative takes the form:

$$\hat{f}_x(x,y) = f_x^r(x,y) = \sum_n \sum_m f_{n,m} h_x(x - n\Delta, y, -m\Delta) \tag{5.62}$$

Formally, the problem is solved by adoption of a criterion whose optimisation uniquely defines the reconstruction/interpolation function $h(x,y)$. One such criterion is as follows:

$$\sum_n \sum_m \left(f^r(n\Delta, m\Delta) - f_{n,m} \right)^2 + \lambda \int_x \int_y \left(f_{xx}^r(x,y) \right)^2 dxdy \tag{5.63}$$

The first term is a measure of the distance between the original samples and the samples of the reconstructed signal. The second term is the signal energy of the second derivative of the reconstructed signal. Due to the double differentiation, the noise in this derivative will be enhanced strongly. As such the second term is a measure of the amount of noise in the reconstruction.

The parameter λ controls the trade off between *noise suppression* and *resolution* or *scale*. The PSF $h(x,y)$ that minimises equation (5.63) defines the optimal reconstruction filter needed for finding the first derivative. Obviously, this PSF depends on the particular choice of the parameter λ.

For 1-dimensional signals, it has been shown that a Gaussian filter is close to the optimal reconstruction filter [Torre and Poggio 1986]. Figure 5.19 illustrates this. Figure 5.19a shows a continuous signal containing a single step function. The signal is contaminated with noise. A sampled version of the signal is also shown. The reconstruction of the signal with a Gaussian filter with standard deviation $\sigma_p = 0.8\Delta$ is depicted in figure 5.19b. Differentiation of the reconstructed signal yields an estimate of the first derivative (figure 5.19c). Finally, resampling applied to the estimated derivative completes the procedure. The whole process is equivalent to a discrete convolution of the input sequence f_n with the sampled version of the first derivative of the Gaussian:

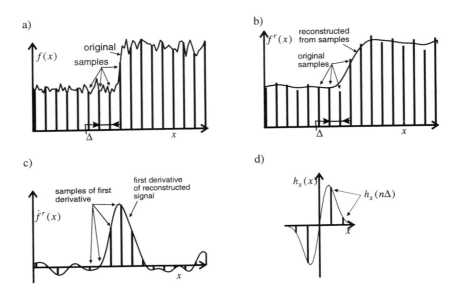

Figure 5.19 Gauss based derivatives
a) Continuous signal with its samples
b) Signal reconstructed from its samples
c) First derivative of reconstructed signal with its sampled version
d) Impulse response to obtain first derivative of reconstructed signal

$$f_n = \sum_i f_i h_x((n-i)\Delta) \quad \text{with:} \quad h_x(x) = \frac{x}{\sigma_p^3 \sqrt{2\pi}} \exp\left(-\frac{x^2}{2\sigma_p^2}\right)$$

The PSF associated with the convolution is shown in figure 5.19d. The standard deviation σ_p is a monotonically increasing function of the parameter λ. For instance, if σ_p is chosen large, then noise suppression is favored above resolution.

Also in two dimensions, the Gauss function is often used as an approximation of the reconstruction filter. The PSF of the reconstruction takes the form:

$$h(x,y) = \frac{1}{2\pi\sigma_p^2} \exp\left(-\frac{x^2+y^2}{2\sigma_p^2}\right) \tag{5.64}$$

For the determination of the derivatives the following convolution masks are appropriate:

$$\begin{aligned}
\frac{\partial}{\partial x} &\rightarrow h_x(x,y) = \frac{-x}{2\pi\sigma_p^4} \exp\left(-\frac{x^2+y^2}{2\sigma_p^2}\right) \\
\frac{\partial^2}{\partial x^2} &\rightarrow h_{xx}(x,y) = \frac{x^2-\sigma_p^2}{2\pi\sigma_p^6} \exp\left(-\frac{x^2+y^2}{2\sigma_p^2}\right) \\
\nabla^2 &\rightarrow h_{xx}(x,y) + h_{xx}(x,y) = \frac{x^2+y^2-2\sigma_p^2}{2\pi\sigma_p^6} \exp\left(-\frac{x^2+y^2}{2\sigma_p^2}\right)
\end{aligned} \tag{5.65}$$

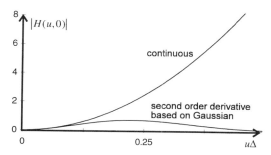

Figure 5.20 Magnitude of the transfer function of a second derivative operation and one based on Gaussian filtering

The convolution of an image with these masks can be regarded as a cascade of a Gaussian low-pass filtering followed by a differential operation. Effectively this is equivalent to a band-pass filter with transfer functions:

$$\frac{\partial}{\partial x} \xrightarrow{FT} H_x(u,v) = 2\pi j u \exp\left(-\frac{\sigma_p^2(u^2+v^2)}{2}\right)$$

$$\frac{\partial^2}{\partial x^2} \xrightarrow{FT} H_{xx}(u,v) = -4\pi^2 u^2 \exp\left(-\frac{\sigma_p^2(u^2+v^2)}{2}\right)$$

$$\nabla^2 \xrightarrow{FT} H_{xx}(u,v) + H_{xx}(u,v) = -4\pi^2(u^2+v^2)\exp\left(-\frac{\sigma_p^2(u^2+v^2)}{2}\right)$$
(5.66)

As an example, figure 5.20 shows the transfer function of a second derivative operation based on Gaussian filtering. The sensitivity to noise will be much lower compared with the difference operation. This is due to the narrow bandwidth of the filter. The bandwidth is inversely proportional to the standard deviation σ_p.

The convolution kernels given in (5.65) are depicted in figure 5.21. Discrete representations are obtained by sampled versions. Since the 2-dimensional Gaussian is separable in x and y, the operators can be implemented quite effectively. Alternatively, with the expressions in (5.66), the filtering can also be accomplished in the frequency domain. Figure 5.21 presents the results of the filters applied to the image in figure 5.2a.

Example 5.2 Edge detection

The capability of estimating the derivatives of an image paves the way to the utilisation of all tools available from differential geometry. Edge detection is an example. The purpose of this operation is to find all locations in the image plane where the irradiance show a step-like transition, an edge; see section 3.3. The detection of these edges is useful since often the edges correspond to object boundaries, shadow transitions and other characteristic points in the scene. The gradient magnitude of the irradiance at a point (x,y) is the slope of the tangent plane at (x,y). As such it is a geometrical property that is strongest on edges. Figure 5.22b is a graphical representation of the estimated gradient magnitude of the image in figure 5.22a. The estimate is based on a Gaussian filter with $\sigma_p = 2\Delta$. The image is sampled on a 128×128-grid. It can be seen that the gradient magnitude is large at locations corresponding to object boundaries.

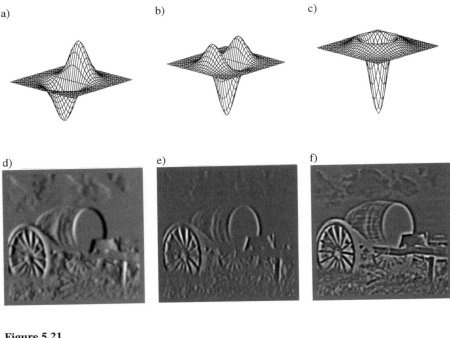

Figure 5.21
a) PSF for first derivative
d) First derivative
b) PSF for second derivative
e) Second derivative
c) PSF for Laplacian
f) Laplacian

Thresholding the gradient magnitude image would result in a map of the image plane containing detected *edge pixels* (or briefly *edges*). However, the map would merely consist of configurations of edge pixels that are several pixels wide. This is caused by the Gaussian low-pass filtering. A better localisation of detected edges is obtained if we use the Laplacian of the image too (figure 5.22c). For straight edge segments, the locations at which the Laplacian of the image is exactly zero (i.e. the zero crossings) often coincide with the locus of maximum gradient. Therefore, the zero crossings of the Laplacian of the image are often used as an indicator of candidate edges. The zero crossings of figure 5.22c are shown in figure 5.22d. Note that the configuration of zero crossings is exactly one pixel wide. Masking the gradient magnitude image with these zero crossings yields an image as shown in figure 5.22e. Finally, thresholding results in a map of detected edge, the configuration of which is exactly one pixel wide. Edge detection based on zero crossings of the Laplacian of the Gaussian filtered image was proposed by Marr and Hildreth [1980]. Hence, the name *Marr-Hildreth operation* is used to denote this method.

In most applications, it is desirable to resolve the detailed structures up to some fixed scale. For instance, in figure 5.22a the edges of interest are the ones corresponding to the two object boundaries. If the task of the vision system is to recognise the object, then finer edge structures (e.g. due to a change of the direction of the surface normal) are less important. The parameter σ_p controls the *scale* on which the operator is working. Increasing σ_p yields an edge map with less detailed structures. This is shown in figure 5.22f and 5.22g, where σ_p is 3Δ and 5Δ, respectively.

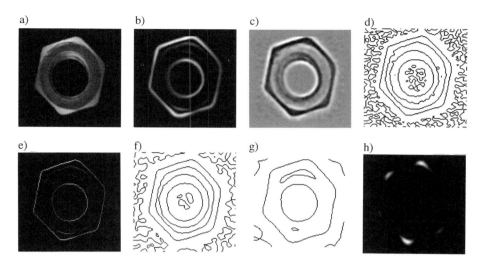

Figure 5.22 Applications of Gaussian based differential operators
a) Original
b) Gradient magnitude
c) Laplacian $\sigma_p = 2\Delta$
d) Zero crossings of c
e) Gradient masked with d
f) Zero crossings $\sigma_p = 3\Delta$
g) Zero crossings $\sigma_p = 5\Delta$
h) Corner enhancement

Example 5.3: Corner enhancement

Another application based on differential geometry is the enhancement of corners. A corner in an image is the location where two straight edge segments (with different directions) meet. The following differential operation enhances the corners in an image [Haar Romeny et al 1991]:

$$g(x,y) = -f_{xx}(x,y)f_y^2(x,y) + 2f_x(x,y)f_y(x,y)f_{xy}(x,y) - f_{yy}(x,y)f_x^2(x,y) \quad (5.67)$$

Figure 5.22h gives an illustration. As in edge detection, the scale of the operation must be selected appropriately.

5.3 CORRELATION TECHNIQUES

Correlation is a technique that can be used to detect and/or to localise a known shape in an image. Suppose that the appearance of an object (or part of an object) within the image is exactly known. Furthermore, suppose that the image of this "object of interest" is given by $h(x-\xi, y-\eta)$. The parameters (ξ, η) define the position of the object within the image plane. Then, the observed image $f(x,y)$ can be thought of as a composition of $h(x,y)$ and the background $\underline{n}(x,y)$ of the scene: $f(x,y) = h(x-\xi, y-\eta) + \underline{n}_1(x,y)$. It is understood that the background contains a deterministic part (i.e. the image of other objects in the scene) and a random part describing the noise. Of course, this model holds true only if the scene really does contain the object of interest. If this is not the case, then the observed image is written as: $f(x,y) = \underline{n}_2(x,y)$. The vision task is to check if the scene contains the object, and if so, to estimate its position within the image plane.

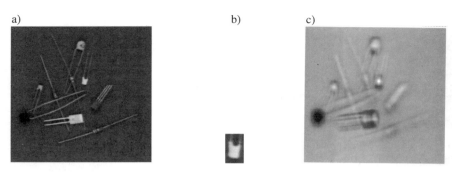

Figure 5.23 Correlation
a) Original image   b) Template   c) Squared Euclidean distance

One way to accomplish this task is to shift a template of the object across the image plane, to match this template locally with the image data, and to select the position $(\hat{\xi},\hat{\eta})$ of the template for which the match is best. A particular choice of the measure of match is the *Euclidean distance* between the template and the observed image.

$$d^2(\xi,\eta) = \iint_{x,y} (f(x,y) - h(x-\xi, y-\eta))^2 \qquad d(\hat{\xi},\hat{\eta}) = \min\{d(\xi,\eta)\} \qquad (5.68a)$$

Note that, upon substitution of integration variables, $d(x,y)$ can be written as:

$$d^2(\xi,\eta) = \iint_{x,y} (f(x+\xi, y+\eta) - h(x,y))^2 \qquad (5.68b)$$

In the discrete case, $d(\xi,\eta)$ corresponds to $d_{n,m} = d(n\Delta, m\Delta)$. Finite summations replace the integrals. Hence:

$$d_{n,m}^2 = \sum_{k=-K}^{K} \sum_{l=-L}^{L} (f_{n+k,m+l} - h_{k,l})^2 \qquad (5.68c)$$

The ranges of summations are determined by the constants K and L. These constants must be chosen in such a way that the template $h_{k,l}$ fully embeds the image of the object of interest.

Figure 5.23 presents an example. Figure 5.23a is the image of a number of electronic components. Figure 5.23b is the template of a certain capacitor. The image $d_{n,m}^2$ is shown in figure 5.23c. For the sake of visibility the contrast in figure 5.23c is inverted. That is, white corresponds to zero, black corresponds to a positive real number. Furthermore, the magnification used in figure 5.23b is two times the magnification used in figures 5.23a and b.

Note that the image in figure 5.23c is suitable to *detect* objects. For that purpose it suffices to threshold the image. Possibly all pixels below a certain threshold correspond to the object that is searched for. The image is also suitable to *localise* objects. For that purpose it suffices to discard all pixels in the image that are not a local minimum. One way to accomplish this task is the so-called *non-local maximum suppression*. A realisation of this algorithm is given in listing 5.7.

Listing 5.7 non-local maximum suppression

```
#define NONMAX     -1          /* identifier used to label pixels              */

void nlm_suppression(f,g,N,M) /* labels all pixels that are not a local maximum
                                  according to 4-neighborhood adjacency         */
double **f;                    /* Pointer to the input image, declared as a
                                  2-dimensional array of doubles.               */
double **g;                    /* Pointer to the output image                   */
int N,M;                       /* number of rows and columns, respectively      */
{
    int i,j;
    for (i=1;i<N-1;i++)
    for(j=1;j<M-1;j++)
    if ((f[i][j]<=f[i-1][j]) ||
        (f[i][j]<=f[i+1][j]) ||
        (f[i][j]<=f[i][j-1]) ||
        (f[i][j]<=f[i][j+1]))  g[i][j]=NONMAX;
    else
                g[i][j]=f[i][j];
}
```

It is interesting to expand the expression in equation (5.68c):

$$\begin{aligned}d_{n,m}^2 &= \sum_{k=-K}^{K}\sum_{l=-L}^{L}\left(f_{n+k,m+l}-h_{k,l}\right)^2 \\ &= \underbrace{\sum_{k=-K}^{K}\sum_{l=-L}^{L}h_{k,l}^2}_{\text{template energy}} - 2\underbrace{\sum_{k=-K}^{K}\sum_{l=-L}^{L}f_{n+k,m+l}h_{k,l}}_{\text{cross correlation}} + \underbrace{\sum_{k=-K}^{K}\sum_{l=-L}^{L}f_{n+k,m+l}^2}_{\text{local image energy}}\end{aligned} \qquad (5.68\text{d})$$

The first term on the right hand side does not depend on the observed image. It is the *norm* (sometimes called *energy*[5]) of the template. Since the term does not depend on (n,m) it can be omitted in further calculations without affecting the quality of the detection and the localisation.

The last term solely depends on the observed image. This term represents the local energy of the image calculated in a $(2K+1)\times(2L+1)$ neighborhood surrounding each pixel (n,m). Generally, the term depends on (n,m). Therefore, it cannot be omitted without degrading the quality of the detection and localisation. In some applications, however, the image is wide-sense stationary (see section 3.2), implying that on an average the local energy is constant everywhere. In that case the third term can be omitted too. If so, the second term remains. This term is called the *(cross) correlation* between image and template. In fact, correlation is equivalent to convolution with the reflected template (see table 2.4). Figure 5.24 shows the correlation of the original image in figure 5.23 with the template. Note that in this example the observed image is not stationary. Consequently, the performance of the detection/localisation will be poor. In example 3.2 the image is such that the stationarity assumption holds. Therefore, in that example, template matching is successful.

Another situation in which template matching often fails is when the image of the object of interest is not perfectly known. Many factors may contribute to this lack of knowledge: unknown factors in the illumination, the reflection, the imaging

---

[5]The term "energy" stems from the analogy to electrical signals. From a physical point of view, the term is not appropriate here. An image is a distribution of radiant energy. Hence, squaring an image has not the physical connotation of "energy".

geometry, the shape, size and orientation of the object, etc. In many occasions this lack of knowledge hinders the utilisation of correlation techniques. However, the influence of some unknown factors can be neutralised quite easily. For instance, if a constant background level of the image is unknown, i.e. $f(x,y) = h(x-\xi, y-\eta) + \underline{n}_1(x,y) + \underline{b}$, then the influence of $\underline{b}$ can be neutralised by normalisation of the (local) average grey level within the template and the image. For this purpose define:

$$\overline{h} = \frac{1}{(2K+1)(2L+1)} \sum_{k=-K}^{K} \sum_{l=-L}^{L} h_{k,l}$$

$$\overline{f_{n,m}} = \frac{1}{(2K+1)(2L+1)} \sum_{k=-K}^{K} \sum_{l=-L}^{L} f_{n+k,m+l}$$

(5.70)

A measure of match can be formulated between the fluctuating part in $h_{n,m}$ and $f_{n,m}$. That is, between $h_{n,m} - \overline{h}$ and $f_{n,m} - \overline{f_{n,m}}$. For instance, if stationarity is assumed it suffices to calculate the cross correlation normalised with respect to background level:

$$\sum_{k=-K}^{K} \sum_{l=-L}^{L} (f_{n+k,m+l} - \overline{f_{n,m}})(h_{k,l} - \overline{h})$$

(5.71)

Another nuisance parameter that sometimes hampers the detection and/or localisation is the gain $\underline{a}$, i.e. $f(x,y) = \underline{a}h(x-\xi, y-\eta) + \underline{n}_1(x,y)$. The parameter $\underline{a}$ can be eliminated by normalisation of the local energy. Define:

$$\overline{h^2} = \frac{1}{(2K+1)(2L+1)} \sum_{k=-K}^{K} \sum_{l=-L}^{L} h_{k,l}^2$$

$$\overline{f_{n,m}^2} = \frac{1}{(2K+1)(2L+1)} \sum_{k=-K}^{K} \sum_{l=-L}^{L} f_{n+k,m+l}^2$$

(5.72)

Invariance with respect to the parameter $\underline{a}$ is obtained by division of $f_{n,m}$ by $(\overline{f_{n,m}^2})^{1/2}$. Note that due to this normalisation the first and third term in equation (5.68c) essentially become constants, so that only the second term is important. The result is the following measure of match:

Figure 5.24 The cross correlation between image and template

$$r_{n,m} = \frac{\sum_{k=-K}^{K}\sum_{l=-L}^{L} f_{n+k,m+l} h_{k,l}}{\sqrt{\overline{f_{n,m}^2}} \sqrt{\overline{h^2}}} \qquad (5.73)$$

If gain and background are unknown, then normalisation with respect to both parameters must be achieved:

$$r_{n,m} = \frac{\sum_{k=-K}^{K}\sum_{l=-L}^{L} (f_{n+k,m+l} - \overline{f_{n,m}})(h_{k,l} - \overline{h})}{\sqrt{\overline{(f_{n,m} - \overline{f_{n,m}})^2}} \sqrt{\overline{(h - \overline{h})^2}}} \qquad (5.74)$$

This last measure is called *cross correlation coefficient* because of its analogy with the normalised correlation coefficient defined between two random variables.

Examples of the various measures of match are depicted in figure 5.25. The last measure is sensitive to noise. This is essentially true in regions with poor contrasts. In these regions the signal is amplified strongly so as to reach unit variance.

5.4 MORPHOLOGICAL OPERATIONS

The study of form and structure of biological organisms is called morphology. This branch of science has evolved into a theory called *mathematical morphology*. This theory describes geometrical structures in two and three dimensions. As such, morphology has become widely applicable as a methodology in image analysis. It provides methods to build image operations with which the spatial structure of objects in images can be modified. These operations are generally referred to as *morphological operations*.

Mathematical morphology is based on *set theory*. The sets of interest can be defined in Euclidean spaces such as $\mathbb{R}^2$ or $\mathbb{R}^3$. For instance, the geometry of an object in 3-dimensional space can be represented as a set of points. In image analysis this object is observed at certain positions in the 2-dimensional image

a)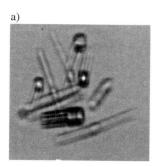
b)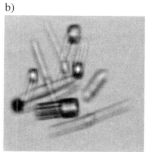
c)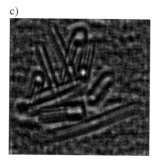

Figure 5.25 Normalisation
a) Normalisation with respect to background level
b) Normalisation with respect to gain
c) Normalisation with respect to background level and gain

plane, i.e. at $A \subset \mathbb{R}^2$. Moreover, since most image analysis technologies are based on discretisation on an orthogonal grid $\mathbb{Z}^2$ (the elements of which are enumerated by pairs of integers), the representation of these positions becomes $A \subset \mathbb{Z}^2$. This last representation is the domain of *digital morphology*. Note, that if there are more objects, we may want to associate a set with each distinguishable object, e.g. A, B, C, etc.

Another representation of the image of an object, equivalent to the set representation $A \subset \mathbb{Z}^2$, is that of a *binary image* or a *bitmap*. A binary image is a function defined in $\mathbb{Z}^2$ that maps each pixel in $\mathbb{Z}^2$ onto the domain $\{0, 1\}$. To each element in A we assign the label "1". Likewise, to each pixel not contained in A we assign the label "0":

$$a_{n,m} = \begin{cases} 1 & \text{if} \quad (n,m) \in A \\ 0 & \text{if} \quad \text{elsewhere} \end{cases} \qquad (5.75)$$

As an example, consider the scene depicted in figure 5.26a. This scene consists of a number of electronic components. Taken together, these objects correspond to one subset of the image plane. In this example, an estimate of the subset can be obtained by thresholding the grey level image. The result is the binary image shown in figure 5.26b. In the image, white areas correspond to objects, generally called the *foreground*. This set reflects the spatial structure of the scene. The complement of the set, the black area, is called the *background*.

This section discusses a class of operations defined on binary images. The operations of interest are *spatial invariant neighborhood operations*. As in the case of discrete convolution, each output pixel depends on the corresponding local neighborhood of the input image; see figure 5.7. Unlike convolution however, in binary operations the input and output pixels take values "0" or "1" only. In the following sections, some definitions are introduced. Next, some basic operations are discussed, followed by sections showing how to use these operations for extraction of shape primitives.

Mathematical morphology does not confine itself to binary operations. The concepts from set theory can be generalised to grey level imagery. This chapter concludes with a section introducing this generalisation briefly.

a)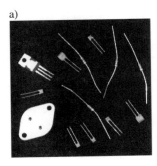
b)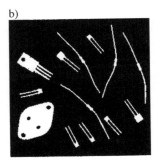

Figure 5.26 Objects represented by foreground and background
a) Image of a scene
b) Thresholded image

5.4.1 Definitions

In this section, the word *pixel* refers to an element from an orthogonal grid, e.g. $a \in \mathbb{Z}^2$. The grid is considered to be infinite. Hence, the finiteness of the image plane is not taken into account. Since local neighborhoods are studied, the neglect of finiteness introduces perturbations only near the borders of the image plane.

A collection of pixels is called a *region*: $A = \{a_i | i = 1, 2, \cdots\} \subset \mathbb{Z}^2$. The *complement* of a region A, denoted by A^c, is the set of all pixels not contained in A. Hence, if A is the foreground, then its complement A^c is the background. If two or more sets A, B, C, etc. are involved, the usual operations from set theory apply:

- union: $\quad A \cup B \quad$ all pixels that belong to A or B (or to both).
- intersection: $\quad A \cap B \quad$ all pixels that belong to both A and B.
- difference: $\quad A - B = A \cap B^c \quad$ pixels that belong to A but not to B.

Consequently, all well-known properties from set theory also apply to our sets. For instance, the operations $\cup$ and $\cap$ are commutative and associative, i.e.:

$$A \cup B = B \cup A \qquad (A \cup B) \cup C = A \cup (B \cup C)$$

$$A \cap B = B \cap A \qquad (A \cap B) \cap C = A \cap (B \cap C)$$

Moreover, the operations $\cup$ and obey the following distributive laws:

$$(A \cup B) \cap C = (A \cap C) \cup (B \cap C) \qquad (A \cap B) \cup C = (A \cup C) \cap (B \cup C)$$

A well-known property of sets is the duality principle (DeMorgan's law):

$$A \cup B = (A^c \cap B^c)^c$$

The duality principle is stated more generally as: "any theorem involving a family of sets can be converted automatically into another theorem by replacing all sets by their complements, all unions by intersections and all intersections by unions".

In "real analysis" a *metric space* is a set equipped with a *distance measure*, see appendix A.2. In a vector space like $\mathbb{R}^2$ one normally uses the Euclidean distance. Let $a = [a_1 \ a_2]$ and $b = [b_1 \ b_2]$, then:

$$D_e(a,b) = \sqrt{(a_1 - b_1)^2 + (a_2 - b_2)^2} \qquad (5.76)$$

In digital morphology, using an orthogonal grid $\mathbb{Z}^2$ more convenient metrics are the city-block distance $D_4(a,b)$ and the *chessboard distance* $D_8(a,b)$, defined as:

$$\begin{aligned} D_4(a,b) &= |a_1 - b_1| + |a_2 - b_2| \\ D_8(a,b) &= \max(|a_1 - b_1|, |a_2 - b_2|) \end{aligned} \qquad (5.77)$$

If the city-block distance is adopted, the sets consisting of pixels with an equal distance to (for instance) the origin are diamonds. Likewise, if the chessboard distance is used, these sets become squares:

$$D_4(.,.): \begin{matrix} & & 2 & & \\ & 2 & 1 & 2 & \\ 2 & 1 & 0 & 1 & 2 \\ & 2 & 1 & 2 & \\ & & 2 & & \end{matrix} \qquad D_8(.,.): \begin{matrix} 2 & 2 & 2 & 2 & 2 \\ 2 & 1 & 1 & 1 & 2 \\ 2 & 1 & 0 & 1 & 2 \\ 2 & 1 & 1 & 1 & 2 \\ 2 & 2 & 2 & 2 & 2 \end{matrix}$$

In metric spaces, the ε-*neighborhood* $N(a)$ of a point a is the collection of points such that the distance $D(a,b)$ from each point b in $N(a)$ to a is smaller than a constant ε. In an orthogonal grid, using the city-block distance, the neighborhood with ε = 1 consists of the central pixel a and the set of horizontal and vertical direct neighbors of a. This neighborhood is denoted by $N_4(a)$. The four direct neighbors are generally referred to as the *4-neighbors* of a. If the chessboard distance is used, the neighborhood $N_8(a)$ consists of a and the set of horizontal, vertical and diagonal direct neighbors. The eight direct neighbors are the *8-neighbors* of a:

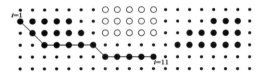

$$N_4(a) = \{a,t,v,w,y\} \qquad N_8(a) = \{a,s,t,u,v,w,x,y,z\} \qquad (5.78)$$

If the city-block distance is used, then a pixel a is *adjacent* to a pixel b, whenever $b \in N_4(a)$. If the chessboard distance is used, a is adjacent to b if $b \in N_8(a)$. The relation "adjacency" is *symmetric*, i.e. if a is adjacent to b, then b is adjacent to a. However, adjacency is *not transitive*: if a is adjacent to b, and b is adjacent to c, then this does not imply that a is adjacent to c. Two pixels are *connected* if they are adjacent to each other. Two pixels a and b are *4-connected*, if $b \in N_4(a)$. The definition of *8-connected* is likewise. Two regions A and B are *connected* if one or more pixels from A are adjacent to one or more pixels from B. As an example, in the following sets:

the regions denoted by ● and ○ are 8-connected. However, these regions are not 4-connected.

A *path* is a collection of, say, M distinct pixels $\{a_i | i = 1, \cdots, M\}$ such that for $i = 2, \cdots, M$ each pixel is a neighbor of its predecessor: $a_i \in N(a_{i-1})$. For instance, in the example given above, a path exists between pixel a_1 and a_{11} (if 8-connectivity is used). Note that this path is not unique. If in a region A a path exists between two pixels a and b, then a and b are said to be *connected in A*.

A *connected component* of a region A is a subset of A in which all pixels are connected in A, and all other pixels in A are not connected to that subset. In the example given above, the region ● has two connected components provided that 8-connectivity is used. If 4-connectivity is used, the region has three connected components.

The definition of 4-connectivity introduces a complication if it is applied to both the background and the foreground. Consider the configuration shown below. If 4-connectivity applies to both sets, each set consists of two connected components. If the components A and D are separated, one would expect that B and

C are connected. However, under the definition of 4-connectivity this is not the case. If 8-connectivity applies to both sets, A and D are connected and so are B and C. These situations are not part of what we expect connectivity to mean. A solution to bypass this problem is to apply 4-connectivity to the background and 8-connectivity to the foreground, or the other way round; see below.

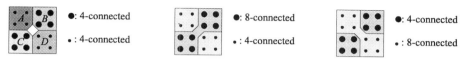

5.4.2 Basic operations

Besides set theoretic operations, like union and intersection, we can add operations known from vector spaces. Especially the *vector addition* $a+b$ and the *vector negation* $-a$ are of particular interest here. These two vector operations lead to definitions of the *translation* and the *reflection* of a region. With the *translation* A_t of a region A by a vector $t \in \mathbb{Z}^2$ is meant the set:

$$A_t = \{a+t \mid \text{for all } a \in A\} \tag{5.79}$$

The *reflection* of A is denoted by $\bar{A}$ and is defined as:

$$\bar{A} = \{-a \mid \text{for all } a \in A\} \tag{5.80}$$

The operations are illustrated in figure 5.27. The example is taken from a subimage of figure 5.26b. This subimage is depicted in figure 5.27a. The figure is enlarged to show the details at pixel level.

Dilation and erosion

The primary operations in mathematical morphology are the *dilation* and the *erosion*. In both operations the image associated with a set is probed with another set of known shape called *structuring element*. For 2-dimensional digital images, structuring elements are subsets of $\mathbb{Z}^2$. In the sequel, the variable B will be used to denote structuring elements. It is important to keep in mind that B is defined with respect to the origin of a co-ordinate system. We can probe an image with B by translating B by all pixels in $\mathbb{Z}^2$.

Dilation of a set A by B relates to the following question: "At which positions of B does B hit A?" The dilated set, denoted $A \oplus B$, is the positions for which this question is answered affirmatively:

$$A \oplus B = \{t \mid B_t \cap A \neq \emptyset\} \tag{5.81}$$

Examples of dilations are given in figure 5.28a, b and c. The original is shown in figure 5.27a. In figure 5.28a the structuring element is a diamond corresponding to the 4-neighborhood N_4 of the origin. The operation is called "4-neighborhood dilation". The foreground grows in four directions by one pixel. Small holes in the foreground will be filled, as with "line-like" holes provided that they are at most two pixels thick and non-diagonally oriented. Thin concavities of the contour are

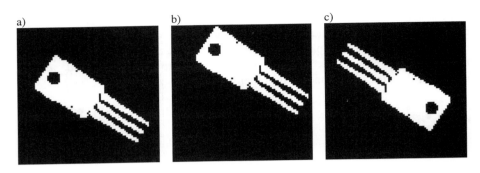

Figure 5.27. Translation and reflection
a) Original set.   b) Translated.   c) Reflected about the centre.

also removed. As such, the operation smoothes the contour. Furthermore, small gaps between two connected components will be bridged causing the fusion of the two components.

8-neighborhood dilation is shown in figure 5.28b. Here, the structuring element is the 8-neighborhood N_8 of the origin. The object grows isotropicly. The removal of concavities and holes is more pronounced compared with 4-neighborhood dilation.

The last example of dilation (figure 5.28c) shows a structuring element that does not contain the origin. The two pixels in B are diagonally oriented with a distance between them that matches the distance between two neighboring leads of the transistor depicted in the original. As a consequence, the dilation replaces each lead with two parallel bars. Since there are three leads, there would be six bars. However, the bars in the middle coincide with other bars, so that only four bars remain.

Erosion can be regarded as the opposite of dilation. Erosion of a set A by the structuring element B answers the following question: "At which positions of B does A fit B?" More precisely, the erosion $A \ominus B$ is defined as:

$$A \ominus B = \{t | B_t \subset A\} \tag{5.82}$$

Figures 5.28d and e show the results of 4-neighborhood and 8-neighborhood erosions. Here, the structuring elements are N_4 and N_8. The foreground shrinks by one pixel. Small or thin objects will vanish, as with protrusions of the contour. Furthermore, an object with a narrow neck will be split into two connected components provided that the width of the neck is at most two pixels.

The structuring element in figure 5.28f consists of two diagonally oriented pixels with a distance that matches the two outer leads of the transistor. This pattern fits the object in the central part of the encapsulation of the transistor and right in the middle between the two outer leads.

Properties

It is straightforward to prove the following properties:

a. commutative laws: $A \oplus B = B \oplus A$. $A \ominus B = B \ominus A$
b. associative law: $A \oplus (B \oplus C) = (A \oplus B) \oplus C$
c. decomposition: $(A \ominus B) \ominus C = A \ominus (B \oplus C)$

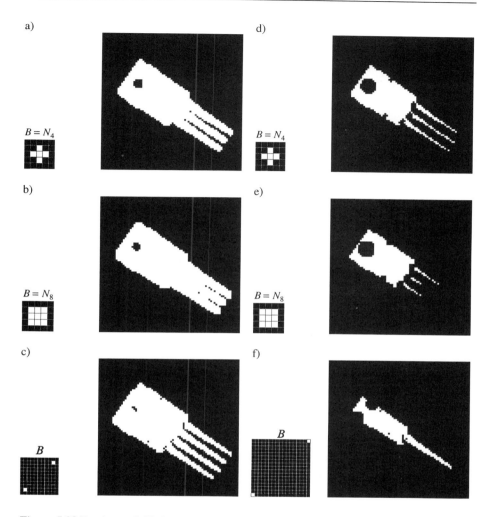

Figure 5.28 Erosion and dilation
a) 4-neighborhood dilation
b) 8-neighborhood dilation
c) Dilation by B
d) 4-neighborhood erosion
e) 8-neighborhood erosion
f) Erosion by B

d. distributive laws: $(A \cup B) \oplus C = (A \oplus C) \cup (B \oplus C)$ $(A \cap B) \ominus C = (A \ominus C) \cap (B \ominus C)$
e. duality: $(A \ominus B)^c = A^c \oplus \breve{B}$ $(A \oplus B)^c = A^c \ominus \breve{B}$ (5.83)
f. translation: $A_t \oplus B = (A \oplus B)_t$ $A \ominus B_t = (A \ominus \breve{B})_t$
g. increasingness: $A \subseteq C \Rightarrow A \oplus B \subseteq C \oplus B$ $A \subseteq C \Rightarrow A \ominus B \subseteq C \ominus B$
h. (anti-)extensivity: $\mathbf{0} \in B \Rightarrow A \subseteq A \oplus B$ $\mathbf{0} \in B \Rightarrow A \ominus B \subseteq A$

The associative law and the decomposition are important, since it states that the dilation (erosion) of a set by a large structuring element can be decomposed into a cascade of dilations (erosions) by smaller structuring elements. The principle of duality states that a dilation can be converted into an erosion, and vice versa. The property of increasingness says that dilation and erosion preserves the order between sets, i.e. if $A \subseteq C$, then so does their dilated (eroded) versions. An operation

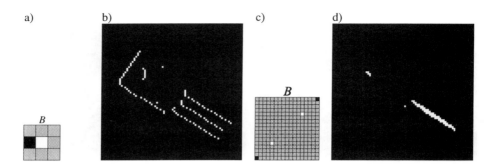

Figure 5.29 Hit-and-miss transforms
a) Structuring element consisting of B_{HIT} (depicted white) and B_{MISS} (depicted black)
b) Result of the hit-and-miss transform of figure 5.27a by B
c) Structuring element suitable to extract the central lead of the transistor
d) Result of application of c)

is extensive, if its output contains the input. The operation is anti-extensive, if the reverse holds true. Dilation and erosion are extensive and anti-extensive, respectively, provided that the structuring element contains the origin.

Hit-and-miss transform

This morphological operation involves the use of two structural elements. One has to hit the foreground; the other has to hit the background. The operation, denoted HMT(.,.), is expressed as the intersection of the foreground eroded by a structuring element B_{HIT} and the background eroded by B_{MISS}.

$$\mathrm{HMT}(A, B) = (A \ominus B_{HIT}) \cap (A^c \ominus B_{MISS}) \tag{5.84}$$

The structural elements are contained in a single composite structuring element $B = \{B_{HIT}, B_{MISS}\}$. The intersection $B_{HIT} \cap B_{MISS}$ must be empty, because otherwise the hit-and-miss transform would always result in the empty set. Therefore, the set can be represented graphically in a single object by the convention that pixels in B_{HIT} are displayed white, pixels in B_{MISS} are black, and remaining pixels (i.e. "don't care's") are shaded. Figure 5.29 provides an example. The structuring element shown in figure 5.29a extracts all foreground pixels that have a left-sided neighboring background pixel; i.e. the left-sided boundary points of the object. Figure 5.29c extracts the central lead of the transistor. In fact, the result is the intersection between figure 5.28f and the complement of figure 5.28c. Note that this extraction is not invariant to orientation and scale of the object.

Opening and closing

In general, two different sets may have equal erosions. Therefore, the erosion operation is irreversible. That is, one cannot recover a set from its eroded version. However, one can determine the smallest set that has a given erosion. The operator that recovers that smallest set is the dilation. Together the operations erosion and dilation are called *opening*:

$$A \circ B = (A \ominus B) \oplus \bar{B} \tag{5.85}$$

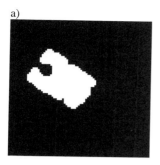 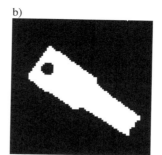

Figure 5.31 Opening and closing
a) Opening by a disk with radius 2
b) Closing by a disk with radius 2

The opening of A by B defines those pixels in A that are covered by some translation of B that itself is entirely contained in A. This can be visualised as follows. Consider the set A as the map of a harbor (see figure 5.30). The complement A^c corresponds to the quay. The top-view of a boat corresponds to the structuring element B. Let us assume that the boat is pointing northwards, and that the boat is not able to turn. Then, the erosion of the harbor is the set of all points that are within reach from the centre of the boat. The opening of the harbor is the collection of all points that are within reach somewhere from the boat.

Opening a set with a disk-like structuring element eliminates small islands and sharp peaks. Application of such an opening to the image of the transistor yields the set shown in figure 5.31a. Note that the leads are removed because these structures cannot contain the disk. Furthermore, the hole has been linked to the background.

The operation *closing* is defined as:

$$A \bullet B = (A \oplus B) \ominus \bar{B} \tag{5.86}$$

The operation is dual to opening in the sense that closing the foreground is equivalent to opening the background. As such all properties of the foreground of an opened set also apply to the background of a closed set. The result of closing the image of the transistor by a disk-like structure is shown in figure 5.31b. It can be seen that all concavities that could not hold the disk are filled.

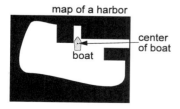

Figure 5.30 The opening is the set of all points within reach of the boat

Listing 5.8 2-dimensional binary dilation

```
void dilation(a,b,c,N,M,K,L)    /* 2-dimensional binary dilation
                                   pixels near the edge of the image are discarded. */
int **a, **b, **c;              /* Pointers to the input image a and the output image c,
                                   respectively, and to the structuring element b.
                                   These images are declared as 2-dimensional arrays of
                                   integers. The background is represented by "0".
                                   The foreground is a "1".                         */
int N,M;                        /* number of rows and columns of the images a and c */
int K,L;                        /* The number of rows and columns of the structuring
                                   element b is 2K+1 and 2L+1, respectively.        */
{
    int n,m;                    /* indices in image.                                */
    int k,l;                    /* indices in structuring element.                  */
    int dk,dl;                  /* number of rows and columns in structuring element*/
    dk=2*K+1;
    dl=2*L+1;
    for (n=K;n<N-K;n++)
    for (m=L;m<M-L;m++)
    {   c[n][m] = 0;
        for (k=0;k<dk;k++)
        for (l=0;l<dl;l++)
            c[n][m] ||= ((b[k][l]) && (a[n-k-K][m-l-L]));
    }
}
```

The following properties holds true for the opening and closing operations:

a. increasingness: $A \subseteq C \Rightarrow A \circ B \subseteq A \circ B$ $A \subseteq C \Rightarrow A \bullet B \subseteq A \bullet B$
b. (anti-)extensivity: $A \circ B \subseteq A$ $A \subseteq A \bullet B$ (5.87)
c. idempotence: $(A \circ B) \circ B = A \circ B$ $(A \bullet B) \bullet B = A \bullet B$

An operation is *idempotent* if repeating the operation induces no further changes.

Implementations

The dilation and erosion operations defined in (5.81) and (5.82) can be implemented in various ways. One dichotomy is to distinguish parallel implementations from sequential implementations. A further categorisation can be based on the type of data representation that is used, for instance: bitmaps, boundary representations, queues, etc. In a parallel implementation, each image pixel is processed independently. In principle, all pixels could be processed at the same time provided that for each pixel one processor is available. The implementation requires one memory for the input image and one memory for the output image. Hence, the processing cannot be done in-place. In sequential implementations the image plane is scanned in a well defined order. When a new output pixel is processed, input pixels are used together with output pixels that are already available. Sometimes this type of implementation is called "serial" or "recursive".

A parallel implementation of the dilation operation given in (5.81) is obtained by the following (equivalent) representation of the dilation operation:

$$A \oplus B = \bigcup_{b \in B} A_{-b}$$

This representation can be transformed into bitmap representation in which "logical 1" denotes foreground (see equation 5.75):

$$C = A \oplus B \iff c_{n,m} = \bigcup_i \bigcup_j a_{n-i,m-j} \cap b_{i,j}$$

Listing 5.8 provides a realisation of this implementation in pseudo-C. The implementation of the erosion operation can be done similar.

Hit-and-miss transforms are often implemented by using a look-up table (LUT) to perform the search for a specific configuration. For instance, in a 3×3-neighborhood centered at a pixel, let the index I be defined as:

a_0	a_1	a_2
a_3	a_4	a_5
a_6	a_7	a_8

$$I = \sum_{i=0}^{8} 2^i a_i$$

This index is an integer with a range of [0,511]. There exists a one-to-one correspondence between the index I and the state of the neighborhood. Therefore, the hit-and-miss transform could be implemented by a LUT that transforms the state of the neighborhood to the appropriate output state of the central pixel. A realisation of this implementation is given in listing 5.9. Note that the calculation of the index is equivalent to convolution.

5.4.3 Morphological operations

The operations discussed in the preceding section are elementary in the sense that a large number of task-oriented procedures can be built with these operations. These tasks range from noise suppression to shape description of connected components. This section discusses noise removal operations, boundary extraction, the skeleton of a region, the convex hull and region filling techniques. The following section gives some examples of applications.

Noise removal

If a binary image is corrupted by noise appearing as pixel variations that are not spatially correlated, morphological operations may be helpful to eliminate these variations. Uncorrelated noise often manifests itself as isolated foreground or background pixels. These isolated pixels can be easily detected by a hit-and-miss

Listing 5.9 2-dimensional hit-and-miss transform

```
void hit_and_miss(a,c,N,M,LUT)     /* 2-dimensional hit-and-miss transform. Pixels
                                      near the edge of the image are discarded      */
int **a, **c;                      /* Pointers to the input image a and the output
                                      image c, respectively. These images are
                                      declared as 2-dimensional arrays of integers.
                                      The background is represented by "0".
                                      The foreground is a "1".                      */
int *LUT;                          /* Look-up-table consisting of 512 entries corresponding
                                      the 512 different states of a 3x3 neighborhood */
int N,M;                           /* number of rows and columns of the images a and c */
{
    int n,m;                       /* indices in image.                              */
    int i,j;                       /* indices in neighborhood.                       */
    int kernel[3][3]={{1,   2,   4},  /* convolution kernel                          */
                      {8,  16,  32},
                      {64,128,256}};
    int index;
    for (n=1;n<N-1;n++)
    for (m=1;m<M-1;m++)
    {   index = 0;
        for (i=0;i<3;i++)
        for (j=0;j<3;j++)
        {   index += kernel[i][j]*a[n-i-1][m-j-1];
        }
        c[n][m] = LUT[index];
    }
}
```

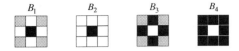

Figure 5.32 Structuring elements to detect isolated pixels

transform. Figure 5.32 shows structuring elements that are suitable for that purpose. B_1 and B_2 are elements that extract isolated background pixels. The elements B_3 and B_4 are appropriate for isolated foreground pixels. The difference between B_1 and B_2 is that B_1 employs 4-connectivity, whereas B_2 uses 8-connectivity. This difference also applies to B_3 and B_4.

Once an isolated pixel has been detected, the removal of it is simple. For instance, when 8-connectivity applies to the background, an isolated background pixel is deleted by the operation:

$$A \cup \mathrm{HMT}(A, B_2) \qquad (5.88)$$

Likewise, an isolated foreground pixel (4-connectivity) is removed by:

$$A - \mathrm{HMT}(A, B_3) \qquad (5.89)$$

If both foreground and background isolated pixels have to be removed, a cascade of the two operations will do: $C \cup \mathrm{HMT}(C, B_2)$ with $C = A - \mathrm{HMT}(A, B_3)$. The result of this operation applied to figure 5.27a is depicted in figure 5.33a. Note that some spurious background pixels that would have been deleted if 4-connectivity had been applied are retained because they have one or more diagonal background neighbours.

Another type of noise removal is based on voting. For instance, the *majority voting operation* turns a foreground pixel into a background pixel whenever the majority of its neighbourhood belongs to the background. Similar, a background pixel becomes foreground whenever the majority of its neighborhood says so. An example is provided in figure 5.33b. Here, most spurious background pixels seem to be removed.

a)
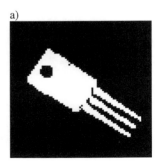

b)
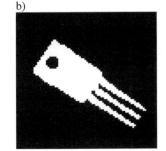

Figure 5.33 Noise removal
a) Isolated pixel removal
b) Majority voting

Figure I RGB-image of a printed circuit board

Figure II RG-image. The blue channel has been suppressed

Figure III The red, green and blue channels shown individually

Figure IV RGB-representation of the image from Figure I shown in a 3-dimensional feature space

Figure V RG-representation of the image from Figure I shown in a 2-dimensional feature space

Boundary extraction

In a metric space, an *interior point* of a set is a point the neighborhood of which entirely belongs to the set. For instance, if 4-connectivity is used, the set of interior points of A is the erosion $A \ominus N_4$. Consequently, all points of A that are not interior points have at least one direct neighbor that belongs to A^c. These points are called *inner boundary points*. The points are defined by:

$$IB(A) = A - (A \ominus N) \tag{5.90}$$

If 4-connectivity is chosen (i.e. $N = N_4$), the resulting inner boundary consists of 8-connected paths. Alternatively, in case of 8-connectivity ($N = N_8$) of A, the resulting inner boundary consists of 4-connected paths. Figure 5.34 illustrates this.

Complementary to the inner boundary, the *outer boundary* of a set A is defined as all points not belonging to A, but with at least one neighbor in A. This is expressed as:

$$OB(A) = A^c \cap (A \oplus N) \tag{5.91}$$

Here too, 8-connectivity of A leads to 4-connected paths, and vice versa. Again, illustrations are given in figure 5.34.

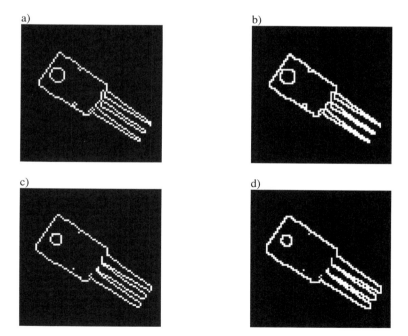

Figure 5.34 Boundaries
a) Inner boundary ($N = N_4$)
b) Inner boundary ($N = N_8$)
c) Outer boundary ($N = N_4$)
d) Outer boundary ($N = N_8$)

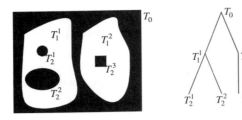

Figure 5.35 The homotopy of a set.

Homotopy preserving thinning

Thinning a set A by a structuring element $B = \{B_{HIT}, B_{MISS}\}$ is an operation defined by:

$$A \otimes B \equiv A - \text{HMT}(A, B) \tag{5.92}$$

An important application of the thinning operation is to obtain a *skeleton*. A skeleton of a set A is a subset of A with the following desirable properties:

1. A pixel in the skeleton is the centre of a maximal-sized disk that is contained in the original set.
2. The skeleton preserves the homotopy of the original set.

The *homotopy* of a (bounded) set A is a tree the root T_0 of which corresponds to the component of A^c that is connected to the infinite; see figure 5.35. The first branches T_1^i correspond to the components of A connected to T_0. The second branches T_2^i correspond to the holes of the components T_1^i, etc.

In an Euclidean continuous space, the set of all points that have property 1 automatically fulfils property 2. However, in $\mathbb{Z}^2$ with the city-block distance or the chessboard distance, this is not the case. Therefore, in order to retain property 2 we can only fulfil property 1 approximately.

A skeleton is obtained by repeated application of a thinning operation, i.e. $A^0 = A$, $A^{i+1} = A^i \otimes B$. The iteration stops as soon as no changes occur (idempotence). However, a homotopy preserving skeltonise operation cannot be realised by a single 3×3-thinning operation. The 3×3-neighborhood is too small to ensure connectivities and yet to retain the pixels with central positions. A possibility is to apply a sequence of 3×3-thinning operations, each one working in a particular direction. If a 4-connected skeleton is desired, a sequence of 8 operations per iteration suffices:

$$A^{i+1} = \left[\left\{ \left(A^i \otimes B_1 \right) \otimes B_2 \right\} \otimes \cdots \otimes B_8 \right] \tag{5.93}$$

The structuring elements are given in figure 5.36a. The result of the operation, applied to the (smoothed) set shown in figure 5.33b, is given in figure 5.36b. If 8-connected skeltons are desired, a sequence of 12 different thinnings. The 12 structuring elements, together with the resulting skeleton, are given in figure 5.36c and d. Note that the 8-connected skeleton is more regular and more symmetric than the 4-connected skeleton.

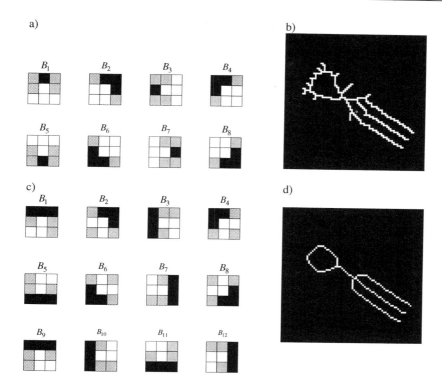

Figure 5.36. Skeletons obtained from figure 5.33b
a) Structuring elements for 4-connected skeleton
b) 4-connected skeleton
c) Structuring elements for 8-connected skeleton
d) 8-connected skeleton

Pruning

The 4-connected skeleton in figure 5.36b shows many tiny spurs. These spurs may be considered as unwanted artefacts caused by small irregularities in the contour of the original set. The artefacts need to be cleaned up. Pruning is a thinning operation that removes the end points of a skeleton. A hit-and-miss transform with the structural elements given in figure 5.37a detects all (8-connected) end points. Consequently, the sequence of thinnings (equation 5.93) by these elements can be used to prune the spurs. Figure 5.37b shows the result of 6 iterations of the pruning operation applied to figure 5.36b. In this example, all artefacts are removed.

Connectivity conversions

A 4-connected path can be easily converted into an 8-connected path by application of the skeltonise operation depicted in figure 5.36c. The reverse, i.e. the conversion of an 8-connected path into a 4-connected path, is obtained by a so-called *diagonal fill*. This operation adds a minimal amount of pixel to the foreground so as to guarantee 4-connectivity between all subsets that are 8-connected.

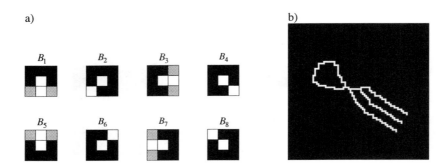

Figure 5.37 Pruning
a) Structural elements performing a pruning operation
b) Iterative application to figure 5.35b

Applied in the hit-and-miss transform, the structuring elements given in figure 5.38a detect all 8-connected structures that are not 4-connected. A union of the original set with these points would assure 4-connectivity. However, such a union would add more points than required. To prevent this, a sequence of hit-and-miss transforms and unions is needed. For that purpose, define a *thickening operation* ¤ as:

$$A \,¤\, B = A \cup \text{HMT}(A, B) \qquad (5.94)$$

With this notation, the diagonal fill operation can be defined as:

$$\bigl(((A \,¤\, B_1) \,¤\, B_2) \,¤\, B_3\bigr) \,¤\, B_4 \qquad (5.95)$$

After application of the first two thickening operations, the hit-and-miss transforms by B_3 and B_4 yield empty sets. Therefore, the last two thickenings are superfluous. The result of diagonal fill applied to figure 5.36d is given in figure 5.38b.

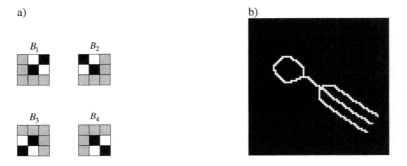

Figure 5.38 Diagonal fill
a) Structural elements
b) Application to figure 5.36d

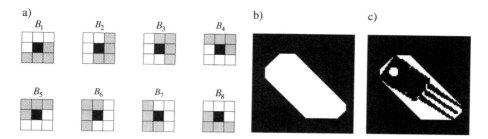

Figure 5.39 Convex hull
a) Structuring elements
b) Convex hull of the image in figure 5.27a
c) Convex deficiency

Convex hull

A set is called *convex* if the straight line between any two points of the set is contained in the set. The *convex hull* of a set is the smallest convex set that contains the original set. The *convex deficiency* of a set is the difference between the convex hull of a set and the original set. A digital approximation of the convex hull consisting of the intersection of half-planes, the normals of which are multiples of 45°, can be obtained by iterative applications of a sequence of eight thickenings. The structuring elements are given in figure 5.39a. The convex hull corresponding to the set in figure 5.27a is shown in figure 5.39b. The convex deficiency is shown in figure 5.39c. Note that the convex hull as defined according to the structuring elements in figure 5.39a is not rotationally invariant.

Reconstruction by dilation

Reconstruction (also called *region filling* or *propagation*) is an operation that involves three input sets. The sets are called the *seed* S, the *mask* M, and the structuring element B. The operation consists of an iterative sequence of *conditional dilations*. Each iteration is a dilation $S^i \oplus B$ followed by an intersection with M:

$$S^{i+1} = (S^i \oplus B) \cap M \qquad \text{with:} \qquad S^0 = S \qquad (5.96)$$

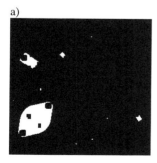
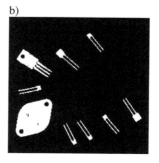

Figure 5.40 Conditional dilation
a) Erosion of figure 5.26b
b) Reconstructed

The iteration stops when no changes occur anymore, i.e. when $S^{i+1} = S^i$. We will denote this reconstruction by $R_B(S,M)$.

The result of the reconstruction is that all components from the set M that are (partly) covered by one or more points from the seed S are fully retrieved. Components of M that contain no points from the seed will be fully suppressed. Figure 5.40a shows the original image of figure 5.26b eroded by a disk with radius 2 (city-block distance). Hence, in this image, all components that cannot contain this disk are fully removed. If this image is used as the seed in a reconstruction operation conditionally to the original set (the mask), all connected components that are sufficiently large are retrieved; see figure 5.40b.

5.4.4 Applications

We conclude this section by two examples in which concepts from digital morphology are applied. The first example shows the ability of morphology to discriminate between components of different size. The second example concerns the relationship between morphology and distances.

Example 5.4 Sifting

Suppose that we have to separate a heap of apples into three categories: small, medium and big. The obvious way to accomplish this is to "sift" the apples through sieves of small and medium sizes. Such a sifting operation can also be applied to sets. A sifting operation must be anti-extensive (the output is a subset of the input set) and increasing (sifting a subset of a heap yields a set that is less than or equal to the result of sifting the whole heap). Furthermore, a sifting operation should be idempotent: no changes occur when a sifted set is sifted again with the same sieve.

The sieve that we use in this example is based on the *thickness* of each connected component. This quantity is the number of (iterative) erosions by (for instance) an 8-neighborhood needed to remove the component. For that purpose, consider the following iterative operation:

$$\underbrace{\left(\left(\left(A \ominus N_8\right) \ominus N_8\right) \cdots \ominus N_8\right)}_{t \text{ times}}$$

The next step is to restore the original components. This can be accomplished by a reconstruction by dilation. The whole operation is defined as:

$$A^t = R_{N_8}(C, A) \quad \text{with:} \quad C = \underbrace{\left(\left(\left(A \ominus N_8\right) \ominus N_8\right) \cdots \ominus N_8\right)}_{t \text{ times}}$$

The resulting set consists of components the thickness of which are all larger than t. As an example, the image in figure 5.40a results from a sequence of two 8-neighborhood erosions. Figure 5.40b shows the reconstructed set (denoted by A^2). Components with thickness less than three (noise pixels and the resistors) are deleted.

Using a number of sift operations, we can separate components based on their thickness. For that purpose we need to calculate the set differences between the set sifted by sieves with increasing sizes, i.e. $A^1 - A^2$, $A^2 - A^3$, etc. For instance, the resistors in the original set (figure 5.26b) are extracted by: $A^1 - A^2$ (figure 5.41a), the capacitors are $A^2 - A^5$ (figure 5.41b). The remaining components A^5 correspond to transistors (figure 5.41c).

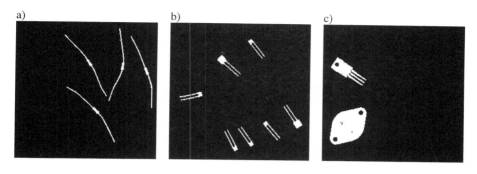

Figure 5.41 Sifting
a) Resistors ($A^1 - A^2$)   b) Capacitors ($A^2 - A^5$)   c) Transistors (A^5)

Example 5.5 Distance transforms

In section 5.4.1 we defined a *path* between two pixels a and b as a sequence of pixels $(a_1, a_2, \cdots, a_M)$ such that $a_1 = a$, $a_M = b$ and each pixel a_i is a neighbor of its predecessor a_{i-1}. By definition the *length* of a path is the number of arcs in it, i.e. $M-1$. The *distance* between two points a and b is the smallest length of all paths linking a to b. In this section we discuss some techniques related to finding the shortest path from one point to another or from one point to a certain set.

We consider a set A. To each point $a \in A$ the *distance transform* $d_A(a)$ is a function that assigns a number that is the minimal distance between a and any point of A^c. Figure 5.42 gives an example taken from geography. Figure 5.42a shows a (digital) map of The Netherlands. The country is surrounded by the Northsea, Belgium and Germany. Furthermore, the map shows us some lakes (in our terminology: "holes"). In figure 5.42b the distance transform is depicted together with the boundary. The grey tone of a point in this image corresponds to the minimal distance of that point to either the Northsea, to Belgium, to Germany or to a lake.

The distance transform can be calculated by a sequence of (iterative) erosions. For that purpose, define: $A^1 = A$ and $A^{i+1} = A^i \ominus B$. The structuring element B should match the distance metric at hand, i.e. N_8 in case of the chessboard distance; N_4 in case of the city-block distance. Then, the distance transform at a point a equals the number of iterations in which a is retained. That is, $d_A(a) = i$ with i selected such that $a \in A^i$ and $a \notin A^{i+1}$.

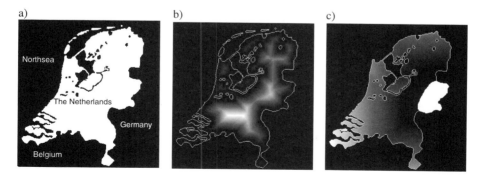

Figure 5.42 Distance transforms
a) Map of the Netherlands   b) Distance to border   c) Distance to eastern region

The operation requires a number of erosions that (in the worst case) equals half of the number of rows (or columns) of the corresponding bitmap. In practice, a recursive implementation in which the image plane is scanned in two passes is much faster. An example of such an implementation is given in listing 5.10. Modifications of this implementation are known for the calculation of the distance transform with other metrics [Borgefors 1984].

Once the distance transform is available, it is easy to find the shortest path from any point $a \in A$ to one of the boundary points of A. Starting from the initial point $a_1 = a$ we seek a neighbor the distance transform of which is less than the first point, i.e. $a_2 \in N(a_1)$ and $d_A(a_2) < d_A(a_1)$. The next points in the path are found similar: $a_{i+1} \in N(a_i)$ with $d_A(a_{i+1}) < d_A(a_i)$. This search continues until a point a_M is reached for which $d_A(a_M) = 0$. Note that the shortest path does not need to be unique.

The *geodesic distance* in a set A between two points $a, b \in A$, denoted by $d_A(a,b)$, is the length of the shortest path between a and b fully contained in A. With b fixed the geodesic distance to any other point a in A can be calculated with the following sequence of conditional dilations:

$$B^0 = \{b\} \quad \text{and} \quad B^{i+1} = (B^i \oplus N) \cap A$$

The geodesic distance at a point a to b in A equals the number of iterations for which a is not reached, i.e. $d_A(a,b) = i$ with i selected such that $a \notin B^i$ and $a \in B^{i+1}$. The shortest path between b and a can be found in a fashion similar to the method pointed out above.

The *geodesic distance in a set A between a point $a \in A$ and a subset $Y \subset A$*, denoted by $d_A(a, Y)$, is the length of the shortest path between a and any point of Y. With fixed Y this distance can be calculated with the procedure stated above:

$$Y^0 = Y \quad \text{and} \quad Y^{i+1} = (Y^i \oplus N) \cap A$$

Here too, the geodesic distance equals the number of iterations for which a is not reached: $d_A(a, Y) = i$ with i such that $a \notin Y^i$ and $a \in Y^{i+1}$. The procedure is illustrated in figure 5.42c. The white area corresponds to the eastern region of the country. In the remaining part of the country the geodesic distance to this region is depicted as grey tones.

Listing 5.10 recursive 4-neighborhood distance transform

```
void 4_n_dist_transform(a,d,N,M)   /* recursive 4-neighborhood distance transform   */
int **a, **d;                      /* Pointers to the input image a and the output
                                      image d, respectively. These images are declared
                                      as 2-dimensional arrays of integers. The
                                      background is represented by "0". The foreground
                                      is a "1".                                       */
                                   /* number of rows and columns of the images a and c.*/
int N,M;
{
    int n,m;                       /* indices in image.                              */
    for (n=0;n<N;n++)
    for (m=0;m<M;m++) d[n][m] = 0;
    for (n=0;n<N;n++)              /* scan the image left-to-right, top-down         */
    for (m=0;m<M;m++)
    {   if (a[n][m]==0) continue;
        if ((n==0)||(m==0)) d[n][m] = 1;
        else d[n][m] = 1 + MIN(d[n-1][m],d[n][m-1]);
    }
    for (n=N-1;n>=0;n--)           /* scan the image in reverse direction            */
    for (m=M-1;m>=0;m--)
    {   if (a[n][m]==0) continue;
        if ((n==N-1)||(m==M-1)) d[n][m] = 1;
        else d[n][m] = MIN(d[n][m],d[n+1][m]+1,d[n][m+1]+1);
    }
}
```

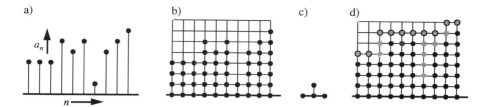

Figure 5.43 1-dimensional grey scale dilation
a) 1-dimensional discrete signal
b) Its umbra
c) Structuring element
d) Dilated umbra

5.4.5 Grey scale morphology

The obvious way to extend the morphological operations to grey scale imagery is by finding an appropriate representation of grey level images with sets. One such a representation is as follows. For any (digital) grey level image $a_{n,m}$, a set representation $A \subset \mathbb{Z}^3$ exists defined as $A = \{(n,m,t) \in \mathbb{Z}^3 | t = a_{n,m}\}$. This representation can be seen as a surface defined on a 2-dimensional plane $\mathbb{Z}^2$. The so-called *umbra* of the set A is another representation of the image $a_{n,m}$. Loosely speaking, the umbra of a surface in $\mathbb{Z}^3$ is the set of points lying on that surface or lying below that surface: $A = \{(n,m,t) \in \mathbb{Z}^3 | t \leq a_{n,m}\}$. The concept of an umbra is illustrated in figure 5.43. In this example, the image a_n (figure 5.43a) is 1-dimensional. Therefore, the umbra (figure 5.43b) is a subset of $\mathbb{Z}^2$. The original signal is seen as the top surface of the umbra. So, one can easily convert one representation into the other.

Once the umbra of a signal has been formed, the definition of the operations *dilation* and *erosion* is straightforward. For instance, suppose that a structuring element $B \subset \mathbb{Z}^3$ is given. Then, the dilation of the umbra $U[A]$ of a set A is defined as (equation 5.81): $U[A] \oplus B = \{t | B_t \cap U[A] \neq \emptyset\}$. As an example, figure 5.43d shows the dilation of figure 5.43b by the structuring element shown in figure 5.43c. The points added to the umbra are grey shaded. The top surface of the dilated umbra (in figure 5.43d, those points that have a black outline) corresponds to the signal representation of the dilated signal.

The umbra of a signal (or grey level image) is a representation that is useful to define the dilation. However, such a representation is not computationally efficient. It can be shown that the following operation is equivalent to the dilation defined above:

$$a_{n,m} \oplus b_{n,m} = \max_{k,l \in D(B)} \{a_{n-k,m-l} + b_{k,l}\} \tag{5.97}$$

The 2-dimensional function $b_{k,l}$ corresponds to the top surface of the structuring element B. The domain $D(B)$ of the kernel consists of those coordinates (k,l) for which the top surface of B is not empty. In the 1-dimensional example given in figure 5.43, the function b_k would be as follows: $b_{-1} = 0$, $b_0 = 1$ and $b_1 = 0$.

128   IMAGE BASED MEASUREMENT SYSTEMS

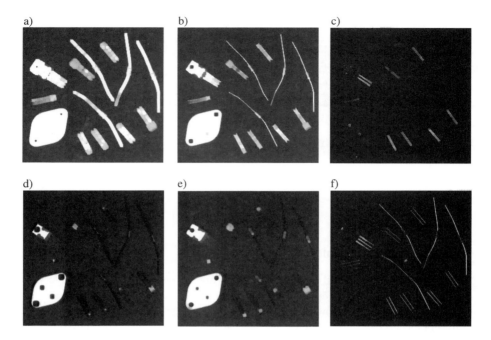

Figure 5.44 Grey level morphology
a) Dilation
b) Closing
c) Arithmetic difference between a and b
d) Erosion
e) Opening
d) Arithmetic difference between d and c

The dilation operation given in (5.97) has a complexity comparable with the convolution operation (equation 5.21). The summation and the multiplications of the convolution are replaced with a maximum-operation and additions, respectively.

Application of the erosion operation (equation 5.82) to the umbra of a grey level image leads to the following equivalent image operation:

$$a_{n,m} \ominus b_{n,m} = \min_{k,l \in D(B)} \{a_{n-k,m-l} - b_{k,l}\} \qquad (5.98)$$

Once the definitions of dilation and erosion for grey level images are established, operations like *opening* and *closing* follows immediately from equations (5.85) and (5.86). Generalisations of the hit-and-miss transform and other binary operations are less trivial.

Morphological grey level operations are useful in a number of applications, varying from noise smoothing to image feature extraction and texture based image segmentation. Figure 5.44 presents some examples. Figure 5.44a shows the image from figure 5.26a dilated by a rectangle-function, the size of which is 5×5. Figure 5.44b results from a closing operation, i.e. an erosion of figure 5.44a by the same rectangle function. In this image small holes, together with the space between the leads of the capacitors and the transistor, are filled. This filling can be made explicit by calculating the arithmetic difference between the closing of the image and its original (figure 5.44c).

Opposite to closing, an opening operation by a 3×3 rectangle function would remove the leads so as to extract the bodies of the components. The erosion and the opening are shown in figures 5.44d and 5.44e, respectively. The leads of the components are extracted by the arithmetic difference between the original and its opening (figure 5.44f).

REFERENCES

Borgefors, G., *Distance Transformations in Arbitrary Dimensions*, CVGIP, Vol. 27, 321-345, 1984.

Gonzalez, R.E. Woods: *Digital Image Processing*: Addison-Wesley, Massachusetts: 1992.

Haar Romeny, B.M. ter, Florack, L.M.J., Koenderink J.J., Viergever M.A., *Scale Space: Its Natural Operators and Differential Invariants*, Second Quinqennial Review 1986-1991, Dutch Society for Pattern Recognition & Image Processing, 1991.

Haralick, R.M. and Shapiro, L.G., *Computer and Robot Vision, Vol. I*, Addison-Wesley, Massachusetts, 1992.

Jain, A.K., *Fundamentals of digital image processing*, Prentice-Hall International, Englewood Cliffs, NJ, 1989.

Marr, D. and Hildreth, E., *Theory of Edge Detection*, Proc. Royal Society London, Vol. B207, 187-217, 1980.

Pratt, W.K., *Digital Image Processing*, Second Edition. J Wiley & Sons, New York 1991.

Rosenfeld, A. Kak: *Digital Picture Processing, Vol. I and II:* Academic Press, New York: 1982

Torre, V. and Poggio, T.A., *On Edge Detection*, IEEE Tr PAMI, Vol. 8, 147-163, March 1986.

6

STATISTICAL PATTERN CLASSIFICATION AND PARAMETER ESTIMATION

The word *pattern* is used in two different contexts. The word is used in the sense of "archetype". In Latin, the word "pater" refers to the word "father". In this context a pattern is something like a "partridge" according to which other objects are made. In the second context, the word "pattern" refers to a set of elements that are arranged more or less regularly. Here, "pattern" is near-synonymous with the word "structure".

Pattern classification is the process which assigns classes to signals that are derived from objects using a sensory system; see figure 6.1. The objects which we have in mind are part of a physical world that - as a whole - form the environment of a measurement system. In this world, objects are grouped according to a criterion that is related to the *physical processes* that originated the objects. We assume that there are a number of different physical processes. Each process defines a class. The task of the measurement system is to recover the class of each object. For that purpose the system has a sensory subsystem with which physical properties from the objects are sensed and converted into a signal. The signal is used to assign a class to an object.

The signal made available by the sensory system may have various forms. We may think of a time-dependent signal, either discrete or continuous. We may also think of a set of time-dependent signals, arranged in a time-dependent vector. In this chapter we will restrict the discussion to a signal consisting of a finite number of measurements, arranged in a finite-dimensional, static vector: the *measurement vector*.

Hopefully, the sensory subsystem is designed such that the measurement vector conveys the information needed to classify all objects correctly. If this is the case, the measurement vectors from all objects behave according to a regular *pattern*. Here we have the word "pattern" in its second context. Ideally, the physical properties are chosen such that all objects from one class cluster together in the measurement space without overlapping the clusters formed by other classes.

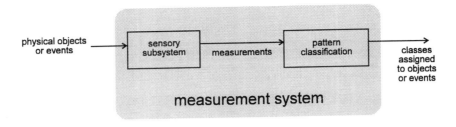

Figure 6.1 Pattern classification

Example 6.1 Classification of chickens

In a chicken farm, chickens are grouped according to the criterion "male" and "female". In order to assign one of these two classes to a chicken (whose gender is unknown) we can measure some physical properties of this chicken, for instance, its weight and its length. The task of the pattern classifier is to decide which class to assign based on these two measurements.

For that purpose, the measurement space is divided into two compartments. To all objects with measurements falling in one compartment the class "male" is assigned. Measurements in the other compartment are labeled "female". In figure 6.2, a measurement space is depicted in which the two compartments are separated by a linear *decision boundary*. In this space also some measurements from both classes are shown. Such a graph is called a *scatter diagram*. It can be seen that in this example the measurements from different classes form their own clusters. However, the clusters also show a small overlapping zone. Therefore, classification in this measurement space will not be completely without classification errors.

In a sense, *parameter estimation* is an extension of pattern classification. Here too, we have a measurement system equipped with a sensory subsystem that senses physical quantities from objects in the environment (figure 6.3). However, objects are now characterised by parameters rather than classes. It is the task of an estimator to transform the signal into an estimate of the parameters of each object. The main difference between classification and parameter estimation is that classes are discrete variables taken from a finite-dimensional set whereas parameters are (usually) continuous variables taken from a normed linear space (see appendix A).

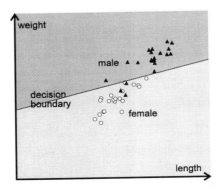

Figure 6.2 Patterns in the measurement space (fictitious data)

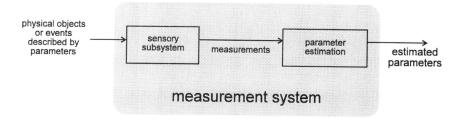

Figure 6.3 Parameter estimation

Example 6.2 Measurement of the diameter of yarn

For the purpose of quality control, the production of yarn in a spinning-mill must be monitored continuously. Suppose that this can be accomplished by constantly measuring the diameter of the yarn. One way to do that is to use a line-scan CCD camera aligned perpendicular to the axial direction of the yarn (figure 6.4b). This sensory system produces a sequence of 1-dimensional scans of the yarn.

Figure 6.4b shows a realisation $f(y)$ of one of such scans. This signal depends on many factors: the illumination and the optical properties of the yarn, the geometry of the optical system, the point spread function, noise sources, and - most important - also on the diameter of the yarn. The task of the parameter estimator is to estimate the diameter of the yarn from this signal. As indicated in the figure, this estimate could be obtained by simply thresholding the signal and measuring the width of the resulting pulse. However, this method is satisfactory only when the illumination is well-conditioned, and all degradations of the signal are negligible. If these conditions are not met, the method is either biased, or it may even fail. A more robust solution is achieved by (statistically) modeling the relationship between all possible diameters d and the signal $f(y)$. Once this model has been established, it is used to estimate the actual diameter from an observed signal.

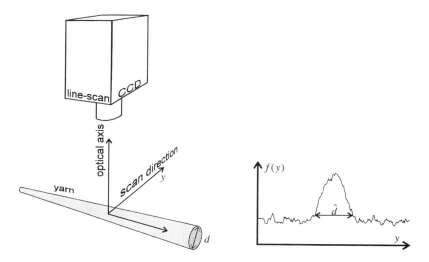

Figure 6.4 Measurement of the diameter of the yarn
a) Sensory system
b) Signal from which the diameter has to estimated (fictitious data)

This chapter addresses the problem of how to build pattern classifiers and parameter estimators. This is done within a Bayesian-theoretic framework (section 6.1 and 6.2). Within this framework one assumption is that the statistics of all measurements and variables is fully known. The remaining sections of the chapter relate to situations in which this knowledge is not fully available, but where a set of examples is given instead. In these situations, we must learn from the examples so as to arrive at a method to classify objects which are not in the set of examples.

6.1 DECISION MAKING AND CLASSIFICATION

Probability theory is a solid base for pattern classification design [Devijver and Kittler 1982]. Essential in this approach is that the pattern-generating mechanism can be represented within a probabilistic framework. Figure 6.5 shows such a framework. Starting point is an experiment (appendix B.1) defined by a set $\Omega = \{\omega_1, \cdots, \omega_K\}$ of K outcomes. The Borel field A of the experiment consists of events $\{\emptyset, \omega_1, \cdots, \omega_K, \Omega\}$. A probability $P(\omega_k)$, $P(\emptyset)$, and $P(\Omega)$ exists for each event. Note that the Borel field is chosen such that events are mutually exclusive, i.e. the probability of an event with two or more outcomes is not defined, except for $P(\Omega)$.

The outcome of the experiment affects a physical process which originates an object. We assume that the physical process is such that the class of the object is fully determined by the outcome. Therefore, class and outcome are in fact identical. The probability $P_k = P(\omega_k)$ of having a class ω_k is called *prior probability*. It is the knowledge we have about the class of an object before measurements are available. Since the number of outcomes is K, we have:

$$\sum_{k=1}^{K} P_k = 1 \tag{6.1}$$

For each object the sensory system eventually brings forth a measurement vector $\underline{x}$ with dimension N. The sensory system must be designed such that this vector conveys class information. Since a class is identified with the outcome of an experiment, this vector is random. In addition, all measurements are subject to some degree of randomness due to all kind of physical limitations, e.g. quantum limitations, thermal noise. This randomness is taken into account by the probability density of $\underline{x}$.

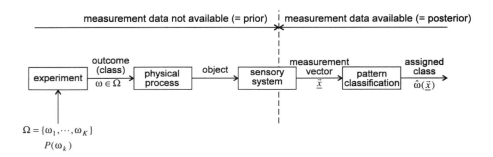

Figure 6.5 Statistical pattern classification

We have to distinguish *conditional probability densities* $p_{\underline{x}|\omega_k}(\underline{x}|\omega_k)$ from the *unconditional probability density* $p_{\underline{x}}(\underline{x})$. The former is the probability density of the measurement vector $\underline{x}$ coming from objects with class ω_k. The latter is the probability density of $\underline{x}$ coming from any object, regardless of the class. Since the classes are mutually exclusive, the unconditional density can be derived from the unconditional densities by weighting these densities with the prior probabilities:

$$p_{\underline{x}}(\underline{x}) = \sum_{k=1}^{K} p_{\underline{x}|\omega_k}(\underline{x}|\omega_k) P_k \qquad (6.2)$$

The pattern classifier converts the measurement vector into a class assigned to the object. In fact, since $\underline{x}$ is an N-dimensional vector this assignment is a function that maps $\mathbb{R}^N$ onto Ω, i.e. $\hat{\omega}(\underline{x})$: $\mathbb{R}^N \to \Omega$. As the pattern classifier decides which class to assign to the object, the function $\hat{\omega}(\underline{x})$ is also called the *decision function*.

Example 6.3 Classification of apples

In agricultural industries, quality control is highly relevant. Often this control is based on visual inspection. As an example, suppose that a particular group of apples may suffer from two different diseases. Furthermore, suppose that these diseases are visible at the skin of an apple by two characteristic features: one parameter that describes the shape of the contour of the apple, and another that describes the texture of the skin. We assume that a vision system is available equipped with a visual front-end processor and a pattern classifier. The visual front-end processor measures both parameters.

Based on these two parameters the pattern classifier has to decide whether the apple is healthy or not. Suppose that the number of known diseases is two. Then we have the following classes: ω_1="healthy", ω_2="disease A" and ω_3="disease B". We have a 3-class problem ($K=3$) with a 2-dimensional measurement space ($N=2$).

Figure 6.6a shows a scatter diagram in the measurement space with examples of apples from all three classes. Each conditional density is represented by a contour plot (see appendix B.2). The unconditional probability density is depicted in figure 6.6b. For the sake of clearness the density is represented by several contour plots.

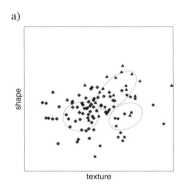
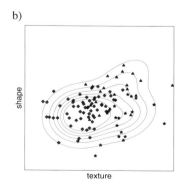

Figure 6.6 A 3-class problem with 2-dimensional measurement vector
a) 2-dimensional measurement space with scatter diagram (3 classes) and conditional probability densities. Measurement vectors are presented by the following marks:
  healthy apple: ♦ , disease A: ★, and disease B: ▲
b) Measurement space with unconditional probability density

6.1.1 Bayes classification

As mentioned above it cannot always be guaranteed that a classification is without mistakes. Most often, an erroneous assignment of a class to an object causes some damage. The damage depends on the application at hand. For instance, if apples have to be classified, then the damage of having a healthy apple being labeled "disease A" or "disease B" is the price of one healthy apple. The damage of a diseased apple being labeled "healthy" may be much more (e.g. the price of a whole heap of apples).

A *Bayes classifier* is a pattern classifier that is based on the following two prerequisites:

- The damage involved when an object is wrongly classified can be quantified as a *cost*.
- The expectation of the cost (known as *risk*) is acceptable as an optimisation criterion.

If the application at hand meets these two conditions, then the development of an optimal pattern classification is straightforward.

The damage is quantified by a *cost function* $C(\hat{\omega}_i|\omega_j)$: $\Omega \times \Omega \to \mathbb{R}^+$. This function expresses the cost which is involved when the true class of an object is ω_j and the decision of the classifier is $\hat{\omega}(\underline{x}) = \hat{\omega}_i$. Since there are K classes, the function $C(\hat{\omega}_i|\omega_j)$ is fully specified by a $K \times K$-matrix. Therefore, sometimes the cost function is called *cost matrix*.

Aforementioned concepts (prior probabilities, conditional probability densities and cost function) are sufficient to design optimal classifiers. For that purpose, another probability has to be derived, called the *posterior probability* $P(\omega_j|\underline{x} = \vec{x})$. It is the probability that an object belongs to class ω_j given that the measurement vector associated with this object is $\vec{x}$. According to Bayes' theorem for conditional probabilities (appendix B.2) we have:

$$P(\omega_j|\underline{x} = \vec{x}) = \frac{p_{\underline{x}|\omega_j}(\vec{x}|\omega_j)P_j}{p_{\underline{x}}(\vec{x})} \tag{6.3}$$

If an arbitrary classifier assigns a class $\hat{\omega}_i$ to a measurement vector $\vec{x}$ coming from an object the true class of which is ω_j, then a cost $C(\hat{\omega}_i|\omega_j)$ is involved. The posterior probability of having such an object is $P(\omega_j|\underline{x} = \vec{x})$. Therefore, the expectation of the cost - denoted by $R(\hat{\omega}_i|\underline{x} = \vec{x})$ - is:

$$R(\hat{\omega}_i|\underline{x} = \vec{x}) = E\{C(\hat{\omega}_i|\omega_j)|\underline{x} = \vec{x}\} = \sum_{j=1}^{K} C(\hat{\omega}_i|\omega_j)P(\omega_j|\underline{x} = \vec{x}) \tag{6.4}$$

This quantity is called *conditional risk*. It expresses the average cost of the assignment $\hat{\omega}_i$ to an object whose measurement vector is $\vec{x}$.

From (6.4) it follows that the conditional risk of a decision function $\hat{\omega}(\vec{x})$ is $R(\hat{\omega}(\vec{x})|\underline{x} = \vec{x})$. The overall cost can be found by averaging the conditional cost over all possible measurement vectors:

$$R = E\{R(\hat{\omega}(\underline{x})|\underline{x} = \vec{x})\} = \int_{\vec{x}} R(\hat{\omega}(\vec{x})|\underline{x} = \vec{x}) p_{\underline{x}}(\vec{x}) d\vec{x} \tag{6.5}$$

The integral extends over the entire measurement space. The quantity R is the *average risk* (or briefly the *risk*) associated with the pattern classifier $\hat{\omega}(\vec{x})$.

The second prerequisite mentioned above states that the optimal classifier is the one with minimal risk R. From the assumption that the cost is non-negative, it follows that the conditional risk is non-negative too. Consequently, the risk in equation (6.5) is the integral of a non-negative function. Therefore, the decision function that minimises the (average) risk is the same as the one that minimises the conditional risk:

$$\hat{\omega}(\vec{x}) = \hat{\omega}_k \quad \text{such that:} \quad R(\hat{\omega}_k | \vec{x} = \vec{x}) \leq R(\hat{\omega}_i | \vec{x} = \vec{x}) \quad i,k = 1,\cdots,K \quad (6.6)$$

This can be expressed more briefly by:

$$\hat{\omega}(\vec{x}) = \underset{\omega \in \Omega}{\operatorname{argmin}} \{ R(\omega | \vec{x} = \vec{x}) \} \quad (6.7)$$

The expression argmin{} gives the element from Ω that minimises $R(\omega | \vec{x} = \vec{x})$. Substitution of (6.3) and (6.4) yields:

$$\hat{\omega}(\vec{x}) = \underset{\omega \in \Omega}{\operatorname{argmin}} \left\{ \sum_{j=1}^{K} C(\omega | \omega_j) P(\omega_j | \vec{x} = \vec{x}) \right\}$$

$$= \underset{\omega \in \Omega}{\operatorname{argmin}} \left\{ \sum_{j=1}^{K} C(\omega | \omega_j) \frac{p_{\vec{x}|\omega_j}(\vec{x}|\omega_j) P_j}{p_{\vec{x}}(\vec{x})} \right\} \quad (6.8)$$

$$= \underset{\omega \in \Omega}{\operatorname{argmin}} \left\{ \sum_{j=1}^{K} C(\omega | \omega_j) p_{\vec{x}|\omega_j}(\vec{x}|\omega_j) P_j \right\}$$

Pattern classification according to (6.8) is called *Bayes classification* or *minimum risk classification*.

Example 6.3 Classification of apples (continued)

We assume that apples with class: ω_1="healthy", ω_2="disease A", and ω_3="disease B" occur with relative frequencies 3:1:1. Hence the prior probabilities are $P_1 = 0.6$, $P_2 = 0.2$, and $P_3 = 0.2$, respectively. Suppose that the cost function is chosen to be:

	ω_1	ω_2	ω_3
$\hat{\omega}_1$	0	1	1
$\hat{\omega}_2$	1	0	1
$\hat{\omega}_3$	1	1	0

The assumption here is that no cost is involved when the classification is correct, and that unit cost is involved when an object is erroneously classified. The resulting pattern classifier (together with the scatter diagram of figure 6.6a) is shown in figure 6.7a. It can be seen that the decision boundary between the class of healthy apples (diamonds) and the other classes is in favour of the healthy apples. The reason for this is that the prior probability of healthy apples is greater than the prior probabilities of the other classes. The pattern classifier combines prior knowledge with measurement data.

It may be argued that in this application the cost given above is not very realistic. As mentioned above the cost which occurs when a diseased apple is treated as a healthy apple may be much more than the cost of a healthy apple that is thrown away. Therefore, we may want to raise the cost of an erroneous classification of a diseased apple. For instance:

	ω_1	ω_2	ω_3
$\hat{\omega}_1$	0	5	5
$\hat{\omega}_2$	1	0	1
$\hat{\omega}_3$	1	1	0

The corresponding decision function is shown in figure 6.7b. The boundary between compartments is shifted in favour of the diseases: raising the cost of misclassifications of diseased apples results in a classifier with low probability to misclassify diseased apples. However, this is at the cost of a number of healthy apples that will be thrown away without need. Nevertheless, on an average the classifier is most economic.

6.1.2 Uniform cost function and minimum error rate

The cost function applied in the Bayes classifier depicted in figure 6.7a is called the *uniform cost function*. Unit cost is assumed when an object is misclassified, zero cost when the classification is correct. This can be written as:

$$C(\hat{\omega}_i|\omega_j) = 1 - \delta(i,j) \qquad \text{with:} \qquad \delta(i,j) = \begin{cases} 1 & \text{if } i = j \\ 0 & \text{elsewhere} \end{cases} \qquad (6.9)$$

$\delta(i,j)$ is the Kronecker-delta function. With this cost function the conditional risk given in (6.4) simplifies to:

$$R(\hat{\omega}_i|\underline{x} = \vec{x}) = \sum_{j=1, j \neq i}^{K} P(\omega_j|\underline{x} = \vec{x}) = 1 - P(\omega_i|\underline{x} = \vec{x}) \qquad (6.10)$$

Minimisation of this risk is equivalent to maximisation of the posterior probability $P(\omega_i|\underline{x} = \vec{x})$. Therefore, with uniform cost function the Bayes decision function (6.8) becomes the *maximum posterior probability classifier* (MAP-classifier):

$$\hat{\omega}(\vec{x}) = \underset{\omega \in \Omega}{\operatorname{argmax}} \{P(\omega|\underline{x} = \vec{x})\} \qquad (6.11a)$$

a)
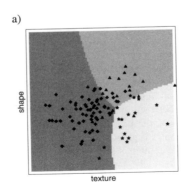

b)
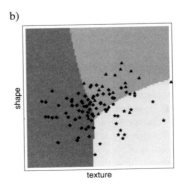

Figure 6.7 Bayes classification
a) With uniform cost function
b) With increased cost of misclassification of ★ and ▲

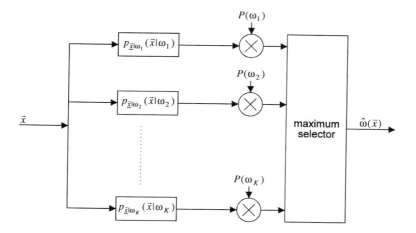

Figure 6.8 Bayes decision function with uniform cost function (MAP-classification)

Application of Bayes' theorem for conditional probabilities and cancellation of irrelevant terms yield a classification, equivalent to a MAP-classification, but fully in terms of prior probabilities and conditional probability densities:

$$\hat{\omega}(\vec{x}) = \underset{\omega \in \Omega}{\operatorname{argmax}} \{ p_{\vec{x}|\omega}(\vec{x}|\omega) P(\omega) \} \qquad (6.11b)$$

The functional structure of this decision function is given in figure 6.8.

Suppose that a class $\hat{\omega}_i$ is assigned to an object whose measurement vector $\vec{x}$ is known. The probability of having a correct classification is $P(\hat{\omega}_i | \vec{x} = \vec{x})$. Consequently, the probability of having a classification error is $1 - P(\hat{\omega}_i | \vec{x} = \vec{x})$. Therefore, the conditional risk given in (6.10) can be given the interpretation of *conditional probability of classification error* $e(\vec{x})$. With arbitrary decision function $\hat{\omega}(\vec{x})$ this probability is:

$$e(\vec{x}) = 1 - P(\hat{\omega}(\vec{x}) | \vec{x} = \vec{x}) \qquad (6.12)$$

In case of a uniform cost function $\hat{\omega}(\vec{x})$ is given by (6.11). Hence:

$$e(\vec{x}) = 1 - \max_{\omega \in \Omega} \{ P(\omega | \vec{x} = \vec{x}) \} \qquad (6.13)$$

The overall probability of classification error is called *error rate* E. The probability is found by averaging the conditional probability of error over all possible measurement vectors:

$$E = \mathrm{E}\{e(\vec{x})\} = \int_{\vec{x}} e(\vec{x}) p_{\vec{x}}(\vec{x}) d\vec{x} \qquad (6.14)$$

The integral extends over the entire measurement space.

The error rate is often used as a characterisation of the performance of a classifier. With uniform cost function, risk and error rate are equal. Therefore, a Bayes classification with uniform cost function is equivalent to a classification with minimum error rate. The minimum error rate is denoted E_B. Of course, the phrase "minimum" relates to a context with fixed sensory subsystem. The performance of an optimal classification with one sensory subsystem may be less than the performance of a non-optimal classification with another sensory subsystem.

6.1.3 Gaussian random vectors

A further development of Bayes classification with uniform cost function requires the specification of the conditional probability densities. This section discusses the case in which these densities are Gaussian. Suppose that the measurement vectors coming from an object with class ω_j are Gaussian distributed with expectation vector $\bar{\mu}_j$ and covariance matrix $\mathbf{C}_j$ (see appendix B.3):

$$p_{\bar{x}|\omega_j}(\bar{x}|\omega_j) = \frac{1}{\sqrt{(2\pi)^N |\mathbf{C}_j|}} \exp\left(\frac{-(\bar{x}-\bar{\mu}_j)^t \mathbf{C}_j^{-1}(\bar{x}-\bar{\mu}_j)}{2}\right) \tag{6.15}$$

N is the dimension of the measurement vector.

Substitution of (6.15) in (6.11b) gives the following minimum error rate classification:

$$\hat{\omega}(\bar{x}) = \underset{\omega_i \in \Omega}{\operatorname{argmax}}\left\{\frac{1}{\sqrt{(2\pi)^N |\mathbf{C}_i|}} \exp\left(\frac{-(\bar{x}-\bar{\mu}_i)^t \mathbf{C}_i^{-1}(\bar{x}-\bar{\mu}_i)}{2}\right) P_i\right\} \tag{6.16a}$$

We can take the logarithm of the function between braces without changing the result of the argmin{} function. Furthermore, all terms not containing $\hat{\omega}_i$ are irrelevant. Therefore, (6.16a) is equivalent to:

$$\hat{\omega}(\bar{x}) = \underset{\omega_i \in \Omega}{\operatorname{argmax}}\left\{-\log|\mathbf{C}_i| + 2\log P_i - (\bar{x}-\bar{\mu}_i)^t \mathbf{C}_i^{-1}(\bar{x}-\bar{\mu}_i)\right\}$$
$$= \underset{\omega_i \in \Omega}{\operatorname{argmax}}\left\{-\log|\mathbf{C}_i| + 2\log P_i - \bar{\mu}_i^t \mathbf{C}_i^{-1}\bar{\mu}_i + 2\bar{x}^t \mathbf{C}_i^{-1}\bar{\mu}_i - \bar{x}^t \mathbf{C}_i^{-1}\bar{x}\right\} \tag{6.16b}$$

Hence, the expression of a minimum error rate classification with Gaussian measurement vectors takes the form of:

$$\hat{\omega}(\bar{x}) = \underset{\omega_i \in \Omega}{\operatorname{argmax}}\left\{w_i + \bar{x}^t \bar{w}_i + \bar{x}^t \mathbf{W}_i \bar{x}\right\} \tag{6.16c}$$

with:

$$w_i = -\log|\mathbf{C}_i| + 2\log P_i - \bar{\mu}_i^t \mathbf{C}_i^{-1} \bar{\mu}_i$$
$$\bar{w}_i = 2\mathbf{C}_i^{-1}\bar{\mu}_i \tag{6.16d}$$
$$\mathbf{W}_i = \mathbf{C}_i^{-1}$$

This classification partitions the measurement space in a number of compartments. The decision boundaries between these compartments are segments of quadratic hypersurfaces in the N-dimensional space. To see this, it suffices to examine the boundary between the compartments of two different classes, e.g. ω_i and ω_j. According to (6.16c) the boundary between the compartments associated with these classes - if it exists - must satisfy the following equation:

$$w_i + \vec{x}^t\vec{w}_i + \vec{x}^t\mathbf{W}_i\vec{x} = w_j + \vec{x}^t\vec{w}_j + \vec{x}^t\mathbf{W}_j\vec{x} \tag{6.17a}$$

or:

$$w_i - w_j + \vec{x}^t(\vec{w}_i - \vec{w}_j) + \vec{x}^t(\mathbf{W}_i - \mathbf{W}_j)\vec{x} = 0 \tag{6.17b}$$

Clearly, the solution of this equation is a quadratic (hyper)surface. If $K > 2$, then pieces of this surface are not part of the solution because of interference by compartments from other classes. A decision function that satisfies equation (6.16c) is called a *quadratic decision function*.

As an example we consider the classifications shown in figure 6.7. In fact, the probability densities shown in figure 6.6a are Gaussian. Therefore, the decision boundaries shown in figure 6.7 must be quadratic curves, i.e. segments of ellipses, hyperbolae, parabolae, circles or straight lines.

Class independent covariance matrices

In this subsection, we discuss the case in which the covariance matrices do not depend on the classes, i.e. $\mathbf{C}_i = \mathbf{C}$ for all $\omega_i \in \Omega$. This situation occurs when the measurement vector of an object equals the (class-dependent) expectation vector corrupted by noise $\vec{x} = \vec{\mu}_i + \vec{n}$. The noise is assumed to be class-independent with covariance matrix $\mathbf{C}$. Hence, all class information is brought forth by expectation vectors.

The quadratic decision function of (6.16a) degenerates into:

$$\begin{aligned}\hat{\omega}(\vec{x}) &= \underset{\omega_i \in \Omega}{\operatorname{argmax}}\left\{2\log P_i - (\vec{x}-\vec{\mu}_i)^t \mathbf{C}^{-1}(\vec{x}-\vec{\mu}_i)\right\} \\ &= \underset{\omega_i \in \Omega}{\operatorname{argmin}}\left\{-2\log P_i + (\vec{x}-\vec{\mu}_i)^t \mathbf{C}^{-1}(\vec{x}-\vec{\mu}_i)\right\}\end{aligned} \tag{6.18a}$$

Since the matrix $\mathbf{C}$ is self-adjoint and positive definite (appendix A.8) the quantity $(\vec{x}-\vec{\mu}_i)^t\mathbf{C}^{-1}(\vec{x}-\vec{\mu}_i)$ can be regarded as a distance measure between the vector $\vec{x}$ and the expectation $\vec{\mu}_i$. The distance measure is called the *Mahalanobis distance*. The decision function of (6.18a) decides for the class whose expectation vector is nearest to the observed measurement vector (with a correction factor $-2\log P_i$ to account for prior knowledge). Hence the name *minimum Mahalonobis distance classifier*.

The decision boundaries between compartments in the measurement space are linear (hyper)planes. This follows from (6.16c) and (6.16d):

$$\hat{\omega}(\vec{x}) = \underset{\omega_i \in \Omega}{\operatorname{argmax}}\left\{w_i + \vec{x}^t\vec{w}_i\right\} \tag{6.18b}$$

where:

$$\begin{aligned}w_i &= 2\log P_i - \vec{\mu}_i^t \mathbf{C}^{-1}\vec{\mu}_i \\ \vec{w}_i &= 2\mathbf{C}^{-1}\vec{\mu}_i\end{aligned} \tag{6.18c}$$

Hence, the equations of the decision boundaries are $w_i - w_j + \vec{x}^t(\vec{w}_i - \vec{w}_j) = 0$. A decision function which has the form of (6.18b) is called a *linear decision function*.

Figure 6.9 gives an example of a 3-class problem and a 2-dimensional measurement space. A scatter diagram with the contour plots of the conditional probability densities are given (figure 6.9a), together with the compartments of the minimum Mahalanobis distance classification (figure 6.9b).

Minimum distance classification

A further simplification occurs when the measurement vector equals the class-dependent vector $\vec{\mu}_i$ corrupted by class-independent white noise with covariance matrix $\mathbf{C} = \sigma_n^2 \mathbf{I}$.

$$\hat{\omega}(\vec{x}) = \underset{\omega_i \in \Omega}{\operatorname{argmin}} \left\{ -2\log P_i + \frac{\|(\vec{x} - \vec{\mu}_i)\|^2}{\sigma_n^2} \right\} \tag{6.19}$$

The quantity $\|(\vec{x} - \vec{\mu}_i)\|$ is the normal (Euclidean) distance between $\vec{x}$ and $\vec{\mu}_i$. The classifier corresponding to (6.19) decides for the class whose expectation vector is nearest to the observed measurement vector (with a correction factor $-2\sigma_n^2 \log P_i$ to account for prior knowledge). Hence the name *minimum distance classifier*. As with the minimum Mahalanobis distance classifier, the decision boundaries between compartments are linear (hyper)planes. The plane separating the compartments of two classes ω_i and ω_j is given by:

$$\sigma_n^2 \log \frac{P_i}{P_j} + \frac{1}{2}\left(\|\vec{\mu}_j\|^2 - \|\vec{\mu}_i\|^2\right) + \vec{x}^t(\vec{\mu}_i - \vec{\mu}_j) = 0 \tag{6.20}$$

The solution of this equation is a plane perpendicular to the line segment connecting $\vec{\mu}_i$ and $\vec{\mu}_j$. The location of the hyperplane depends on the factor $\sigma_n^2 \log(P_i/P_j)$. If $P_i = P_j$, the hyperplane is the perpendicular bisector of the line segment (see figure 6.10).

Figure 6.11 gives an example of the decision function of a minimum distance classification.

a)
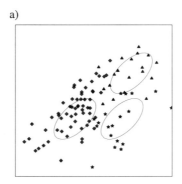

b)
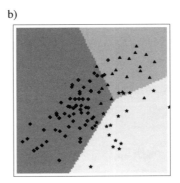

Figure 6.9 Minimum Mahalanobis classification
a) Scatter diagram with contour plots of conditional probability densities
b) Decision function

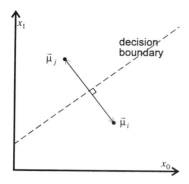

Figure 6.10 Decision boundary of a minimum distance classifier

Class-independent expectation vectors

Another interesting situation is when class-information is solely brought forth by differences between covariance matrices. In that case, the expectation vectors do not depend on the class: $\vec{\mu}_i = \vec{\mu}$. Hence, the central parts of the conditional probability densities overlap. Whatever the covariance matrices are, the probability of making the wrong decision will never be zero. The decision function takes the form of:

$$\hat{\omega}(\vec{x}) = \underset{\omega_i \in \Omega}{\operatorname{argmax}}\left\{-\log|\mathbf{C}_i| + 2\log P_i - (\vec{x}-\vec{\mu})'\mathbf{C}_i^{-1}(\vec{x}-\vec{\mu})\right\} \tag{6.21}$$

If the covariance matrices are of the type $\sigma_i^2 \mathbf{I}$, the decision boundaries are concentric circles or (hyper)spheres. Figure 6.12a gives an example of such a classification. If the covariance matrices are rotated versions of one "elliptic" prototype, the decision boundaries are hyperbolae. If the prior probabilities are equal, these hyperbolae degenerate into a number of linear planes or (if $N=2$) linear lines. An example is given in figure 6.12b.

a)

b)

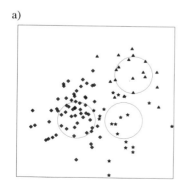
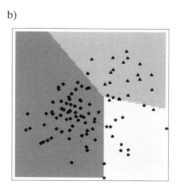

Figure 6.11 Minimum distance classification
a) Scatter diagram with contour plots of conditional probability densities
b) Decision function

6.1.4 The 2-class case

In the 2-class case ($K = 2$) the Bayes classfication with uniform cost function reduces to the test (see equation 6.11):

$$p_{\bar{x}|\omega_1}(\bar{x}|\omega_1)P_1 > p_{\bar{x}|\omega_2}(\bar{x}|\omega_2)P_2 \qquad (6.22a)$$

If the test fails, it is decided for ω_2, otherwise for ω_1. We write symbolically:

$$p_{\bar{x}|\omega_1}(\bar{x}|\omega_1)P_1 \underset{\omega_2}{\overset{\omega_1}{\gtrless}} p_{\bar{x}|\omega_2}(\bar{x}|\omega_2)P_2 \qquad (6.22b)$$

Rearrangement gives:

$$\frac{p_{\bar{x}|\omega_1}(\bar{x}|\omega_1)}{p_{\bar{x}|\omega_2}(\bar{x}|\omega_2)} \underset{\omega_2}{\overset{\omega_1}{\gtrless}} \frac{P_2}{P_1} \qquad (6.22c)$$

Regarded as a function of ω the conditional probability density $p_{\bar{x}|\omega}(\bar{x}|\omega)$ is called the *likelihood function of* ω. Therefore, the ratio:

$$L(\bar{x}) = \frac{p_{\bar{x}|\omega_1}(\bar{x}|\omega_1)}{p_{\bar{x}|\omega_2}(\bar{x}|\omega_2)}$$

is called the *likelihood ratio*. With this definition the classification becomes a simple likelihood ratio test:

$$L(\bar{x}) \underset{\omega_2}{\overset{\omega_1}{\gtrless}} \frac{P_2}{P_1} \qquad (6.22d)$$

The test is equivalent to a threshold operation applied to $L(\bar{x})$ with threshold P_2/P_1.

a) b)

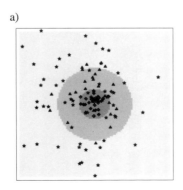
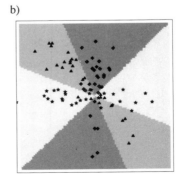

Figure 6.12 Classification of objects with equal expectation vectors
a) Concentric conditional probability densities
b) Eccentric conditional probability densities; see text

STATISTICAL PATTERN CLASSIFICATION AND PARAMETER ESTIMATION

In case of Gaussian distributed measurement vectors, it is convenient to replace the likelihood ratio test with a *log-likelihood ratio test*, see equation 6.16:

$$\Lambda(\underline{\bar{x}}) \underset{\omega_2}{\overset{\omega_1}{\gtrless}} T \quad \text{with:} \quad \Lambda(\underline{\bar{x}}) = \log(L(\underline{\bar{x}})) \quad \text{and} \quad T = \log\left(\frac{P_2}{P_1}\right) \tag{6.23}$$

For Gaussian distributed vectors, the log-likelihood ratio is:

$$\Lambda(\underline{\bar{x}}) = -\frac{1}{2}\left(\log|\mathbf{C}_1| - \log|\mathbf{C}_2| + (\underline{\bar{x}} - \underline{\bar{\mu}}_1)^t \mathbf{C}_1^{-1}(\underline{\bar{x}} - \underline{\bar{\mu}}_1) - (\underline{\bar{x}} - \underline{\bar{\mu}}_2)^t \mathbf{C}_2^{-1}(\underline{\bar{x}} - \underline{\bar{\mu}}_2)\right) \tag{6.24}$$

which is an expression much simpler than the likelihood ratio. When the covariance matrices of both classes are equal ($\mathbf{C}_1 = \mathbf{C}_2$) the expression becomes even more simple:

$$\Lambda(\underline{\bar{x}}) = \left(\underline{\bar{x}} - \frac{1}{2}(\underline{\bar{\mu}}_1 + \underline{\bar{\mu}}_2)\right)^t \mathbf{C}^{-1}(\underline{\bar{\mu}}_1 - \underline{\bar{\mu}}_2) \tag{6.25}$$

In a 2-class decision problem there are two types of error. Suppose that $\hat{\omega}(\underline{\bar{x}})$ is the result of a decision based on the measurement $\underline{\bar{x}}$. The true class of an object is either ω_1 or ω_2. Then the following four states may occur [Kreyszig 1970]:

	ω_1	ω_2
$\hat{\omega}_1$	correct decision I	type II error
$\hat{\omega}_2$	type I error	correct decision II

The performance of the classification is fully defined when the probabilities of the two types of error are specified. Usually these probabilities are given conditionally with respect to the true classes, i.e $P(\hat{\omega}_1|\omega_2)$ and $P(\hat{\omega}_2|\omega_1)$. The overall probability of having an error of type I can be derived from the prior probability using Bayes' theorem, e.g. $P(\text{type I error}) = P(\hat{\omega}_2, \omega_1) = P(\hat{\omega}_2|\omega_1)P_1$. The probabilities $P(\hat{\omega}_1|\omega_2)$ and $P(\hat{\omega}_2|\omega_1)$ follow from (6.23):

$$P(\hat{\omega}_2|\omega_1) = P(\Lambda(\underline{x}) < T|\omega = \omega_1)$$
$$P(\hat{\omega}_1|\omega_2) = P(\Lambda(\underline{x}) > T|\omega = \omega_2) \tag{6.26}$$

In general, it is difficult to find analytical expressions for $P(\hat{\omega}_1|\omega_2)$ and $P(\hat{\omega}_2|\omega_1)$. In the case of Gaussian measurement vectors with $\mathbf{C}_1 = \mathbf{C}_2$, expression (6.26) can be further developed. Equation (6.25) shows that $\Lambda(\underline{\bar{x}})$ is linear in $\underline{\bar{x}}$. Since $\underline{\bar{x}}$ is Gaussian, so is $\Lambda(\underline{\bar{x}})$; see appendix B.3. The posterior distribution of $\Lambda(\underline{\bar{x}})$ is fully specified by its conditional expectation and its variance. As $\Lambda(\underline{\bar{x}})$ is linear in $\underline{\bar{x}}$, these parameters are obtained easily:

$$E\{\Lambda(\underline{\bar{x}})|\omega_1\} = \left(E\{\underline{\bar{x}}|\omega_1\} - \frac{1}{2}(\underline{\bar{\mu}}_1 + \underline{\bar{\mu}}_2)\right)^t \mathbf{C}^{-1}(\underline{\bar{\mu}}_1 - \underline{\bar{\mu}}_2)$$
$$= \left(\underline{\bar{\mu}}_1 - \frac{1}{2}(\underline{\bar{\mu}}_1 + \underline{\bar{\mu}}_2)\right)^t \mathbf{C}^{-1}(\underline{\bar{\mu}}_1 - \underline{\bar{\mu}}_2) \tag{6.27a}$$
$$= \frac{1}{2}(\underline{\bar{\mu}}_1 - \underline{\bar{\mu}}_2)^t \mathbf{C}^{-1}(\underline{\bar{\mu}}_1 - \underline{\bar{\mu}}_2)$$

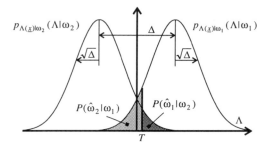

Figure 6.13 The conditional probability densities of the log-likelihood ratio ($C_1 = C_2$)

Likewise:

$$E\{\Lambda(\vec{x})|\omega_2\} = -\frac{1}{2}(\vec{\mu}_1 - \vec{\mu}_2)^t \mathbf{C}^{-1}(\vec{\mu}_1 - \vec{\mu}_2) \qquad (6.27b)$$

and:

$$\mathrm{Var}\{\Lambda(\vec{x})|\omega_1\} = (\vec{\mu}_1 - \vec{\mu}_2)^t \mathbf{C}^{-1}(\vec{\mu}_1 - \vec{\mu}_2)$$
$$= \mathrm{Var}\{\Lambda(\vec{x})|\omega_2\} \qquad (6.27c)$$

The quantity $(\vec{\mu}_1 - \vec{\mu}_2)^t \mathbf{C}^{-1}(\vec{\mu}_1 - \vec{\mu}_2)$ is the Mahalanobis distance between $\vec{\mu}_1$ and $\vec{\mu}_2$. We will denote this measure by Δ. With this notation, the variance of the log-likelihood ratio becomes simply Δ, and the conditional expectations become $\Delta/2$ and $-\Delta/2$. The probability densities are shown in figure 6.13.

Knowing the conditional probabilities it is easy to evaluate the expressions in (6.26). The distribution function of a Gaussian random variable is the *error function* erf(). Hence:

$$P(\hat{\omega}_2|\omega_1) = \mathrm{erf}\left(\frac{T - \frac{1}{2}\Delta}{\sqrt{\Delta}}\right)$$
$$P(\hat{\omega}_1|\omega_2) = 1 - \mathrm{erf}\left(\frac{T + \frac{1}{2}\Delta}{\sqrt{\Delta}}\right) \qquad (6.28)$$

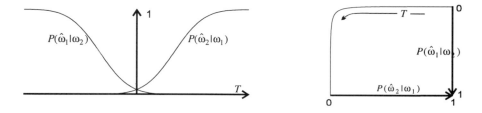

Figure 6.14 Performance of 2-case classification
a) Conditional probability of type I and type II error versus the threshold
b) $P(\hat{\omega}_1|\omega_2)$ and $P(\hat{\omega}_2|\omega_1)$ as a parametric plot of T

Figure 6.14a shows a graph of both probabilities when the threshold T varies. It can be seen that the requirements for T are contradictory. The probability of type I is small if the threshold is chosen small. However, the probability of type II is small if the threshold is chosen large. A balance must be found between both types of errors. This is also shown in figure 6.14b. Here, the probabilities of type II and type I are given as a parametric curve of T. Ideally, both probabilities are zero, but the figure shows that no threshold exists for which this occurs.

As stated before, the overall probability of an error of type I is $P(\text{type I error}) = P(\hat{\omega}_2, \omega_1) = P(\hat{\omega}_2 | \omega_1) P_1$. A similar expression exists for the probability of type II errors. Since the two errors are mutually exclusive, the error rate of the classification is the sum of both probabilities:

$$E = P(\hat{\omega}_2 | \omega_1) P_1 + P(\hat{\omega}_1 | \omega_2) P_2 \qquad (6.29)$$

In the example of figures 6.13 and 6.14 the Mahalonobis distance Δ equals 10. If this distance becomes larger, the probabilities of type II and type I become smaller. Hence, the error rate E is a montonically decreasing function of Δ.

Example 6.4 Detection of weaving-faults

In a weaving mill, it is often important to detect weaving faults in an early stage. One way to accomplish this is to inspect the tissue with a line-can CCD camera aligned perpendicular to the direction of transport (figure 6.15). Suppose that with a proper illumination the weaving faults are visible. Figure 6.15b and c give two realisations of the signal $f(y)$ coming from the CCD. One realisation is from tissue without weaving fault. The other realisation contains a weaving fault at position y_0. The weaving fault appears as a small spike-like signal. Note that both realisations contain a more or less periodic pattern induced by the texture of the tissue. The task of the detection system is to determine whether a fault has occured or not. In fact, this is a two-class pattern classification problem: $\omega_1 = $"No weaving fault", $\omega_2 = $"weaving fault".

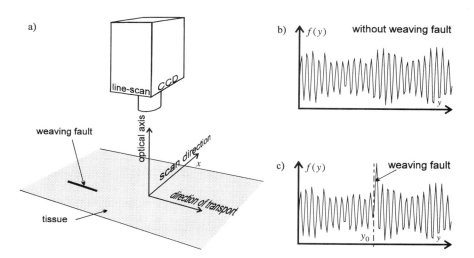

Figure 6.15 Fault detection in textile tissue
a)   Sensory system
b) and c) Measurements from one scan (fictitious data)

For the sake of simplicity we will assume that the position y_0 of a possible fault is known. Later in the chapter we will consider the case in which this assumption is not valid.

The development of a solution starts with formulating a model of the signal measured from the tissue. Suppose that the spike-like signal indicating the presence of a fault is given by $s(y)$. The pattern induced by the texture of the tissue is modeled by a Gaussian, stationary stochastic process $\underline{n}(y)$. The autocovariance function $C_{nn}(y)$ of this process describes the periodic behavior of the texture. We will assume that this function is known. The measurements are given by:

$$\begin{aligned} \omega_1: \quad & \underline{f}(y) = \underline{n}(y) \\ \omega_2: \quad & \underline{f}(y) = \underline{n}(y) + s(y - y_0) \end{aligned} \qquad (6.30)$$

The next step is to convert the space-continuous signal $\underline{f}(y)$ into a finite dimensional measurement vector $\underline{\tilde{x}}$. The obvious way to do this is by sampling. Suppose that the record-length of the scan is Y, and that the number of samples needed to represent this record without too much aliasing is N. Then, the elements of $\underline{\tilde{x}}$ are defined by:

$$\underline{x}_i = \underline{f}\left(\frac{i}{N}Y\right) \qquad \text{with:} \qquad i = 0, \cdots, N-1 \qquad (6.31)$$

Since $\underline{f}(y)$ is Gaussian, so is $\underline{\tilde{x}}$. Clearly, the expectation vector of $\underline{\tilde{x}}$ is class-dependent. If no fault has occured (ω_1), the expectation vector is zero: $\vec{\mu}_1 = \vec{0}$. If a fault has occured (ω_2), the expectation vector $\vec{\mu}_2$ corresponds to $s(y - y_0)$:

$$\mu_{2_i} = s\left(\frac{i}{N}Y - y_0\right) \qquad \text{with:} \qquad i = 0, \cdots, N-1 \qquad (6.32)$$

The random part of $\underline{\tilde{x}}$ comes from the stochastic process $\underline{n}(y)$. Hence, the covariance matrix $\mathbf{C}$ is class-independent. The elements of $\mathbf{C}$ are found as:

$$\mathbf{C}_{i,j} = C_{nn}\left(\frac{|i-j|}{N}Y\right) \qquad (6.33)$$

With that, the log-likelihood ratio in this problem becomes:

$$\Lambda(\tilde{x}) = -\tilde{x}^t \mathbf{C}^{-1} \vec{\mu}_2 + \text{constant} \qquad (6.34)$$

In detection theory, specific names are given to the two types of errors discussed before:

	ω_1	ω_2
$\hat{\omega}_1$	nameless	missed event
$\hat{\omega}_2$	false alarm	detection

The nomenclature is taken from radar technology where the problem is to detect a target. In this field of technology, the parametric curve shown in figure 6.14b is called *receiver operating characteristic (ROC)*. The curve shows the probability of false alarm versus the probability of detection. In fact, the unusual choice of the direction of the axes in figure 6.14b is induced by the conventions of an ROC.

6.1.5 Rejection

In some applications, it is advantageous to provide the classification with a so-called *rejection option*. For instance, in automated quality control in a production process, it might occur that the classification of a few objects is ambiguous. In these cases, it may be beneficial to subject the ambiguous objects to additional manual inspection. Instead of accepting or rejecting the *object*, we reject the *classification*. We may take this into account by extending the range of the decision function by a new element: the *rejection option* ω_0. The range of the decision function becomes: $\Omega^+ = \{\omega_0, \omega_1, \cdots, \omega_K\}$. The decision function itself is a function $\hat{\omega}(\vec{x})$: $\mathbb{R}^N \to \Omega^+$. In order to develop a Bayes classification, the definition of the cost function must also be adapted: $C(\hat{\omega}|\omega)$: $\Omega^+ \times \Omega \to \mathbb{R}^+$. The definition is extended so as to express the cost of the rejection option. $C(\omega_0|\omega_j)$ is the cost of the rejection option while the true class of the object is ω_j.

With these extensions, the decision function of a Bayes classification (equation 6.8) becomes:

$$\hat{\omega}(\vec{x}) = \underset{\hat{\omega} \in \Omega^+}{\operatorname{argmin}} \left\{ \sum_{j=1}^{K} C(\hat{\omega}|\omega_j) P(\omega_j|\underline{x} = \vec{x}) \right\} \tag{6.35}$$

In this expression, a conversion from posterior probabilities to prior probabilities and conditional densities is trivial.

The minimum error rate classification can also be extended with a rejection option. Suppose that the cost of a rejection is C_{rej} regardless of the true class of the object. All other costs are defined by equation (6.9).

We first note that if it is decided for the rejection option, the risk is C_{rej}. If it is not, the conditional risk is $e(\vec{x})$ given by equations (6.10) and (6.13). In that case, the optimal decision is the one which minimises this risk (equation 6.11). Minimisation of C_{rej} and $e(\vec{x})$ yields the following optimal decision function:

$$\begin{aligned}\hat{\omega}(\vec{x}) &= \omega_0 & \text{if: } & C_{rej} < e(\vec{x}) \\ \hat{\omega}(\vec{x}) &= \underset{\omega \in \Omega}{\operatorname{argmax}}\{P(\omega_j|\underline{x} = \vec{x})\} & \text{if: } & C_{rej} \geq e(\vec{x})\end{aligned} \quad \text{with: } e(\vec{x}) = 1 - \underset{\omega \in \Omega}{\max}\{P(\omega|\underline{x} = \vec{x})\} \tag{6.36}$$

The maximum posterior probability $\max\{P(\omega|\underline{x} = \vec{x})\}$ is always greater than $1/K$. Therefore, the minimum probability of error is bounded by $1 - 1/K$. Consequently, in (6.36) the rejection option never wins if $C_{rej} \geq 1 - 1/K$.

The overall probability of having a rejection is called the *rejection rate*. It is found by calculating the fraction of measurements that fall inside the rejection region:

$$R_{rej} = \int_{\{\vec{x}|C_{rej} < e(\vec{x})\}} p_{\underline{x}}(\vec{x}) d\vec{x} \tag{6.37}$$

The integral extends over those regions in the measurement space for which $C_{rej} < e(\vec{x})$. The error rate is found by averaging the conditional error over all measurements except those that fall inside the rejection region:

$$E = \int_{\{\vec{x}|C_{rej} \geq e(\vec{x})\}} e(\vec{x}) p_{\vec{x}}(\vec{x}) d\vec{x} \qquad (6.38)$$

Comparison of (6.38) with (6.14) shows that the error rate of a classification with rejection option is bounded by the error rate of a classification without rejection option.

Example 6.3 Classification of apples (continued)

In the classification of apples discussed in example 6.3 in section 6.1.1 it might be advantageous to manually inspect the apples whose automatic classification is likely to fail. If the cost of manual inspection is assumed to be about 60% of the cost price of a healthy apple, the cost function becomes (compare with the table in section 6.1.1):

	ω_1	ω_2	ω_3
$\hat{\omega}_0$	0.6	0.6	0.6
$\hat{\omega}_1$	0	5	5
$\hat{\omega}_2$	1	0	1
$\hat{\omega}_3$	1	1	0

The corresponding decision function is shown in figure 6.16. The ambiguous regions in the measurement space appear to be a zone in which the conditional densities show a large overlap and for which the cost of a misclassification is pretty high.

An indication of the performance of the classifier can be obtained by making a frequency table of all events shown in figure 6.16:

	ω_1	ω_2	ω_3
$\hat{\omega}_0$	18	5	6
$\hat{\omega}_1$	35	1	0
$\hat{\omega}_2$	7	11	1
$\hat{\omega}_3$	0	3	13

Such a frequency table is called a *confusion matrix*. Ideally, this matrix is diagonal. Off-diagonal elements are the number of times that a particular misclassification occurs. The matrix shows that the probability of having a diseased apple being classified as healthy is very low. This is the result of our particular choice of the cost function. An estimate of the rejection rate, based on the confusion matrix, is $R_{rej} \approx 29/100 = 29\%$. An estimate of the error rate is $E \approx 12/100 = 12\%$.

Figure 6.16 Bayes classification with rejection option (the rejection region is dark shaded)

6.2 PARAMETER ESTIMATION

Figure 6.17 gives a framework in which parameter estimation can be defined. The framework is similar to the one of pattern classification. Starting point is an experiment defined by the set Ω, a Borel field A, and a probability function defined on A. The difference with pattern classification is that the set Ω in parameter estimation is often infinite, whereas the one in pattern classification is always finite. Furthermore, in parameter estimation we assume that each outcome corresponds to a real vector $\underline{\alpha}$ defined in the M-dimensional linear space $\mathbb{R}^M$.

The parameter vector $\underline{\alpha}$ affects a physical object or a physical event. The object (or event) is sensed by a sensory system which brings forth an N-dimensional measurement vector $\underline{x}$. It is the task of the parameter estimator to recover the original parameter vector given the measurement vector. This is done by a function $\hat{\underline{\alpha}}(\underline{x})$: $\mathbb{R}^N \to \mathbb{R}^M$.

By the phrase "physical object" we mean concrete 3-dimensional objects. But the word can also have a more abstract interpretation, i.e. a physical system consisting of interacting elements. In a *dynamic system*, the measurement vector, and eventually also the parameter vector, depend on time. Often, such a system is described in terms of differential equations. This section mainly deals with *static systems* or *quasi-static systems*. At the end of the section an introduction to estimation in dynamic systems will be given.

We assume that the parameter vector has prior probability $p_{\underline{\alpha}}(\underline{\alpha})$. As mentioned above, the dimension of the vector is M. The situation in which the parameter is a single scalar is embedded in this assumption as we can always choose $M = 1$.

The conditional probability density $p_{\underline{x}|\underline{\alpha}}(\underline{x}|\underline{\alpha})$ gives the connection between parameter vector and measurements. The randomness of the measurement vector is, on one hand, due to the stochastic nature of $\underline{\alpha}$, on the other hand, due to physical noise sources elsewhere in the system. The density $p_{\underline{x}|\underline{\alpha}}(\underline{x}|\underline{\alpha})$ is conditional. The overall probability density of $\underline{x}$ is found by averaging the conditional density over the complete set of parameters:

$$p_{\underline{x}}(\underline{x}) = \int_{\underline{\alpha}} p_{\underline{x}|\underline{\alpha}}(\underline{x}|\underline{\alpha}) p_{\underline{\alpha}}(\underline{\alpha}) d\underline{\alpha} \tag{6.39}$$

The integral extends over the entire M-dimensional parameter space $\mathbb{R}^M$.

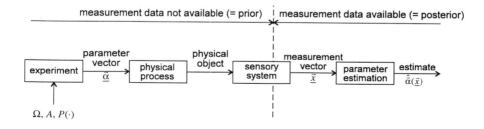

Figure 6.17 Statistical parameter estimation

6.2.1 Bayes estimation

Bayes' optimisation criterion (i.e. minimum risk) applies to statistical parameter estimation provided that two conditions are met [Sorenson 1980]. Firstly, it must be possible to quantify the cost of the damage involved when the estimates differ from the true parameters. Secondly, the expectation of the cost - the risk - is acceptable as an optimisation criterion.

Suppose that the damage is quantified by a cost function $C(\hat{\vec{\alpha}}|\vec{\alpha}): \mathbb{R}^M \times \mathbb{R}^M \rightarrow \mathbb{R}^+$. Ideally, this function represents the true cost. However, in most applications it is difficult to quantify the damage accurately. In addition, it appears that in a wide variety of estimation problems the "optimal" estimator is not very sensitive to a change of cost function. Therefore, it is common practice to choose a cost function whose mathematical treatment is not too complex. Often, the assumption is that the cost function only depends on the difference between estimated and true parameters: the estimation error $\vec{\varepsilon} = \hat{\vec{\alpha}} - \vec{\alpha}$. With this assumption, the following cost functions are well-known (see figure 6.18):

- *quadratic cost function:*

$$C(\hat{\vec{\alpha}}|\vec{\alpha}) = \|\hat{\vec{\alpha}} - \vec{\alpha}\|_2^2 = \sum_{m=0}^{M-1} (\hat{\alpha}_m - \alpha_m)^2 \qquad (6.40a)$$

- *absolute value cost function:*

$$C(\hat{\vec{\alpha}}|\vec{\alpha}) = \|\hat{\vec{\alpha}} - \vec{\alpha}\|_1 = \sum_{m=0}^{M-1} |\hat{\alpha}_m - \alpha_m| \qquad (6.40b)$$

- *uniform cost function:*

$$C(\hat{\vec{\alpha}}|\vec{\alpha}) = \begin{cases} 1 & \text{if } \|\hat{\vec{\alpha}} - \vec{\alpha}\|_2 > \Delta \\ 0 & \text{if } \|\hat{\vec{\alpha}} - \vec{\alpha}\|_2 \leq \Delta \end{cases} \quad \text{with: } \Delta \rightarrow 0 \qquad (6.40c)$$

The first two cost functions are instances of the distance measure based on the l_p-norm (see appendix A.1 and A.2). The third cost function is also a distance measure (appendix A.2).

With the estimate $\hat{\vec{\alpha}}$ and a given measurement vector $\vec{x}$, the *conditional risk* of $\hat{\vec{\alpha}}$ is defined as the expectation of the cost function:

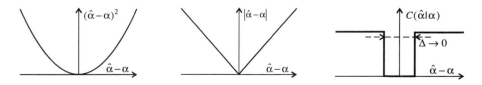

Figure 6.18. Cost functions
a) Quadratic cost
b) Absolute value cost
c) Uniform cost

$$R(\hat{\underline{\alpha}}|\underline{\tilde{x}} = \tilde{x}) = \mathrm{E}\{C(\hat{\underline{\alpha}}|\underline{\alpha})|\underline{\tilde{x}} = \tilde{x}\} = \int_{\tilde{\alpha}} C(\hat{\underline{\alpha}}|\tilde{\alpha}) p_{\underline{\alpha}|\underline{\tilde{x}}}(\tilde{\alpha}|\tilde{x}) d\tilde{\alpha} \qquad (6.41)$$

In Bayes estimation (or minimum risk estimation) the estimate is the parameter vector that minimises the risk:

$$\hat{\underline{\alpha}}(\tilde{x}) = \underset{\tilde{\alpha}}{\mathrm{argmin}} \{R(\tilde{\alpha}|\underline{\tilde{x}} = \tilde{x})\} \qquad (6.42)$$

The minimisation extends over the entire parameter space.

In the 1-dimensional case ($M = 1$), the Bayesian solution with the cost functions given in (6.40) can be given an interpretation in terms of the posterior probability of the parameter:

- The solution with quadratic cost function is the expectation conditioned on $\underline{\tilde{x}} = \tilde{x}$ (often called *conditional mean*):

$$\hat{\alpha}(\tilde{x}) = \mathrm{E}\{\underline{\alpha}|\underline{\tilde{x}} = \tilde{x}\} = \int_{\alpha} \alpha p_{\underline{\alpha}|\underline{\tilde{x}}}(\alpha|\tilde{x}) d\alpha \qquad (6.43\mathrm{a})$$

- The solution with absolute value cost function is the *median* of the posterior probability density:

$$\hat{\alpha}(\tilde{x}) = \hat{\alpha} \quad \text{such that:} \quad P_{\underline{\alpha}|\underline{\tilde{x}}}(\hat{\alpha}|\tilde{x}) = \frac{1}{2} \qquad (6.43\mathrm{b})$$

- The solution with uniform cost function is the *mode* of the posterior probability:

$$\hat{\alpha}(\tilde{x}) = \underset{\alpha}{\mathrm{argmax}} \{p_{\underline{\alpha}|\underline{\tilde{x}}}(\hat{\alpha}|\tilde{x})\} \qquad (6.43\mathrm{c})$$

If the posterior probability density is symmetric, the first two solutions give identical results. If, in addition, the posterior probability is uni-modal, the third solution gives also identical results.

6.2.2 Minimum variance estimation

The solution based on the quadratic cost function (6.40a) is called the *minimum variance estimate*. Other authors use the name *minimum mean square error estimate* (abbreviated to MSE). Substitution of (6.40a) and (6.41) in (6.42) gives:

$$\hat{\underline{\alpha}}_{MSE}(\tilde{x}) = \underset{\tilde{\alpha}}{\mathrm{argmin}} \left\{ \int_{\tilde{\alpha}} (\hat{\underline{\alpha}} - \tilde{\alpha})^t (\hat{\underline{\alpha}} - \tilde{\alpha}) p_{\underline{\alpha}|\underline{\tilde{x}}}(\tilde{\alpha}|\tilde{x}) d\tilde{\alpha} \right\} \qquad (6.44\mathrm{a})$$

Differentiating the function between braces with respect to $\hat{\underline{\alpha}}$ (see appendix A.7), and equating this result to zero yields a system of M linear equations, the solution of which is:

$$\hat{\underline{\alpha}}_{MSE}(\tilde{x}) = \int_{\tilde{\alpha}} \tilde{\alpha} p_{\underline{\alpha}|\underline{\tilde{x}}}(\tilde{\alpha}|\tilde{x}) d\tilde{\alpha} = \mathrm{E}\{\underline{\alpha}|\underline{\tilde{x}} = \tilde{x}\} \qquad (6.44\mathrm{b})$$

The conditional risk of this solution is the sum of the variances of the estimated parameters:

$$R(\hat{\underline{\alpha}}_{MSE}(\vec{x})|\vec{\underline{x}} = \vec{x}) = \int_{\underline{\alpha}} (\hat{\underline{\alpha}}_{MSE}(\vec{x}) - \underline{\alpha})^t (\hat{\underline{\alpha}}_{MSE}(\vec{x}) - \underline{\alpha}) p_{\underline{\alpha}|\vec{\underline{x}}}(\underline{\alpha}|\vec{x}) d\underline{\alpha}$$

$$= \sum_{m=0}^{M-1} \text{Var}\{\underline{\alpha}_m | \vec{\underline{x}} = \vec{x}\}$$

(6.45)

Hence the name minimum variance estimator.

Suppose the measurement vector can be expressed as a linear combination of the parameter vector corrupted by additive Gaussian noise:

$$\vec{\underline{x}} = \mathbf{H}\underline{\alpha} + \vec{\underline{n}} \qquad (6.46)$$

where $\vec{\underline{n}}$ is an N-dimensional random vector with zero expectation and covariance matrix $\mathbf{C}_n$. $\mathbf{H}$ is a $N \times M$-matrix, and $\underline{\alpha}$ is the M-dimensional parameter vector. Let us assume that prior to the measurement we know that $\underline{\alpha}$ has a Gaussian distribution with expectation $\text{E}\{\underline{\alpha}\} = \vec{\mu}_\alpha$ and covariance matrix $\mathbf{C}_\alpha$.

The assumption that $\vec{\underline{n}}$ is Gaussian implies that the conditional probability density of $\vec{\underline{x}}$ equals:

$$p_{\vec{\underline{x}}|\underline{\alpha}}(\vec{x}|\underline{\alpha}) = C_1 \exp\left(-\tfrac{1}{2}(\mathbf{H}\underline{\alpha} - \vec{x})^t \mathbf{C}_n^{-1}(\mathbf{H}\underline{\alpha} - \vec{x})\right) \qquad (6.47)$$

C_1 is a constant that depends neither on $\vec{x}$ nor on $\underline{\alpha}$. Assuming that the prior probability density of $\underline{\alpha}$ is also Gaussian:

$$p_{\underline{\alpha}}(\underline{\alpha}) = C_2 \exp\left(-\tfrac{1}{2}(\underline{\alpha} - \vec{\mu}_\alpha)^t \mathbf{C}_\alpha^{-1}(\underline{\alpha} - \vec{\mu}_\alpha)\right) \qquad (6.48)$$

where C_2 is a constant that does not depend on $\underline{\alpha}$ it follows that:

$$p_{\underline{\alpha}|\vec{\underline{x}}}(\underline{\alpha}|\vec{x}) = \frac{p_{\vec{\underline{x}}|\underline{\alpha}}(\vec{x}|\underline{\alpha}) p_{\underline{\alpha}}(\underline{\alpha})}{p_{\vec{\underline{x}}}(\vec{x})}$$

$$= C_3 \frac{\exp\left(-\tfrac{1}{2}(\mathbf{H}\underline{\alpha} - \vec{x})^t \mathbf{C}_n^{-1}(\mathbf{H}\underline{\alpha} - \vec{x})\right) \exp\left(-\tfrac{1}{2}(\underline{\alpha} - \vec{\mu}_\alpha)^t \mathbf{C}_\alpha^{-1}(\underline{\alpha} - \vec{\mu}_\alpha)\right)}{p_{\vec{\underline{x}}}(\vec{x})} \qquad (6.49)$$

$$= C_4 \frac{\exp\left(-\tfrac{1}{2}\left(\underline{\alpha}^t(\mathbf{H}^t\mathbf{C}_N^{-1}\mathbf{H} + \mathbf{C}_\alpha^{-1})^{-1}\underline{\alpha} - 2\underline{\alpha}^t(\mathbf{H}^t\mathbf{C}_n^{-1}\vec{x} + \mathbf{C}_\alpha^{-1}\vec{\mu}_\alpha)\right)\right)}{p_{\vec{\underline{x}}}(\vec{x})}$$

C_3 and C_4 are constants that do not depend on $\underline{\alpha}$. The last equality in equation (6.49) shows that the posterior probability density of $\underline{\alpha}$ is Gaussian. Its expectation and covariance matrix, conditioned on $\vec{x}$, are:

$$\hat{\underline{\alpha}}_{MSE}(\vec{x}) = \text{E}\{\underline{\alpha}|\vec{\underline{x}} = \vec{x}\} = \left(\mathbf{H}^t\mathbf{C}_n^{-1}\mathbf{H} + \mathbf{C}_\alpha^{-1}\right)^{-1}\left(\mathbf{H}^t\mathbf{C}_n^{-1}\vec{x} + \mathbf{C}_\alpha^{-1}\vec{\mu}_\alpha\right)$$

$$\mathbf{C}_{\underline{\alpha}|\vec{\underline{x}}=\vec{x}} = \left(\mathbf{H}^t\mathbf{C}_n^{-1}\mathbf{H} + \mathbf{C}_\alpha^{-1}\right)^{-1} \qquad (6.50)$$

The risk equals the trace of the conditional covariance matrix.

The minimum variance estimate consists of two terms. The first term is linear in $\vec{x}$. It represents the information coming from the measurement. The second term is linear in $\bar{\mu}_\alpha$. It represents the prior knowledge. To show that this interpretation is correct it is instructive to see what happens at the extreme ends: either no information from the measurement, or no prior knowledge.

The measurements are useless if the matrix $\mathbf{H}$ is virtually zero, or if the noise is too large, i.e. $\mathbf{H}'\mathbf{C}_n^{-1}$ is too small. The estimate given in (6.50) shows that in that case the second term prevails. In the limit, the estimate becomes $\bar{\mu}_\alpha$ with covariance matrix $\mathbf{C}_\alpha$, i.e. the estimate is purely based on prior knowledge.

On the other hand, if the prior knowledge is weak, i.e. the variances of the parameters are very large, the inverse covariance matrix $\mathbf{C}_\alpha^{-1}$ becomes irrelevant. In the limit, the estimate becomes $(\mathbf{H}'\mathbf{C}_n^{-1}\mathbf{H})^{-1}\mathbf{H}'\mathbf{C}_n^{-1}\bar{x}$. In this solution, the prior knowledge, i.e. $\bar{\mu}_\alpha$, is completely ruled out.

6.2.3 MAP estimation

If the uniform cost function is chosen, the conditional risk (6.41) becomes:

$$R(\hat{\bar{\alpha}}|\underline{\bar{x}} = \bar{x}) = \int_{\bar{\alpha}} p_{\bar{\alpha}|\bar{x}}(\bar{\alpha}|\bar{x})d\bar{\alpha} - p_{\bar{\alpha}|\bar{x}}(\hat{\bar{\alpha}}|\bar{x})\Delta$$

$$= 1 - p_{\bar{\alpha}|\bar{x}}(\hat{\bar{\alpha}}|\bar{x})\Delta \qquad (6.51)$$

The estimate which now minimises the risk is called the maximum a posterior (MAP) estimate:

$$\hat{\bar{\alpha}}_{MAP}(\bar{x}) = \underset{\bar{\alpha}}{\operatorname{argmax}}\left\{p_{\bar{\alpha}|\bar{x}}(\bar{\alpha}|\bar{x})\right\} \qquad (6.52)$$

This solution equals the mode of the posterior probability. It can also be written entirely in terms of prior probability densities and conditional probabilities:

$$\hat{\bar{\alpha}}_{MAP}(\bar{x}) = \underset{\bar{\alpha}}{\operatorname{argmax}}\left\{\frac{p_{\bar{x}|\bar{\alpha}}(\bar{x}|\bar{\alpha})p_{\bar{\alpha}}(\bar{\alpha})}{p_{\bar{x}}(\bar{x})}\right\} = \underset{\bar{\alpha}}{\operatorname{argmax}}\left\{p_{\bar{x}|\bar{\alpha}}(\bar{x}|\bar{\alpha})p_{\bar{\alpha}}(\bar{\alpha})\right\} \qquad (6.53)$$

This expression is a similar to the one of MAP-classification (equation 6.11b).

If the measurement vector is linear in $\bar{\alpha}$ and corrupted by additive Gaussian noise, as given in equation (6.46), the posterior probability density is Gaussian (equation 6.49). The mode of the density coincides with the expectation. Thus, in this case, MAP-estimation and the MSE-estimation coincide.

Example 6.2 Measurement of the diameter of yarn (continued)

We return to the problem of estimating the diameter (or radius) of yarn produced in a spinning mill. The measurement set-up is as indicated by figure 6.4a. A model of the set-up is depicted in figure 6.19a.

With the assumption that the direction of the illumination of the yarn is along the optical axis, and that yarn is a Lambertian diffuser, the projection of the yarn at the target of the CCD camera has the following form (section 2.1 and 2.2):

$$g_\alpha(y) = \begin{cases} C_1\dfrac{\alpha^2 - (y - y_0)^2}{\alpha^2} & \text{if: } (y - y_0) < \alpha \\ C_0 & \text{elsewhere} \end{cases} \quad \text{with: } d = 2\alpha \qquad (6.54)$$

C_0 and C_1 are radiometric constants. y_0 is the position of the centre of the yarn. In this example, all these constants are assumed to be known. α is the radius of the yarn.

In the sensory system, image blurring is modeled with a line spread function $h(y)$. The noise $\underline{n}(x)$ in the measurements is stationary, additive and Gaussian with autocovariance function $C_{nn}(y)$. The resulting signal is given as:

$$\underline{f}(y) = h(y) * g_{\underline{\alpha}}(y) + \underline{n}(y) \tag{6.55}$$

Sampling a record of this signal in an interval with length Y and with sampling period Δy yields the measurement vector $\underline{\bar{x}}$:

$$\underline{x}_i = \underline{f}\left(\frac{i\Delta y}{Y}\right) \quad \text{with:} \quad i = 0, \cdots, N-1 \tag{6.56}$$

The measurement vector conditioned on the parameter $\underline{\alpha}$ is Gaussian with expectation vector $\bar{\mu}_x(\underline{\alpha})$ and covariance matrix $\mathbf{C}_n$:

$$\mu_x(\underline{\alpha})_i = m_{\underline{\alpha}}\left(\frac{i\Delta y}{Y}\right) \quad \text{with:} \quad m_{\underline{\alpha}}(y) = h(y) * g_{\underline{\alpha}}(y)$$
$$\mathbf{C}_{n_{i,j}} = C_{nn}(|i-j|\Delta y) \tag{6.57}$$

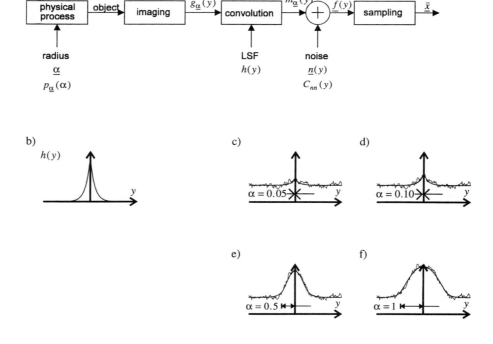

Figure 6.19 Measurement of the radius of yarn
a) Model of the imaging process
b) Line Spread Function used in c-f
c-f) Realisations with and without noise and with different instances of the radius

In general, the relation between $\underline{\alpha}$ and $\vec{\mu}_x(\underline{\alpha})$ is non-linear. If the width of the yarn is larger than the width of the LSF, the width of the image of the yarn is proportional to its radius. In figure 6.19, this is the case when approximately $\alpha > 0.5$ (see figure 6.19 e and f). If the width of the yarn falls below the resolution imposed by the LSF, the width of the image is invariant. In this case, the height of the image is proportional to the radius. Figure 6.19c and d are examples.

Suppose that the prior knowledge of the parameter $\underline{\alpha}$ is modelled with a Gaussian distribution with expectation μ_α and standard deviation σ_α. The MAP-estimate of $\underline{\alpha}$ follows from (6.53):

$$\hat{\alpha}_{MAP}(\vec{x}) = \underset{\alpha}{\operatorname{argmax}} \{ p_{\vec{x}|\underline{\alpha}}(\vec{x}|\alpha) p_{\underline{\alpha}}(\alpha) \}$$
$$= \underset{\alpha}{\operatorname{argmax}} \{ \log(p_{\vec{x}|\underline{\alpha}}(\vec{x}|\alpha)) + \log(p_{\underline{\alpha}}(\alpha)) \} \quad (6.58)$$

Upon substitution of the Gaussian densities and elimination of irrelevant terms we have:

$$\hat{\alpha}_{MAP}(\vec{x}) = \underset{\alpha}{\operatorname{argmax}} \left\{ -\frac{(\alpha - \mu_\alpha)^2}{\sigma_\alpha^2} - (\vec{x} - \vec{\mu}_x(\alpha))^t \mathbf{C}_n^{-1} (\vec{x} - \vec{\mu}_x(\alpha)) \right\} \quad (6.59)$$

The analytical maximisation of the function between braces is difficult because of the complex relation between $\underline{\alpha}$ and $\vec{\mu}_x(\underline{\alpha})$. However, a maximisation of the function using numerical techniques should be quite easy. This is illustrated in figure 6.20 which shows a graph of the logarithm of $p_{\vec{x}|\underline{\alpha}}(\vec{x}|\tilde{\alpha}) p_{\underline{\alpha}}(\tilde{\alpha})$ versus α. In this example $\vec{x}$ is as shown in figure 6.19c. The graph in figure 6.20 shows that $p_{\vec{x}|\underline{\alpha}}(\vec{x}|\tilde{\alpha}) p_{\underline{\alpha}}(\tilde{\alpha})$ is a smooth function of α. Hence, finding the maximum can be accomplished easily by standard numerical methods (e.g. Newton-Raphson, or steepest ascent).

The MAP-estimate of the radius using the measurement vector shown in figure 6.19c is the position in figure 6.20 indicated by $\hat{\alpha}_{MAP}$. The estimate is about 0.09 whereas the true parameter is $\alpha = 0.05$. The prior knowledge used is that $\underline{\alpha}$ is a Gaussian with expectation $\mu_\alpha = 0.5$ and standard deviation $\sigma_\alpha = 0.2$

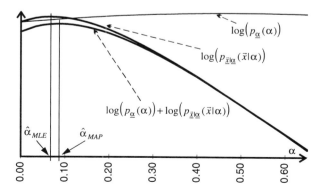

Figure 6.20 The optimisation function $\log(p_{\underline{\alpha}}(\alpha)) + \log(p_{\vec{x}|\underline{\alpha}}(\vec{x}|\alpha))$, used in the MAP-estimation of the radius of yarn

6.2.4 Maximum likelihood estimation

In many practical situations the prior knowledge needed in MAP-estimation is not available. Therefore, an estimator which does not depend much on prior knowledge is desirable. One attempt in that direction is given by a method referred to as *maximum likelihood estimation* (ML-estimation). The method is based on the observation that in MAP estimation written in the logarithm form (6.58) the peak of the first term $\log(p_{\underline{x}|\underline{\alpha}}(\vec{x}|\vec{\alpha}))$ is often in an area of $\vec{\alpha}$ in which the second term $\log(p_{\underline{\alpha}}(\vec{\alpha}))$ is almost flat. This holds true especially if not much prior knowledge is available. In these cases, in finding the MAP-estimate, the $\log(p_{\underline{\alpha}}(\vec{\alpha}))$-function does not affect the position of the maximum very much. Discarding the term, and maximising the function $\log(p_{\underline{x}|\underline{\alpha}}(\vec{x}|\vec{\alpha}))$ solely, gives the ML-estimate. Furthermore, instead of maximising $\log(p_{\underline{x}|\underline{\alpha}}(\vec{x}|\vec{\alpha}))$ one can equally well maximise the conditional probability density $p_{\underline{x}|\underline{\alpha}}(\vec{x}|\vec{\alpha})$:

$$\hat{\vec{\alpha}}_{MLE}(\vec{x}) = \mathrm{argmax}_{\vec{\alpha}}\left\{p_{\underline{x}|\underline{\alpha}}(\vec{x}|\vec{\alpha})\right\} = \mathrm{argmax}_{\vec{\alpha}}\left\{\log\left(p_{\underline{x}|\underline{\alpha}}(\vec{x}|\vec{\alpha})\right)\right\} \tag{6.58}$$

Regarded as a function of $\vec{\alpha}$ the conditional probability density is called the *likelihood function*. Hence the name "maximum likelihood estimation".

If the measurement vector $\underline{x}$ is linear in $\underline{\alpha}$ and corrupted by additive Gaussian noise, as given in equation (6.46), the likelihood of $\underline{\alpha}$ is given in (6.47). The parameter vector which maximises this density is:

$$\hat{\vec{\alpha}}_{MLE}(\vec{x}) = \left(\mathbf{H}^t \mathbf{C}_n^{-1} \mathbf{H}\right)^{-1} \mathbf{H}^t \mathbf{C}_n^{-1} \vec{x} \tag{6.60}$$

Example 6.2 Measurement of the diameter of yarn (continued)

In the estimation problem concerning the radius (or diameter) of yarn, a maximum likelihood estimate can be derived from equation (6.58) and (6.59):

$$\begin{aligned}\hat{\alpha}_{MLE}(\vec{x}) &= \mathrm{argmax}_{\alpha}\left\{\log\left(p_{\underline{x}|\underline{\alpha}}(\vec{x}|\alpha)\right)\right\} \\ &= \mathrm{argmax}_{\alpha}\left\{-(\vec{x}-\vec{\mu}_x(\alpha))^t \mathbf{C}_n^{-1}(\vec{x}-\vec{\mu}_x(\alpha))\right\}\end{aligned} \tag{6.61}$$

A graph of this optimisation function is shown in figure 6.20. The function finds its maximum near $\alpha = 0.06$. The true parameter is 0.05. In fact, in this example the estimation error of the ML-estimate (no prior knowledge) is less than the error in the MAP-estimate. The reason is that with the prior knowledge assumed ($\mu_\alpha = 0.5$, $\sigma_\alpha = 0.2$) the true parameter (0.05) is very unlikely. Therefore, the use of the prior knowledge tends to pull the estimate towards the vicinity of the prior expectation.

6.2.5 Least squares fitting and ML-estimation

Least squares fitting is a technique in which the measurement process is modeled as:

$$\underline{x} = \vec{h}(\vec{\alpha}) + \underline{n} \tag{6.62}$$

where $\vec{h}()$ is a deterministic vector function, and $\underline{n}$ is additive noise. The purpose of least squares fitting is to find the parameter vector $\vec{\alpha}$ which "best" fits the measurements $\underline{x}$.

Suppose that $\hat{\underline{\alpha}}$ is an estimate of $\underline{\alpha}$. This estimate is able to "predict" the deterministic part of $\underline{x}$, but it cannot predict the noise. The *prediction* of the estimate $\hat{\underline{\alpha}}$ *is* given by $\vec{h}(\hat{\underline{\alpha}})$. The *residuals* $\underline{\vec{\varepsilon}}$ are defined as the difference between observed and predicted measurements:

$$\underline{\vec{\varepsilon}} = \underline{\vec{x}} - \vec{h}(\hat{\underline{\alpha}}) \tag{6.63}$$

Figure 6.21 gives an overview of the variables and the processes introduced so far.

An optimisation criterion frequently used in the fitting process is the *sum of the squares of the residuals*:

$$J(\hat{\underline{\alpha}}) = \|\underline{\vec{\varepsilon}}\|_2^2 = \sum_{i=0}^{N-1} \varepsilon_i^2 = \sum_{i=0}^{N-1} (x_i - h_i(\hat{\underline{\alpha}}))^2 \tag{6.64}$$

The parameter vector which best fits the measurements is defined as the one that minimises the criterion function:

$$\hat{\underline{\alpha}}_{LSE}(\underline{\vec{x}}) = \underset{\underline{\alpha}}{\operatorname{argmin}} \{J(\underline{\vec{\alpha}})\} \tag{6.65}$$

Finding this vector is called *least squares fitting* or *least squares estimation* (LSE).

Note that in least squares fitting the parameter vector is modeled as a deterministic, but unknown vector. In contrast with that, the estimators discussed above consider the parameter vector to be random. Apart from this subtle distinction, ML-estimation and LSE-estimation are equivalent provided that the measurement model is given as in (6.62) and that $\underline{\vec{n}}$ is Gaussian with covariance matrix $C_n = I$. In that case, the likelihood function of $\underline{\alpha}$ becomes:

$$p_{\underline{\vec{x}}|\underline{\alpha}}(\underline{\vec{x}}|\underline{\vec{\alpha}}) = \frac{1}{\sqrt{(2\pi)^N}} \exp\left(-\frac{(\underline{\vec{x}} - \vec{h}(\underline{\vec{\alpha}}))^t (\underline{\vec{x}} - \vec{h}(\underline{\vec{\alpha}}))}{2}\right) \tag{6.66}$$

The vector that maximises the likelihood function is:

$$\hat{\underline{\alpha}}_{MLE}(\underline{\vec{x}}) = \hat{\underline{\alpha}}_{LSE}(\underline{\vec{x}}) = \underset{\underline{\alpha}}{\operatorname{argmin}}\{(\underline{\vec{x}} - \vec{h}(\underline{\vec{\alpha}}))^t(\underline{\vec{x}} - \vec{h}(\underline{\vec{\alpha}}))\} = \underset{\underline{\alpha}}{\operatorname{argmin}}\{\|\underline{\vec{x}} - \vec{h}(\underline{\vec{\alpha}})\|_2^2\} \tag{6.67}$$

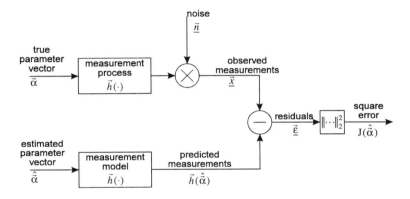

Figure 6.21 Observed and predicted measurements in least squares fitting

For example, if the measurement model is linear ($\underline{\tilde{x}} = \mathbf{H}\underline{\tilde{\alpha}} + \underline{\tilde{n}}$) and $\mathbf{H}$ is an $N \times M$-matrix having a rank $M < N$, then it can be proven that:

$$\hat{\underline{\alpha}}_{LSE}(\tilde{x}) = \left(\mathbf{H}^t\mathbf{H}\right)^{-1}\mathbf{H}^t\tilde{x} \tag{6.68}$$

Provided that the noise is white, i.e. $\mathbf{C}_n = \mathbf{I}$, this solution is indeed equivalent to the ML-estimator in (6.60).

6.2.6 Simultaneous estimation and classification

Some applications require simultaneous estimation and classification. This is the case in example 6.4, the weaving fault detection problem given in section 6.1.4. In that example, it was presumed that the position y_0 of a possible fault is known. Of course, this is a simplification. A fault may occur anywhere. A model, more realistic in this application, is that the prior probability density of the position is uniform along the width Y of the material. The task of the detection/estimation system is to check whether there is a weaving fault, and if so, to find the position. This is a simultaneous estimation and detection problem.

The conditional probability density of the measurement vector in the example in section 6.1.4 takes the form:

$$p_{\underline{\tilde{x}}|\underline{y}_0,\omega}(\tilde{x}|y_0,\omega) = \begin{cases} p_{\underline{\tilde{n}}}(\tilde{x}) & \text{if } \omega = \omega_1 \quad (\text{no fault, noise only}) \\ p_{\underline{\tilde{n}}}(\tilde{x} - \underline{\tilde{\mu}}_2(y_0)) & \text{if } \omega = \omega_2 \quad (\text{fault at } y_0 + \text{noise}) \end{cases} \tag{6.69}$$

$\underline{\tilde{n}}$ is the vector resulting from the noise $n(y)$; see equation (6.30) and (6.31). The vector is a zero mean Gaussian random vector with covariance matrix $\mathbf{C}$ given in (6.33). The vector $\underline{\tilde{\mu}}_2(y_0)$ is the expectation of the measurement vector if a fault has occurred at position y_0; see equation (6.32).

A MAP-solution consists of the selection of those arguments that simultaneously maximise the posterior probability density:

$$\hat{\omega}(\tilde{x}) = \underset{\omega \in \{\omega_1,\omega_2\}}{\operatorname{argmax}} \left\{ \underset{y_0 \in Y}{\max} \left\{ p_{\omega, \underline{y}_0|\underline{\tilde{x}}}(\omega, y_0|\tilde{x}) \right\} \right\}$$

$$\hat{y}_0(\tilde{x}) = \underset{y_0 \in Y}{\operatorname{argmax}} \left\{ p_{\omega, \underline{y}_0|\underline{\tilde{x}}}(\omega_2, y_0|\tilde{x}) \right\} \tag{6.70a}$$

The first equation states that a fault is assumed to exist if the posterior probability of a fault occurring at the most likely position exceeds the posterior probability of having no fault. Of course, seeking the position of a fault makes sense only if a fault is assumed to exist. Therefore, in the second equation of (6.70a) the argmax function is restricted to the situation in which a fault exists.

Using Bayes' theorem of conditional probability, rearranging the results, and cancelling irrelevant terms, we have:

$$p_{\underline{\tilde{x}}|\omega_1}(\tilde{x}|\omega_1)P_1 \underset{\omega_2}{\overset{\omega_1}{\gtrless}} \underset{y_0 \in Y}{\max} \left\{ p_{\underline{\tilde{x}}|\omega_2,\underline{y}_0}(\tilde{x}|\omega_2,y_0) \right\} P_2$$

$$\hat{y}_0(\tilde{x}) = \underset{y_0 \in Y}{\operatorname{argmax}} \left\{ p_{\underline{\tilde{x}}|\omega_2,\underline{y}_0}(\tilde{x}|\omega_2,y_0) \right\} \tag{6.70b}$$

STATISTICAL PATTERN CLASSIFICATION AND PARAMETER ESTIMATION 161

In the last line of this equation the knowledge that y_0 is uniform within Y is used. A further development of the detector/estimator involves the substitution of the probability density $p_{\bar{x}|y_0,\omega}(\bar{x}|y_0,\omega)$ given in (6.69). After some straightforward mathematical manipulations of the result of that, we arrive at the conclusion that the MAP-estimator/detector must set $\hat{y}_0$ equal to the position y_0 which maximises:

$$\bar{\mu}_2^t(y_0)C^{-1}\bar{x} \qquad \text{with:} \qquad y_0 \in Y \tag{6.70c}$$

The fault has been detected if $\bar{\mu}_2^t(\hat{y}_0)C^{-1}\bar{x}$ exceeds a suitably chosen threshold. The proof of this is left as an exercise to the reader.

The operation in (6.70c) can have the interpretation of a correlation. To see this it is instructive to assume that the noise is white, i.e. $C = I$. In that case, the operation takes the form of $\bar{\mu}_2^t(y_0)\bar{x}$. Returning back to the continuous domain, replacing $\bar{\mu}_2(y_0)$ with its continuous representation $s(y_0)$ in accordance with (6.31) and (6.32), and replacing the summation with an integral, the operation becomes:

$$\int_\eta s(\eta - y_0) \underline{f}(\eta) d\eta \tag{6.70d}$$

Clearly, this operation is the correlation between the observed signal $f(y)$ and the "template" $s(y)$; see table 2.4, and sections 3.4 and 5.3.

With this interpretation the estimator/detector can be implemented with a convolution, a *non-maximum suppression* and a threshold operation; see figure 6.22. The convolution - with impulse response $s(-y)$ - implements the correlation given in (6.70d). Non-maximum suppression is an operation that sets the signal to zero everywhere, except at the position of its maximum. At this position the signal is retained. As such after non-maximum suppression, the position of a hypothesised fault is seen as the position for which the signal is non-zero. The magnitude of this position is useful as a measure of the posterior probability of having a fault. Thresholding this magnitude completes the process.

If the noise is non-white, the operation in (6.70c) becomes $\bar{\mu}_2^t(\hat{y}_0)C^{-1}\bar{x}$. The behavior of C^{-1} is similar to a decorrelation operation (see appendix B.3.1). It can be implemented as a *noise whitening filter*. As such it can be included in the convolution operation by adapting the impulse response.

6.2.7 Properties of estimators

Figure 6.23 gives an overview of the estimators discussed so far. The estimators considered are all Bayesian. Differences arise from the adoption of different cost functions and from the type of prior knowledge.

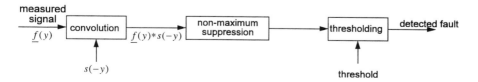

Figure 6.22 Implementation of a 1-dimensional estimator/detector

Adoption of the quadratic cost function leads to minimum variance estimation (MSE-estimation). The cost function is such that small errors are regarded as unimportant, while larger errors are considered more and more serious. The solution is found as the conditional mean, i.e. the expectation of the posterior probability density.

Another possibility is the unit cost function which leads to MAP-estimation. The cost function is such that the damage of small errors and large errors are equally weighted. The solution appears to be the mode of the posterior probability density.

It is remarkable that although these two cost functions differ a lot the solutions are identical provided that the posterior density is uni-modal and symmetric. An example of this occurs when the prior probability density and conditional probability density are both Gaussian. In that case the posterior probability density is Gaussian too.

If the solution space of a minimum variance estimation problem is restricted to the class of linear operations, the resulting class of estimation is called *linear mean square estimation* (LMSE-estimation). As shown above, if all random vectors involved are Gaussian, the solution appears to be linear. Hence, in that particular case, LMSE-, MSE- and MAP-solutions are all identical.

The performance of an estimator is determined by many aspects: the bias, the mean square error, the variance, the efficiency, the stability, the robustness, etc.

An estimator of a parameter vector is said to be *unbiased* if its expectation equals the true vector $E\{\hat{\underline{\alpha}}(\underline{\tilde{x}})|\underline{\alpha} = \tilde{\alpha}\} = \tilde{\alpha}$. Otherwise, the estimator is called *biased*. The difference between expectation and true parameter, $E\{\hat{\underline{\alpha}}(\underline{\tilde{x}})|\underline{\alpha} = \tilde{\alpha}\} - \tilde{\alpha}$, is the *bias*. A small bias is desirable.

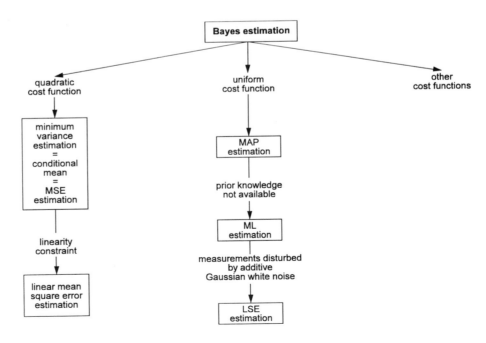

Figure 6.23 Overview of estimation methods within Bayes' family

The mean square error $E\{\|\hat{\bar{\alpha}}(\bar{x})-\bar{\alpha})\|^2\}$ and the variance defined as $\mathrm{Var}\{\hat{\bar{\alpha}}(\bar{x})\} =$
$= E\{\|\hat{\bar{\alpha}}(\bar{x})-\bar{\alpha}\|^2\}-\|bias\|^2$ are other performance measures. The first measure incorporates both the deterministic and the random part of the estimate. The second one depends on the random part only. Of course, it is desirable to have both measures as small as possible.

If $\hat{\bar{\alpha}}(\bar{x})$ is an unbiased estimator of $\bar{\alpha}$ and no other unbiased estimator of $\bar{\alpha}$ has smaller variance than $\hat{\bar{\alpha}}(\bar{x})$, then $\hat{\bar{\alpha}}(\bar{x})$ is called an *efficient* estimate. If $\tilde{\bar{\alpha}}(\bar{x})$ is an unbiased estimate and $\hat{\bar{\alpha}}(\bar{x})$ is an efficient estimator, then the *efficiency* of $\tilde{\bar{\alpha}}(\bar{x})$ in estimating $\bar{\alpha}$ is the ratio of the variances of both estimators:

$$\text{efficiency} = \frac{\mathrm{Var}\{\tilde{\bar{\alpha}}(\bar{x})\}}{\mathrm{Var}\{\hat{\bar{\alpha}}(\bar{x})\}} \tag{6.71}$$

By definition, an unbiased minimum variance estimator is (most) efficient.

Other properties related to the quality of an estimator are *stability* and *robustness*. In this context, stability refers to the property of being insensitive to small random errors in the measurement vector. Robustness is the property of being insensitive to large errors in a few elements of the measurements (outliers). An example of an estimator with neither much stability nor much robustness is a LSE-estimator in which the dimension of the measurement vector is equal or less than the dimension of the parameter vector. In general, enlargement of prior knowledge increases both the stability and the robustness.

In practical situations, one would like to use the performance measures mentioned above to quantify and to compare the different estimators. However, the calculation of these measures is often cumbersome. In these cases, approximations are needed. Alternatively, one could try to find upper and/or lower bounds of the measures. For instance, the variance of an unbiased estimator has a lower bound given by the so-called Cramèr-Rao inequality [Patrick 1972].

6.2.8 Estimation of dynamically changing parameters

A prior probability $p_{\bar{\alpha}}(\bar{\alpha})$ is called *reproducing* with respect to the conditional density $p_{\bar{x}|\bar{\alpha}}(\bar{x}|\bar{\alpha})$, if the posterior density $p_{\bar{\alpha}|\bar{x}}(\bar{\alpha}|\bar{x})$ has the same functional form of $p_{\bar{\alpha}}(\bar{\alpha})$ differing only in the values of the parameters characterising $p_{\bar{\alpha}}(\bar{\alpha})$. An example of a reproducing density is the Gaussian density given in equation (6.48). With a known measurement vector this density reproduces itself in another Gaussian density, the one given in equation (6.49).

In a dynamic estimation process, the parameters to be estimated depend on time. In course of the time, refreshed measurement data become available. In such systems, a reproducing density is most desirable. As an example, consider the estimation of the radius of yarn (example 6.2). In fact, this is a dynamic process since the production of yarn goes on and on. The prior knowledge needed in the estimation given in (6.59) could be derived from previous estimates. If this is the case, and if the prior probability density of the radius reproduces itself, the estimation process can be implemented recursively.

Example 6.2 Measurement of the diameter of yarn (continued)

We can demonstrate the use of recursively implemented estimated processes as follows. Let m be the discrete time-index. This means that refreshed measurements $\underline{\tilde{x}}_m$ become available at time-discrete moments $t_m = m\Delta t$, where Δt is the sampling period. The task is to estimate the parameter $\underline{\alpha}_m = \underline{\alpha}(t_m)$ at each time-discrete moment t_m. We assume that the radius of yarn cannot change rapidly. Therefore, the prior knowledge used in the m-th estimate $\hat{\underline{\alpha}}_m$ can be based on previous estimates $\hat{\underline{\alpha}}_{m-1}$, $\hat{\underline{\alpha}}_{m-2}$, and so on. One of the simplest models of $\underline{\alpha}_m$ which can be used for that purpose is a first order difference equation, i.e. an *autoregressive (AR)* process:

$$\underline{\alpha}_m = \Phi \underline{\alpha}_{m-1} + \underline{w}_m + \beta \tag{6.72}$$

where Φ is a constant satisfying $0 \le \Phi < 1$. This constant determines the correlation between present radius and previous radii. The driving force behind these fluctuating radii is $\underline{w}_m$ modeled as a sequence of uncorrelated, zero mean Gaussian random variables with variance σ_w^2. It can be shown that $\underline{\alpha}_m$ is a sequence of Gaussian random variables with variance: $\sigma_\alpha^2 = \sigma_w^2/(1-\Phi^2)$. The expectation of $\underline{\alpha}_m$ is β. In dynamic systems, $\underline{\alpha}_m$ is called a *state variable*.

Suppose that it is known that $\underline{\alpha}_{m-1} = \alpha_{m-1}$. Then, without using any measurement data, the expectation of $\underline{\alpha}_m$ is:

$$E\{\underline{\alpha}_m | \underline{\alpha}_{m-1} = \alpha_{m-1}\} = \Phi \alpha_{m-1} + \beta \tag{6.73}$$

The best estimate of $\underline{\alpha}_m$ is $\hat{\alpha}_m = \Phi \alpha_{m-1} + \beta$. The estimation error is $\underline{w}_m$ with variance σ_w^2.

The measurement vector $\underline{\tilde{x}}_m$ can be modeled with:

$$\underline{\tilde{x}}_m = \bar{\mu}_x(\underline{\alpha}_m) + \underline{\tilde{n}}_m \tag{6.74}$$

where $\bar{\mu}_x(\underline{\alpha}_m)$ is the expectation vector of $\underline{\tilde{x}}_m$ conditioned on $\underline{\alpha}_m$; see (6.57). The zero mean Gaussian random vector $\underline{\tilde{n}}_m$ represents the randomness in the measurements. We assume that the noise vectors taken at different time-discrete moments are independent. Therefore, the covariance matrix C_n given in equation (6.57) is sufficient to define the noise statistically. Figure 6.24 shows a graphical representation of the model used to describe the physical process and the sensory system.

Conditioned on $\underline{\alpha}_{m-1} = \alpha_{m-1}$ the following (optimal) estimator results from substitution of (6.73) and (6.74) in (6.59):

$$\hat{\alpha}_m = \underset{\alpha}{\operatorname{argmax}} \left\{ -\frac{(\alpha - \Phi \alpha_{m-1} - \beta)^2}{\sigma_w^2} - (\tilde{x}_m - \bar{\mu}_x(\alpha))^t C_n^{-1} (\tilde{x}_m - \bar{\mu}_x(\alpha)) \right\} \tag{6.75}$$

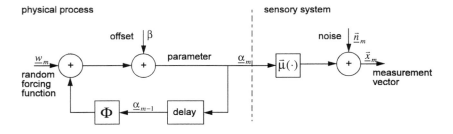

Figure 6.24 Mathematical model of the physical process and the sensory system

As the true parameter α_{m-1} is unknown, this estimator cannot be realised as such. However, we can replace the true parameter with its estimate $\hat{\alpha}_{m-1}$, thereby arriving at:

$$\hat{\alpha}_m = \underset{\alpha}{\operatorname{argmax}}\left\{-\frac{(\alpha-\Phi\hat{\alpha}_{m-1}-\beta)^2}{\sigma_w^2}-(\vec{x}_m-\vec{\mu}_x(\alpha))^t\mathbf{C}_n^{-1}(\vec{x}_m-\vec{\mu}_x(\alpha))\right\} \quad (6.76)$$

In fact, this estimator is a recursive, non-linear filter with the sequence $\vec{x}_m$ as input and $\hat{\alpha}_m$ as output. It can be realised, but it is not optimal. This is due to the uncertainty caused by the usage of the estimate $\hat{\alpha}_{m-1}$ instead of α_{m-1}. This uncertainty has not been incorporated as yet. Unfortunately, the calculation of the variance of $\hat{\underline{\alpha}}_{m-1}$ is difficult.

A simplification occurs if the dependency of $\vec{\mu}_x(\underline{\alpha}_m)$ on $\underline{\alpha}_m$ is linear. In that case the measurement model is:

$$\underline{\vec{x}}_m = \mathbf{H}\underline{\alpha}_m + \underline{\vec{n}}_m \quad (6.76)$$

where $\mathbf{H}$ is an $N\times 1$-matrix. In the estimation problem of the yarn, a linear measurement model is accurate provided that the effective width of the PSF is much larger than the radius of the yarn. See, for instance, figure 6.19c and 6.19d.

With this simplification, the key to an optimal solution of the problem lies in the minimum variance estimator given in equation (6.50). Translated to a 1-dimensional static problem ($\vec{x} = \mathbf{H}\underline{\alpha}+\vec{n}$) the estimator takes the form:

$$\hat{\alpha} = \left(\mathbf{H}'\mathbf{C}_n^{-1}\mathbf{H}+\sigma_\alpha^{-2}\right)^{-1}\left(\mathbf{H}'\mathbf{C}_n^{-1}\vec{x}+\sigma_\alpha^{-2}\mu_\alpha\right)$$
$$\sigma_{\hat{\alpha}}^2 = \left(\mathbf{H}'\mathbf{C}_n^{-1}\mathbf{H}+\sigma_\alpha^{-2}\right)^{-1} \quad (6.77)$$

In order to arrive at a solution applicable in the dynamic case we assume that an estimate of the previous state is available. We use this estimate to predict the parameter at the present state and use this prediction as prior knowledge (μ_α in 6.77). Next we combine this prior knowledge with the current measurement vector to obtain an updated estimate. Along with this prediction and update we also use the second line in (6.77) to update variances.

Suppose that in the previous state an estimate $\hat{\underline{\alpha}}_{m-1}$ has been found with variance $\sigma_{\hat{\alpha},m-1}^2$. According to (6.73) a prediction from the previous state to the present is:

$$\hat{\mu}_{\alpha,m} = \Phi\hat{\underline{\alpha}}_{m-1}+\beta \quad (6.78a)$$

with variance:

$$\sigma_{\hat{\mu},m}^2 = \Phi^2\sigma_{\hat{\alpha},m-1}^2+\sigma_w^2 \quad (6.78b)$$

Combination of this prior knowledge with the measurement vector $\underline{\vec{x}}_m$ in the minimum variance estimator given in (6.77) yields:

$$\hat{\underline{\alpha}}_m = \left(\mathbf{H}'\mathbf{C}_n^{-1}\mathbf{H}+\sigma_{\hat{\mu},m}^{-2}\right)^{-1}\left(\mathbf{H}'\mathbf{C}_n^{-1}\vec{x}_m+\sigma_{\hat{\mu},m}^{-2}\hat{\underline{\mu}}_{\alpha,m}\right) \quad (6.79a)$$

with variance:

$$\sigma_{\hat{\alpha},m}^2 = \left(\mathbf{H}'\mathbf{C}_n^{-1}\mathbf{H}+\sigma_{\hat{\mu},m}^{-2}\right)^{-1} \quad (6.79b)$$

Taken together, equations (6.78) and (6.79) give the solution of the estimation problem. Usually, this solution is given in a compact form using a time-dependent matrix $\mathbf{K}_m$ called the *Kalman gain matrix*. In our case (estimation of a 1-dimensional parameter) the matrix degenerates into a $1 \times N$-matrix defined as follows:

$$\mathbf{K}_m = \left(\mathbf{H}^t \mathbf{C}_n^{-1} \mathbf{H} + \sigma_{\hat{\mu},m}^{-2}\right)^{-1} \mathbf{H}^t \mathbf{C}_n^{-1} \tag{6.80}$$

With that, (6.79) transforms into:

$$\hat{\alpha}_m = \hat{\mu}_{\alpha,m} + \mathbf{K}_m \left(\vec{x}_m - \mathbf{H}\hat{\mu}_{\alpha,m}\right) \tag{6.81a}$$

with variance:

$$\sigma_{\hat{\alpha},m}^2 = (1 - \mathbf{K}_m \mathbf{H}) \sigma_{\hat{\mu},m}^2 \tag{6.81b}$$

Figure 6.25 shows an overview of the estimation process given in (6.78) up to (6.81). An interpretation of this process is as follows. With each prediction of the parameter a predicted measurement vector is associated. The difference between predicted and actual measurement vector is used to correct the predicted parameter.

The structure of the estimator shown in figure 6.25 is that of a so-called *discrete Kalman filter* [Gelb 1974]. The filter is linear. Because of the dependency of $\mathbf{K}_m$ on time it is time-variant. The filter can easily be generalised to the case of an M-dimensional parameter vector instead of a single parameter. The model of the physical process is given then by a linear difference equation of order M:

$$\vec{\alpha}_m = \Phi \vec{\alpha}_{m-1} + \vec{w}_m \tag{6.82}$$

Under the condition that the driving random vector $\vec{w}_m$ and the noise vector $\vec{n}_m$ are Gaussian the Kalman filter gives an optimal (minimum variance) estimate of $\hat{\alpha}_m$. The optimality is founded on the very fact that a Gaussian prior probability density reproduces itself in the posterior probability (provided that the estimator is linear).

The concept of Kalman filtering applies even if the matrices $\mathbf{H}$ and Φ depend on time. Moreover, the assumption that random vectors are stationary is not needed.

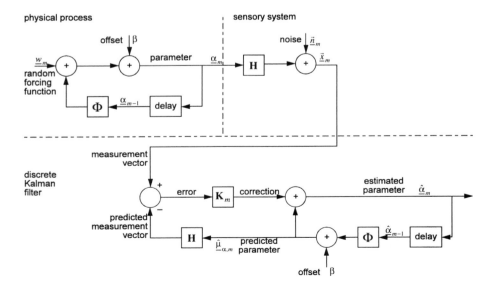

Figure 6.25 System model and discrete Kalman filter

The covariance matrices of these vectors are allowed to be time-dependent. However, if these matrices are constant - as assumed in our example - the Kalman gain matrix $\mathbf{K}_m$ will finally reach a steady state $\mathbf{K}_\infty$.

6.3 SUPERVISED LEARNING

The development of a Bayes classificator or estimator requires the availability of the prior probabilities and the conditional probability densities. One possibility to obtain this knowledge is by proper modeling the physical process and the sensory system. The measurement of the radius of yarn (example 6.2) is based on such an approach. In many other applications modeling the process is very difficult if not impossible. For instance, in the classification of healthy and diseased apples (example 6.3) the visual appearance of diseased apples results from a complex biological process which is very difficult to model.

An alternative to the *model-based* approach is an approach based on *learning from examples*. Suppose that in the apple classification problem a set of samples selected from a population of apples is available. We may assume that the measurement vector of each sample is known. The purpose of learning (also called *training*) is to use the set of samples in order to build a pattern classifier.

The general requirement concerning the set of samples is that the selection occurs randomly. From the given population, each apple has a known probability of being selected. Most simple, this probability is the same for each apple. A second requirement is that the selection of samples is independent. That is, the probability that one member of the population is selected does not depend on the selection of other members of the population.

Usually, the set of samples is called the *training set* (synonyms are: *learning data* and *design set*). Suppose that the number of samples in the training set is N_S. The objects in the training set are enumerated by the symbol $n = 1, \cdots, N_S$. Object n has a measurement vector $\bar{x}_n$.

If the true classes of the objects are unknown, the training set is called *unlabeled*. Finding a decision function with an unlabeled training set is *unsupervised learning*. Almost synonymous to this is *cluster analysis*. If the classes of the objects have overlapping regions in the measurement space, unsupervised learning is difficult. This is the main reason why we will not discuss this topic any further.

Finding a decision function is facilitated if the training set is *labeled*. In that case, the true classes of the objects are known. There are several methods to find the true class of a sample: manual inspection, additional measurements, destructive analysis, etc. Often, these methods are expensive and therefore allowed only if the number of samples is not too large. Suppose the true class of the n-th object is denoted by θ_n. Then, a labeled training set T_S contains elements $(\bar{x}_n, \theta_n)$ each one consisting of a measurement vector and the true class.

$$T_S = \{(\bar{x}_n, \theta_n) | \theta_n \in \Omega; n = 1, \cdots, N_S\} \qquad (6.83)$$

Since the true class of each measurement vector is known, we can split the training set into K disjunct subsets T_k, each subset containing vectors from one particular class:

$$T_k = \{(\vec{x}_n, \theta_n) | \theta_n \in \omega_k ; n = 1, \cdots, N_k\} \quad \text{with:} \quad k = 1, \cdots, K \tag{6.84}$$

It is understood that the numberings of objects in the subsets T_k in (6.84) and the one of objects in T_S (equation 6.83) do not coincide. Since the subsets are disjunct, we have:

$$T_S = \bigcup_{k=1}^{K} T_k \quad \text{and} \quad N_S = \sum_{k=1}^{K} N_k \tag{6.85}$$

The methods used to find a decision function based on a training set can roughly be divided into *parametric learning* and *non-parametric learning*. In the next subsection the former type of learning will be discussed. The remaining subsections deal with various types of non-parametric learning.

6.3.1 Parametric learning

The basic assumption in parametric learning is that the form of the probability densities is known. The process of learning from the samples boils down to finding suitable values of the parameters that describe the densities. This process is analogous to parameter estimation discussed in section 6.2. The difference is that the parameters in section 6.2 describe a physical process whereas the parameters discussed here are parameters of probability densities describing classes of the objects. Moreover, in parametric learning a training set consisting of many measurement vectors is available. These differences are not essential. The concepts used in Bayes estimation also applies to the current estimation problem.

The measurement vectors $\vec{x}_n$ in the training set are repeated realisations of a single random vector $\underline{\vec{x}}$. Alternatively, we may consider the vectors as being produced by a number of random vectors which share the same probability density. In the latter case, each training set T_k consists of N_k random vectors $\underline{\vec{x}}_n$ which are independent. The joint probability density of these vectors - conditioned on the class ω_k and the parameter vector describing the density - is:

$$p_{\underline{\vec{x}}_1, \underline{\vec{x}}_2, \cdots, \underline{\vec{x}}_{N_k} | \omega_k, \underline{\vec{\alpha}}_k}(\vec{x}_1, \vec{x}_2, \cdots, \vec{x}_{N_k} | \omega_k, \vec{\alpha}_k) = \prod_{n=1}^{N_k} p_{\underline{\vec{x}} | \omega_k, \underline{\vec{\alpha}}_k}(\vec{x}_n | \omega_k, \vec{\alpha}_k) \tag{6.86}$$

With that, the complete machinery from Bayes estimation (minimum risk, MSE-estimation, MAP-estimation, ML-estimation) becomes available to find estimators for the parameter vectors $\vec{\alpha}_k$ describing the conditional probability densities. Known concepts to evaluate these estimators (bias, variance, efficiency, etc.) also apply.

Example 6.5 Gaussian distribution, expectation vector unknown

Let us assume that under class ω_k the measurement vector $\underline{\vec{x}}$ is a Gaussian random vector with known covariance matrix $\mathbf{C}_k$ and unknown expectation vector $\vec{\mu}_k$. No prior knowledge is available concerning this unknown vector. The purpose is to find an estimator for $\vec{\mu}_k$.

Since no prior knowledge is assumed, a maximum likelihood estimator seems appropriate (section 6.2.4). Substitution of (6.86) in (6.58) gives the following general expression of a maximum likelihood estimator:

STATISTICAL PATTERN CLASSIFICATION AND PARAMETER ESTIMATION 169

$$\hat{\vec{\mu}}_k = \underset{\vec{\mu}}{\operatorname{argmax}} \left\{ \prod_{n=1}^{N_k} p_{\vec{x}|\omega_k, \vec{\mu}_k}(\vec{x}_n | \omega_k, \vec{\mu}) \right\}$$

$$= \underset{\vec{\mu}}{\operatorname{argmax}} \left\{ \sum_{n=1}^{N_k} \log\left(p_{\vec{x}|\omega_k, \vec{\mu}_k}(\vec{x}_n | \omega_k, \vec{\mu}) \right) \right\}$$
(6.87)

The logarithms transform the product into a summation. This is only a technical matter which facilitates the maximisation.

Knowing that $\vec{x}$ is Gaussian the likelihood of $\vec{\mu}_k$ from a single observation $\vec{x}_n$ is:

$$p_{\vec{x}|\omega_k, \vec{\mu}_k}(\vec{x}_n | \omega_k, \vec{\mu}_k) = \frac{1}{\sqrt{(2\pi)^N |\mathbf{C}_k|}} \exp\left(-\frac{1}{2}(\vec{x}_n - \vec{\mu}_k)' \mathbf{C}_k^{-1} (\vec{x}_n - \vec{\mu}_k)\right)$$
(6.88)

Upon substitution of (6.88) in (6.87), rearrangement of terms, and elimination of irrelevant terms, we have:

$$\hat{\vec{\mu}}_k = \underset{\vec{\mu}}{\operatorname{argmin}} \left\{ \sum_{n=1}^{N_k} (\vec{x}_n - \vec{\mu})' \mathbf{C}_k^{-1} (\vec{x}_n - \vec{\mu}) \right\}$$

$$= \underset{\vec{\mu}}{\operatorname{argmin}} \left\{ \sum_{n=1}^{N_k} \vec{x}_n' \mathbf{C}_k^{-1} \vec{x}_n + \sum_{n=1}^{N_k} \vec{\mu}' \mathbf{C}_k^{-1} \vec{\mu} - 2 \sum_{n=1}^{N_k} \vec{x}_n' \mathbf{C}_k^{-1} \vec{\mu} \right\}$$
(6.89a)

Differentiating the expression between braces with respect to $\vec{\mu}$, (appendix A.7) and equating the result to zero yields the average calculated over the training set:

$$\hat{\vec{\mu}}_k = \frac{1}{N_k} \sum_{n=1}^{N_k} \vec{x}_n$$
(6.89b)

The estimate in (6.89b) is a single realisation of the random variable:

$$\hat{\underline{\vec{\mu}}}_k = \frac{1}{N_k} \sum_{n=1}^{N_k} \underline{\vec{x}}_n$$
(6.89c)

Being a sum of Gaussian random variables, this estimate has a Gaussian distribution too. If the true expectation of $\vec{x}$ is $\vec{\mu}_k$, the expectation of the estimate is:

$$E\{\hat{\underline{\vec{\mu}}}_k\} = \frac{1}{N_k} \sum_{n=1}^{N_k} E\{\underline{\vec{x}}_n\} = \frac{1}{N_k} \sum_{n=1}^{N_k} \vec{\mu}_k = \vec{\mu}_k$$
(6.90)

Hence, the estimation is unbiased. The variance is the trace of the covariance matrix of the estimate. This covariance matrix follows from:

$$\mathbf{C}_{\hat{\mu}} = E\left\{ (\hat{\underline{\vec{\mu}}}_k - \vec{\mu}_k)(\hat{\underline{\vec{\mu}}}_k - \vec{\mu}_k)' \right\} = \frac{1}{N_k} \mathbf{C}_k$$
(6.91)

It can be shown that the estimator of (6.89c) is efficient. No other estimator exists with smaller variance.

Suppose that prior knowledge concerning the expectation is available in the sense that without the availability of a training set the expectation vector is approximately known. This can be modeled by considering the unknown expectation vector as a Gaussian random vector with expectation $\vec{\mu}_{k,0}$ and covariance matrix $\mathbf{C}_{k,0}$. Since prior knowledge is available, an estimation method appropriate for this case is minimum variance estimation (section 6.2.2). With Gaussian distributions this estimation is equivalent to MAP-estimation (section 6.2.3). The estimator takes the form:

$$\hat{\bar{\mu}}_k = \mathbf{C}_{0,k}\left(\mathbf{C}_{0,k}+\frac{1}{N_K}\mathbf{C}_k\right)^{-1}\left(\frac{1}{N_K}\sum_n^{N_K}\bar{x}_n\right) + \frac{1}{N_K}\mathbf{C}_k\left(\mathbf{C}_{0,k}+\frac{1}{N_K}\mathbf{C}_k\right)^{-1}\bar{\mu}_{0,k}$$

$$\mathbf{C}_{\hat{\mu}} = \frac{1}{N_K}\mathbf{C}_k\left(\mathbf{C}_{0,k}+\frac{1}{N_K}\mathbf{C}_k\right)^{-1}\mathbf{C}_{0,k}$$

(6.92)

The proof (based on equation (6.50)) is left as an exercise to the reader. The interpretation of (6.92) is that the estimate consists of a weighted linear combination of the ML-estimate and the prior expectation. The balance between these two contributions is determined by the uncertainties in both contributions. If $N_k \to \infty$ the contribution of the ML-estimator prevails.

If the prior covariance matrix $\mathbf{C}_{k,0}$ is a scaled version of the true covariance matrix: $\mathbf{C}_{0,k} = \gamma \mathbf{C}_k$, then:

$$\hat{\bar{\mu}}_k = \frac{\gamma N_k}{\gamma N_k+1}\left(\frac{1}{N_K}\sum_n^{N_K}\bar{x}_n\right) + \frac{1}{\gamma N_k+1}\bar{\mu}_{0,k}$$

$$\mathbf{C}_{\hat{\mu}} = \frac{\gamma}{\gamma N_k+1}\mathbf{C}_k$$

(6.93)

The conclusion is that the density of $\underline{\mu}_k$ reproduces itself. This property is useful in applications in which at successive stages new measurement vectors become available. The reproduction allows a recursive update of the estimate. This brings us close to Kalman filtering (section 6.2.8)

Example 6.6 Gaussian distribution, covariance matrix unknown

Next, we consider the case where under class ω_k the measurement vector $\underline{\bar{x}}$ is a Gaussian random vector with unknown covariance matrix $\mathbf{C}_k$. For the moment we assume that the expectation vector $\bar{\mu}_k$ is known. No prior knowledge is available The purpose is to find an estimator for $\mathbf{C}_k$.

The maximum likelihood estimate follows from (6.87) and (6.88):

$$\hat{\mathbf{C}}_k = \underset{\mathbf{C}}{\operatorname{argmax}}\left\{\sum_{n=1}^{N_k}\log\left(p_{\underline{\bar{x}}|\omega_k,\bar{\mu}_k}(\bar{x}_n|\omega_k,\mathbf{C})\right)\right\}$$

$$= \frac{1}{N_k}\sum_{n=1}^{N_k}(\bar{x}_n-\bar{\mu}_k)(\bar{x}_n-\bar{\mu}_k)^t$$

(6.94)

The last step in (6.94) is non-trivial. The proof is rather technical and can be found in Patrick [1972]. The probability distribution of the random variables in $\underline{\mathbf{C}}_k$ is called a *Wishart distribution*.

With known expectation vector the estimator is unbiased. The variances of the elements of $\underline{\mathbf{C}}_k$ are:

$$\operatorname{Var}\{\hat{\underline{\mathbf{C}}}_{k_{i,j}}\} = \frac{1}{N_k}\left(\mathbf{C}_{k_{i,i}}\mathbf{C}_{k_{j,j}} + \mathbf{C}_{k_{i,j}}^2\right)$$

(6.95)

If the expectation vector is unknown, the estimation problem becomes more complicated because then we have to estimate the expectation vector as well. It can be deduced that the following estimators for $\mathbf{C}_k$ and $\bar{\mu}_k$ are unbiased:

$$\hat{\vec{\mu}}_k = \frac{1}{N_k} \sum_{n=1}^{N_k} \vec{x}_n$$

$$\hat{\mathbf{C}}_k = \frac{1}{N_k - 1} \sum_{n=1}^{N_k} (\vec{x}_n - \hat{\vec{\mu}}_k)(\vec{x}_n - \hat{\vec{\mu}}_k)^t \qquad (6.96)$$

Note that in the estimation of $\mathbf{C}_k$ in (6.96) the sample mean replaces the unknown expectation. Although this yields an unbiased estimator, it is not a maximum likelihood estimator. However, the estimator is consistent. This means that if $N_K \to \infty$ the variance becomes zero. Therefore, the estimator is of much practical importance.

In classification problems the inverse $\mathbf{C}_k^{-1}$ is often needed, for instance in quantities like: $\vec{x}^t \mathbf{C}_k^{-1} \vec{\mu}$, $\vec{x}^t \mathbf{C}_k^{-1} \vec{x}$, etc. Frequently, $\hat{\mathbf{C}}_k^{-1}$ is used as an estimate. The variance of $\hat{\mathbf{C}}_k$ (equation 6.95) is not much helpful to determine the number of samples needed to calculate $\mathbf{C}_k^{-1}$ accurately. To see this it is instructive to rewrite the inverse as (see appendix A.8):

$$\mathbf{C}_k^{-1} = \mathbf{V}_k \Lambda_k^{-1} \mathbf{V}_k^t \qquad (6.97)$$

where Λ_k is a diagonal matrix containing the eigenvalues of $\mathbf{C}_k$. Clearly, the behavior of $\mathbf{C}_k^{-1}$ is strongly affected by small eigenvalues in Λ_k. In fact, the number of non-zero eigenvalues in the estimate $\hat{\mathbf{C}}_k$ given in (6.94) cannot exceed N_k. If N_k is smaller than the dimension $N+1$, the estimate $\hat{\mathbf{C}}_k$ is not invertible. Therefore, we must require that N_k is much larger than N. As a rule of thumb, the number of samples must be chosen such that $N_k > 5N$ [Kalayeh 1983].

Figure 6.26 is an illustration of the use of parametric learning. The conditional probabilities to be estimated are the one shown in figure 6.6a (Gaussian densities). The parameters are estimated with the samples also shown in figure 6.6. The estimator used is given in (6.96). Figure 6.26 shows the contour plots of the true densities along with the one obtained by estimation.

Example 6.7 Multinomial distribution

As stated in section 6.1, an object is supposed to belong to a class taken from Ω. The prior probability $P(\omega_k)$ is denoted by P_k. There are exactly K classes. Having a labeled training set with N_S elements, the number N_k of elements from the set with class ω_k has a *multinomial distribution*. If $K = 2$ the distribution is *binomial*.

Another example of a multinomial distribution occurs when the measurement vector $\vec{x}$ can only take a finite number of states. For instance, if the sensory system is such that each element in the measurement vector is binary, i.e. either "1" or "0", then the number of states the vector can take is at most 2^N. The number of vectors in the training set with a certain state $\vec{x}$ has a multinomial distribution.

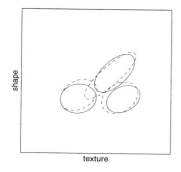

Figure 6.26 Contour plots of true and estimated (dashed) probability densities

These two examples show that in classification problems the estimation of the parameters of a multinomial distribution is of practical importance.

For the moment the estimation of the prior probabilities will serve as an example. Since these probabilities sum up to one, the number of unknown parameters that describe the multinomial distribution is $K-1$. The training set is labeled. Therefore, the number N_k of events with class ω_k occurring in the training set can be regarded as a random variable $\underline{N}_k$ of which a realisation N_k is available. As before, the total number of events in the training set is N_S. Intuitively, the following estimator is appropriate:

$$\hat{\underline{P}}_k = \frac{\underline{N}_k}{N_S} \tag{6.98}$$

In this model, the number N_S is deterministic (we fully control the total number of samples). The variable $\underline{N}_k$ has a binomial distribution with parameters P_k and N_S. The estimator in (6.98) is unbiased. The variance is:

$$\mathrm{Var}\{\hat{\underline{P}}_k\} = \frac{P_k(1-P_k)}{N_S} \tag{6.99}$$

This shows that the estimator is consistent (if $N_S \to \infty$, then $\mathrm{Var}\{\hat{\underline{P}}_k\} \to 0$). The number of required samples follows from the constraint that $\sqrt{\mathrm{Var}\{\hat{\underline{P}}_k\}} \ll P_k$. For instance, if $P_k = 0.01$, then N_S must be greater than about 1000 in order to have reasonable precision.

In many applications, (6.98) as an estimator for prior probabilities suffices. However, if the estimator is used to estimate the probability of a state in an N-dimensional measurement vector, the estimator may fail. A small example can demonstrate this. Suppose the dimension of the vector is $N=10$. Then the total number of states is about 10^3. This implies that most states must have a probability of less than 10^{-3}. The precision with which these probabilities must be estimated must be a fraction of that, say 10^{-5}. The number of samples needed to guarantee this precision is $N_S \approx 10^7$. Needless to say that in many applications such a number of samples is much too expensive. Moreover, with even a slight increase of N the problem becomes much larger.

One way to avoid a large variance is to incorporate more prior knowledge. For instance, without the availability of a training set, it is known beforehand that all parameters are bounded by: $0 \le P_k \le 1$. If nothing further is known, one could "guess" that all states are equally likely. For instance, in the estimation of prior probabilities one could assume that without knowledge of the training set all classes occur with the same frequency, i.e. $P_k = 1/K$. Based on this guess, the estimator takes the form:

$$\hat{\underline{P}}_k = \frac{\underline{N}_k + 1}{N_S + K} \qquad \text{with variance:} \qquad \mathrm{Var}\{\hat{\underline{P}}_k\} = \frac{N_S P_k(1-P_k)}{(N_S + K)^2}$$

It is obvious that the variance is reduced at the cost of a small bias.

One can increase the effect by assuming stronger prior knowledge. This is accomplished most easily by expressing the prior knowledge in terms of a fictitious training set consisting of R_S samples with exactly R_k samples belonging to class ω_k. With that fictitious set the model of the parameter P_k becomes:

$$\hat{\underline{P}}_k = \frac{R_k + 1}{R_S + K} \qquad \text{(prior: training set not available)} \tag{6.100}$$

where it is assumed that the random variable $\underline{R}_k$ has a binomial distribution with parameters R_S and R_k/R_S. It can be proven [Patrick 1972] that with the model of equation (6.100) the minimum variance estimator of P_k becomes:

$$\hat{\underline{P}}_k = \frac{N_k + R_k + 1}{N_S + R_S + K} \quad \text{(posterior: training set available)} \quad (6.101)$$

The variances are:

$$\begin{aligned}
\text{Var}\{\hat{\underline{P}}_k\} &= \frac{R_S P_k (1 - P_k)}{(R_S + K)^2} \quad \text{(prior)} \\
\text{Var}\{\hat{\underline{P}}_k\} &= \frac{N_S P_k (1 - P_k)}{(N_S + R_S + K)^2} \quad \text{(posterior)}
\end{aligned} \quad (6.102)$$

By carefully choosing R_S and R_k one can find a balance between bias and variance.

6.3.2 Parzen estimation

Non-parametric methods are learning methods for which prior knowledge about the functional form of the conditional probability densities are not needed. Parzen estimation is one method in this category. Here, one tries to estimate the densities by functional approximations.

As before, the estimation is based on a labeled training set T_S. We split the training set in K subsets T_k, each having N_k samples all belonging to class ω_k. The goal is to estimate the conditional density $p_{\bar{x}|\omega}(\bar{x}|\omega_k)$ for all $\bar{x}$.

The obvious way to reach that goal is to partition the measurement space into a finite number of disjunct regions R_i, called *bins*, and to count the number of times that a vector falls in one of these bins. The estimated probability density within a bin is proportional to that count. This technique is called *histogramming*. It is useful if the dimension of the measurement space is small. An example in which histogramming can be applied successfully is the estimation of the probability density of grey levels; see sections 3.1 and 5.1.1.

The probability density of a histogram is that of a multinomial distribution. This implies that if the dimension of the measurement space is large, histogramming is likely to fail; see example 6.7.

Parzen estimation is a technique that can be considered as a refinement of histogramming. The first step in the development of the estimator is to consider only one sample from the training set. Suppose that $\bar{x}_j \in T_k$. Then, we are certain that at this position in the measurement space the density is non-zero, i.e. $p_{\bar{x}|\omega}(\bar{x}_j|\omega_k) \neq 0$. Under the assumption that $p_{\bar{x}|\omega}(\bar{x}|\omega_k)$ is continuous over the entire measurement space it follows that in a small neighborhood of $\bar{x}_j$ the density is likely to be non-zero too. However, the further we move away from $\bar{x}_j$, the less we can say about $p_{\bar{x}|\omega}(\bar{x}|\omega_k)$. The basic idea behind Parzen estimation is that the knowledge gained by observation of $\bar{x}_j$ is represented by a function positioned at $\bar{x}_j$ and with an influence restricted to a small vicinity of $\bar{x}_j$. This function is the contribution of $\bar{x}_j$ to the estimate. Summing together the contributions of all vectors in the training set yields the final estimate.

Let $\rho(\bar{x}, \bar{x}_j)$ be a distance measure (appendix A.2) defined in the measurement space. The knowledge gained by the observation $\bar{x}_j \in T_k$ is represented by the function $h(\cdot)$: $\mathbb{R}^+ \to \mathbb{R}^+$. This function must be such that $h(\rho(\vec{x},\vec{x}_j))$ has its maximum at $\vec{x} = \vec{x}_j$, i.e. at $\rho(\vec{x},\vec{x}_j) = 0$. Furthermore, $h(\rho(\cdot,\cdot))$ must be monotonically decreasing as $\rho(\cdot,\cdot)$ increases, and $h(\rho(\cdot,\cdot))$ must be normalised: $\int h(\rho(\vec{x},\vec{x}_j))d\vec{x} = 1$ where the integration extends over the entire measurement space.

The contribution of a single observation is $h(\rho(\vec{x},\vec{x}_j))$. The contributions of all observations are summed together to yield the final Parzen estimate:

$$\hat{p}_{\vec{x}|\omega}(\vec{x}|\omega_k) = \frac{1}{N_k} \sum_{\vec{x}_j \in T_k} h(\rho(\vec{x},\vec{x}_j)) \tag{6.103}$$

The operation is similar to the reconstruction of a continuous signal from discrete samples; see section 4.2.2. In the analogy, the function $h(\rho(\cdot,\cdot))$ corresponds to the interpolation function.

Figure 6.27 gives an example of Parzen estimation in a 1-dimensional measurement space. The true probability density is uniform. The estimation is based on six samples. The interpolation function chosen is a Gaussian function with width σ_P. The distance measure is Euclidean. Figure 6.27a shows an estimation in which σ_P is a fraction of 0.04 of the width of the true density. The width σ_P in figure 6.27b is about 0.2. These graphs illustrate a phenomenon related to the choice of the interpolation function. If the interpolation function is peaked (the influence of a sample is very local), the variance of the estimator is large. But if the interpolation is smooth, the variance decreases, but at the same time the estimator looses resolution. One could state that by changing the width of the interpolation function one can choose between a large bias or a large variance. Of course, both types of errors are reduced by enlargement of the training set.

In the N-dimensional case, various interpolation functions are useful. A popular one is the Gaussian function:

$$\rho(\vec{x},\vec{x}_j) = \sqrt{(\vec{x}-\vec{x}_j)^t \mathbf{C}^{-1}(\vec{x}-\vec{x}_j)}$$

$$h(\rho) = \frac{1}{\sigma_p^N \sqrt{(2\pi)^N |\mathbf{C}|}} \exp\left(-\frac{\rho^2}{2\sigma_p^2}\right) \tag{6.104}$$

The constant σ_p controls the size of the influence zone of $h(\cdot)$. It is chosen smaller as the number of samples in the training set increases. The matrix $\mathbf{C}$ must be

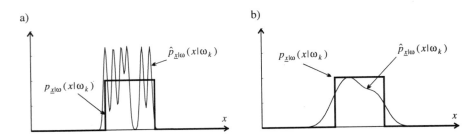

Figure 6.27 Parzen estimation of a uniform density
a) With peaked interpolation function
b) With smooth interpolation function

symmetric and positive definite (appendix A.8). If the training set is very large, the actual choice of $\mathbf{C}$ is less important. If the set is not very large, a suitable choice is the covariance matrix $\hat{\mathbf{C}}_k$ determined according to (6.96).

Figure 6.28 is an illustration of Parzen estimation in a 2-dimensional space. The conditional probability densities and the training set are taken from the apple classification problem (figure 6.6a). Figure 6.28 shows contour plots of both the true densities and the estimated densities (dashed). The width of the interpolation function in figure 6.28b is three times that of figure 6.28a. Comparison of these results with the results from parametric learning (figure 6.26) shows that the latter approximates the true densities better than the results from Parzen estimation. The reason is that parametric learning methods use more prior knowledge.

6.3.3 Nearest neighbor classification

In Parzen estimation, each sample in the training set contributes in like manner to the estimate. The estimation process is "space-invariant". As a consequence, the trade-off which exists between resolution and variance is a global one. A refinement would be to have an estimator with high resolution in regions where the training set is dense, and low resolution in other regions. The advantage of this is that the balance between resolution and variance can be adjusted locally.

Nearest neighbor estimation is a method that implements such a refinement. The method is based on the following observation. Let $R(\vec{x}) \in \mathbb{R}^N$ be a hypersphere with volume V. The centre of $R(\vec{x})$ is $\vec{x}$. If the number of samples in the training set T_k is N_k, then the probability of having exactly $\underline{n}$ samples within $R(\vec{x})$ has a binomial distribution with expectation:

$$E\{\underline{n}\} = N_k \int_{\vec{\xi} \in R(\vec{x})} p_{\vec{x}|\omega}(\vec{\xi}|\omega_k) d\vec{\xi} \approx N_k V p_{\vec{x}|\omega}(\vec{x}|\omega_k) \qquad (6.105)$$

Suppose that the radius of the sphere around $\vec{x}$ is selected such that this sphere contains exactly κ samples. It is obvious that this radius depends on the position $\vec{x}$

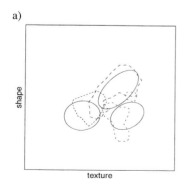
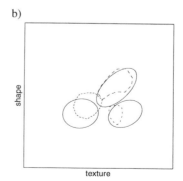

Figure 6.28 Contour plots of true densities and densities obtained with Parzen estimation
a) Peaked interpolation function
b) Smooth interpolation function

in the measurement space. Therefore, the volume will also depend on $\vec{x}$. We have to write $V(\vec{x})$ instead of V. With that, an estimate of the density is:

$$\hat{p}_{\vec{x}|\omega}(\vec{x}|\omega_k) = \frac{\kappa}{N_k V(\vec{x})} \qquad (6.106)$$

The expression shows that in regions where $p_{\vec{x}|\omega}(\vec{x}|\omega_k)$ is large the volume is expected to be small. This is similar to having a small "interpolation zone". If, on the other hand, $p_{\vec{x}|\omega}(\vec{x}|\omega_k)$ is small, the sphere needs to grow in order to collect the required κ samples.

The parameter κ controls the balance between resolution and variance. This is alike the parameter σ_p in Parzen estimation. The choice of κ should be such that:

$$\begin{array}{ll} \kappa \to \infty & \text{as} \quad N_k \to \infty \\ \kappa/N_k \to 0 & \text{as} \quad N_k \to \infty \end{array} \qquad (6.107)$$

A suitable choice is to make κ proportional to $\sqrt{N_k}$.

Nearest neighbor estimation is of particular practical interest. The reason is that it paves the way to a classification technique that directly uses the training set, i.e. without explicitly estimating probability densities. The development of this technique is as follows.

We consider the entire training set $T_S = \cup_k T_k$. The total number of samples is N_S. Estimates of the prior probabilities follow from (6.98):

$$\hat{P}_k = \frac{N_k}{N_S} \qquad (6.108)$$

As before, let $R(\vec{x}) \in \mathbb{R}^N$ be a hypersphere with volume $V(\vec{x})$. In order to classify a vector $\vec{x}$ we select the radius of the sphere around $\vec{x}$ such that this sphere contains exactly κ samples taken from T_S. These samples are called the κ-*nearest neighbors*[1] of $\vec{x}$. Let κ_k denote the number of samples found with class ω_k. An estimate of the conditional density is:

$$\hat{p}_{\vec{x}|\omega}(\vec{x}|\omega_k) \approx \frac{\kappa_k}{N_k V(\vec{x})} \qquad (6.109)$$

Combination of (6.108) and (6.109) in the Bayes classification with uniform cost function (6.11b) produces the following sub-optimal classification:

$$\hat{\omega}(\vec{x}) = \omega_k \quad \text{with:} \quad k = \underset{k=1,\cdots,K}{\operatorname{argmax}}\{\hat{p}_{\vec{x}|\omega}(\vec{x}|\omega_k)\hat{P}_k\} = \underset{k=1,\cdots,K}{\operatorname{argmax}}\left\{\frac{\kappa_k}{N_k V(\vec{x})}\frac{N_k}{N_S}\right\} \qquad (6.110)$$

$$= \underset{k=1,\cdots,K}{\operatorname{argmax}}\{\kappa_k\}$$

The interpretation of this classification is simple. The class assigned to a vector $\vec{x}$ is the class with the maximum number of votes coming from κ samples nearest to $\vec{x}$. In literature, this classification method is known as k-*nearest neighbor rule*

---

[1] In the literature about nearest neighbor classification, it is common to use the symbol k to denote the number of samples in a volume. However, in order to avoid confusion with symbols like ω_k, T_k, etc. we prefer to use κ.

classification (k-NNR, in our nomenclature κ-NNR). The special case in which κ = 1 is simply referred to as *nearest neighbor rule classification* (NNR).

Figure 6.29 is an illustration of the nearest neighbor rules. The training set is the one shown in figure 6.6a. Figure 6.29a shows the decision function according to the NNR. The decision function belonging to the 3-NNR is depicted in figure 6.29b.

The analysis of the performance of κ-nearest neighbor classification is difficult. This holds true especially if the number of samples in the training set is finite. In the limiting case, when the number of samples grows to infinity, some bounds on the error rate can be given. Let the error rate of a Bayes classifier with uniform cost function be denoted by E_B. In section 6.1.2 it has been shown that this error rate is minimal. Therefore, if the error rate of a κ-NNR is denoted E_κ, then a lower error bound is:

$$E_B \leq E_\kappa \qquad (6.111)$$

It can be shown that for the 1-NNR the following upper bound holds:

$$E_1 \leq E_B\left(2 - \frac{K}{K-1}E_B\right) \leq 2E_B \qquad (6.112a)$$

This means that replacing the true probability densities with estimations based on the first nearest neighbor gives an error rate which is at most twice the minimum. Thus, half of the classification information in a dense training set is contained in the first nearest neighbor.

In the two class problem ($K = 2$) the following bound can be proven:

$$\begin{aligned} E_\kappa &\leq E_B + \frac{E_1}{\sqrt{0.5(\kappa-1)\pi}} & \text{if } \kappa \text{ is odd} \\ E_{\kappa-1} &= E_\kappa & \text{if } \kappa \text{ is even} \end{aligned} \qquad (6.112b)$$

The importance of (6.112) is that it shows that the performance of the κ-NNR approximates the optimum as κ increases. Of course, this asymptotic optimality holds true only if the training set is dense ($N_S \to \infty$). Nevertheless, even in the small

a)
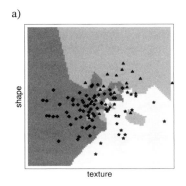
texture

b)
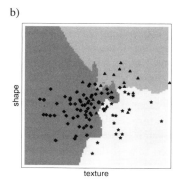
texture

Figure 6.29 Nearest neighbour classification
a) 1-NNR classification
b) 3-NNR classification

sized training set given in figure 6.29 it can be seen that the 3-NNR is superior to the 1-NNR. The topology of the compartments in figure 6.29a is very specific for the given realisation. This is in contrast with the topology of the 3-NNR in figure 6.29b. This topology equals the one of the optimal Bayes classification (figure 6.7). The 3-NNR *generalises* better than the 1-NNR.

Unfortunately, κ-NNR classifications also have some serious disadvantages:

- A distance measure is needed in order to decide which sample in the training set is nearest. Usually the Euclidean distance measure is chosen, but this choice needs not to be optimal. A systematic method to find the optimal measure is not available.
- The optimality is reached only when $\kappa \to \infty$. But since at the same time it is required that $\kappa/N_S \to 0$, this demands that the training set must be extremely large. If this is not the case, κ-NNR classification may be far from optimal.
- If the training set is large, the computational complexity of κ-NNR classification becomes a serious burden.

Many attempts have been made to remedy the last drawback. One approach is to design fast algorithms with suitably chosen data structures (e.g. hierarchical data structures combined with a preordered training set).

Another approach is to preprocess the training set so as to speed up the search for nearest neighbours. There are two principles on which this reduction can be based. In the first place one can try to edit the training set such that the (expensive) κ-NNR can be replaced with the 1-NNR. Secondly, one can try to remove samples that do not affect the classification anyhow.

In the first approach, the strategy is to remove those samples from the training set that when used in the 1-NNR would cause erroneous results. An algorithm that accomplishes this is the so-called *multi-edit algorithm*. The following algorithm is from Devijver and Kittler [1982].

Algorithm 6.1 Multi-edit

Input: a labeled training set T_S.
1. *Diffusion:* make a random partition of the training set T_S into a L disjunct subsets: $T_1', T_2', \cdots, T_L'$ with $T_S = \bigcup_l T_l'$, and $L \geq 3$.
2. *Classification:* classify the samples in T_l' using 1-NNR classification with $T_{(l+1) \mod L}'$ as training set.
3. *Editing:* discard all the samples that were misclassified at step 2.
4. *Confusion:* pool all the remaining samples to constitute a new training set T_S.
5. *Termination:* if the last couple of I iterations produced no editing, then exit with the final training set, else go to step 1.

Output: a subset of T_S.

The subsets created in step 1 are regarded as independent random selections. A minimum of three subsets is required in order to avoid a two-way interaction between two subsets. Because in the first step the subsets are randomised, it cannot be guaranteed that, if during one iteration no changes in the training set occurred, changes in further iterations are ruled out. Therefore, the algorithm does not stop immediately after that an iteration with no changes occurred.

The effect of the algorithm is that ambiguous samples in the training set are removed. This eliminates the need to use the κ-NNR. The 1-NNR can be used instead. An example of an edited training set is given in figure 6.30a. It can be seen that the topology of the resulting decision function does not change the one of the 3-NNR given in figure 6.29b. Hence, multi-editing improves the generalisation property.

Another algorithm that is helpful to reduce the computational cost is *condensing*. This algorithm - also from Devijver and Kittler [1982] - is used to eliminate all samples in the training set that are irrelevant.

Algorithm 6.2 Condensing

Input: a labeled training set T_S.

1. *Initiation:* set up the two new training sets T_{STORE} and $T_{GRABBAG}$; place the first sample of T_S in T_{STORE}, all other samples in $T_{GRABBAG}$.
2. *Condensing:* use 1-NNR classification with the current T_{STORE} to classify a sample in $T_{GRABBAG}$; if classified correctly, the sample is retained in $T_{GRABBAG}$, otherwise it is moved from $T_{GRABBAG}$ to T_{STORE}; repeat this operation for all other samples in $T_{GRABBAG}$.
3. *Termination:* if one complete pass is made through step 2 with no transfer from $T_{GRABBAG}$ to T_{STORE}, or if $T_{GRABBAG}$ is empty, then terminate; else go to step 2.

Output: a subset of T_S.

The effect of this algorithm is that in regions where the training set is overcrowded with samples of the same class most of these samples will be removed. The remaining set will, hopefully, contain samples close to the Bayes decision boundaries.

Figure 6.30b shows that condensing can be successful when applied to a multi-edited training set. The decision boundaries in figure 6.30b are close to those in figure 6.30a, especially in the more important areas of the measurement space. If a non-edited training set is fed into the condensing algorithm, this may result in erroneous decision boundaries, especially in areas of the measurement space where the training set is ambiguous.

6.3.4 Linear discriminant functions

Another learning technique that directly uses the training set to arrive at a classification method is based on *discriminant functions*. These are functions

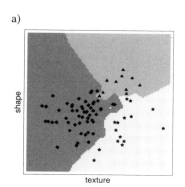
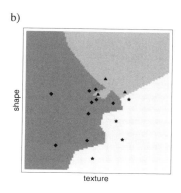

Figure 6.30 Preprocessing the training set
a) 1-NNR classification with edited training set
b) 1-NNR classification with edited and condensed training set

$g_k(\vec{x})$, $k = 1, \cdots, K$ with a given functional form. The functions are used in a classificator as follows:

$$\hat{\omega}(\vec{x}) = \omega_k \quad \text{with:} \quad k = \underset{k=1,\cdots,K}{\operatorname{argmax}} \{g_k(\vec{x})\} \qquad (6.113)$$

Clearly, if $g_k(\vec{x})$ were chosen to be the posterior probabilities $P(\omega_k|\underline{x}=\vec{x})$, the decision function of (6.113) becomes a Bayes decision function with a uniform cost function. Since the posterior probabilities are not known, the strategy is to replace the probabilities with some suitably selected functions. The selection is to be based on a labeled training set.

The prerequisite in this selection process is that a suitable functional form of the functions is known. An assumption often made is that the samples in the training set can be classified correctly with linear decision boundaries. In that case, the discriminant functions take the form of:

$$g_k(\vec{x}) = \vec{w}_k^t \vec{x} + v_k \qquad (6.114)$$

Functions of this type are called *linear discriminant functions*.

Discriminant functions depend on a set of parameters. In equation (6.114) these parameters are the vectors $\vec{w}_k$ and the scalars v_k. In essence, the learning process boils down to a search for parameters such that with these parameters the decision function in (6.113) correctly classifies all samples in the training set.

The basic approach to find the parameters is to define a performance measure that depends on both the training set and the set of parameters. Adjustment of the parameters such that the performance measure is maximised gives the "optimal" decision function; see figure 6.31. This approach is quite similar to parametric learning. The main difference is that with discriminant functions the functional form of the probability densities are unspecified.

Strategies to adjust the parameters may be further categorised into "iterative" and "non-iterative". Non-iterative schemes are found if the performance measure is such that an analytic solution of the optimisation is feasible. For instance, suppose that the set of parameters is denoted by W and that the performance measure is a continuous function $J(W)$ of W. The optimal solution is one which extremises $J(W)$. Hence, the solution must satisfy $\partial J(W)/\partial W = 0$. Examples of performance measures allowing such a mathematical treatment will be given in subsequent sections.

In iterative strategies the procedure to find a solution is numerical. Samples from the training set are fed into the decision function. The classes found are

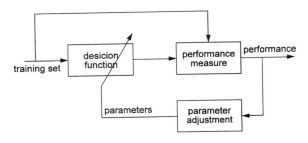

Figure 6.31 Training by means of performance optimisation

compared with the true classes. The result controls the adjustment of the parameters. The adjustment is in a direction which improves the performance. An *epoch* is the period in which all samples in the training set are processed once. With repeated epochs it is hoped that the optimal setting of the parameters is found.

Example 6.8 Perceptron learning

The perceptron is a classification method based on the linear discriminant functions given in equation (6.114). It is convenient to adapt the notation a little so as to arrive at a compact mathematical description:

$$g_k(\vec{x}) = \vec{w}_k^t \vec{x}' \tag{6.115a}$$

$\vec{w}_k$ are $N+1$-dimensional parameter vectors. The vector $\vec{x}'$ is the measurement vector $\vec{x}$ augmented with a constant "1 ":

$$\vec{x}' = \begin{bmatrix} \vec{x} \\ 1 \end{bmatrix} \tag{6.115b}$$

Note that the old vector $\vec{w}_k$ and the scalar v_k in (6.114) are absorbed in the newly defined vector $\vec{w}_k$ used in (6.115a).

For the sake of simplicity we consider the 2-class case ($K=2$) only. With that, the decision function stated in (6.113) is equivalent to a test $g_1(\vec{x}) - g_2(\vec{x}) > 0$. If the test fails, it is decided for ω_2, otherwise for ω_1. The test can be accomplished equally well with a single linear function:

$$g(\vec{x}) = \vec{w}^t \vec{x}' \tag{6.116}$$

defined as $g(\vec{x}) = g_1(\vec{x}) - g_2(\vec{x})$. Figure 6.32 is a graphical representation of a single perceptron. It consists of a weighted summation of the elements of the (augmented) input vector. The sum is thresholded at zero indicating either class ω_1, or ω_2.

The perceptron has been trained as soon as an $N+1$-dimensional weight vector $\vec{w}$ is found such that all vectors in the training set with class ω_1 (i.e. $\vec{x}_i \in T_1$) satisfy: $\vec{w}^t \vec{x}_i > 0$, and all vectors with class ω_2 (i.e. $\vec{x}_j \in T_2$) satisfy: $\vec{w}^t \vec{x}_j < 0$. For that purpose, the perceptron uses the following performance measure:

$$J_{perceptron}(\vec{w}) = \sum_{\vec{x}_j \in T_2} \vec{w}^t \vec{x}_j \, U(\vec{w}^t \vec{x}_j) - \sum_{\vec{x}_i \in T_1} \vec{w}^t \vec{x}_i \, U(-\vec{w}^t \vec{x}_i) \tag{6.117}$$

where $U(\cdot)$ is a Heaviside step-function. The definition of the performance measure is such that $J_{perceptron}(\vec{w}) = 0$ if all vectors are correctly classified; and $J_{perceptron}(\vec{w}) > 0$ if some vectors are misclassified. Furthermore, the measure is continuous in $\vec{w}$. A strategy to find a minimum of $J_{perceptron}(\vec{w})$ is the steepest

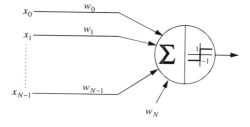

Figure 6.32 The perceptron

descent procedure. This procedure consists of an iterative update of the weight vector in the direction of maximum decrease of $J_{perceptron}(\vec{w})$. This direction is given by the negative of the gradient vector $\nabla J_{perceptron}(\vec{w})$.

$$\nabla J_{perceptron}(\vec{w}) = \sum_{\vec{x}_j \in T_2} \vec{x}_j \, U(\vec{w}^t \vec{x}_j) - \sum_{\vec{x}_i \in T_1} \vec{x}_i \, U(-\vec{w}^t \vec{x}_i) \qquad (6.118)$$

Suppose that during the n-th epoch the weight vector is $\vec{w}_n$. Then, the updated weight vector becomes:

$$\vec{w}_{n+1} = \vec{w}_n - \alpha \nabla J_{perceptron}(\vec{w}_n) \qquad (6.119)$$

where α is a suitably chosen constant. The update occurs after each epoch. Alternately, an update can also take place after that each sample has been processed. Equations (6.118) and (6.119) show that the learning rule with respect to a single sample is very simple:

- If a vector $\vec{x}_i$ from class T_1 is misclassified, then update $\vec{w}$ by $\alpha \vec{x}_i$.
- If a vector $\vec{x}_j$ from class T_2 is misclassified, then update $\vec{w}$ by $-\alpha \vec{x}_j$.
- If a vector is correctly classified, do nothing.

This algorithm is known as *perceptron learning*.

6.3.5 Generalised linear discriminant functions

A limitation of classification based on linear discriminant functions is that the compartments in the measurement space are always convex polyhedrons separated by linear decision boundaries. One way to overcome this limitation is to use *generalised linear discriminant functions*. These functions are of the type:

$$g_k(\vec{x}) = \vec{v}_k^t \, \vec{f}_k(\vec{x}) \qquad (6.120)$$

where $\vec{f}_k(\vec{x})$ are non-linear vector functions with dimension D. The D-dimensional vector $\vec{v}_k$ is a parameter vector. An example of a vector function with dimension $D = 3$ are the *quadratic discriminant functions*:

$$\vec{f}_k(\vec{x}) = \begin{bmatrix} 1 \\ \vec{u}_k^t \vec{x} \\ \vec{x}^t U_k \vec{x} \end{bmatrix} \qquad (6.121)$$

where $\vec{u}_k$ are N-dimensional vectors, and U_k are $N \times N$-matrices. The functional form of this vector function is suitable to classify Gaussian random vectors; see equation (6.16).

As in the linear case, the discriminant functions depend on parameters contained in $\vec{v}_k$, $\vec{u}_k$, and U_k. The parameters have to be tuned so as to optimise the performance. We make this dependency explicit by writing $g_k(\vec{x}, \vec{w}_k)$. The vector $\vec{w}_k$ is defined such that it comprises all parameters. The goal of training is to find $\vec{w}_k$ such that a maximum number of samples in the training set is classified correctly.

The more the complexity of the discriminant functions increases, the more difficult it will be to find analytic solutions of the optimisation problem. Therefore, the solution is often sought numerically in an iterative process. A performance measure suitable for this task is the *squared error* (SQE).

The so-called *target* of the discriminant functions is defined as follows:

$$g_{k,TARGET}(\vec{x}) = \begin{cases} 1 & \text{if } \vec{x} \text{ is from class } \omega_k \\ 0 & \text{otherwise} \end{cases} \qquad (6.122)$$

Ideally, the parameters $\vec{w}_k$ are selected such that for each sample $\vec{x}_n$ in the training set: $g_k(\vec{x}_n, \vec{w}_k) = g_{k,TARGET}(\vec{x}_n)$. Usually, the number of parameters is much smaller than the number of samples N_S. Therefore, the set of equations is overdetermined, and no exact solution exists. Alternatively, we can seek a solution that minimises the SQE measure. This measure is defined as:

$$J_{SQE}(\vec{w}) = \sum_{k=1}^{K} \sum_{n=1}^{N_S} \left(g_k(\vec{x}_n, \vec{w}_k) - g_{k,TARGET}(\vec{x}_n) \right)^2 \qquad (6.123)$$

where $\vec{w}$ is a vector comprising all parameter vectors $\vec{w}_1, \vec{w}_2, \cdots, \vec{w}_K$. The gradient vector of the SQE measure - provided that it exists - is $\nabla J_{SQE}(\vec{w})$. With that, the steepest descent procedure becomes:

$$\vec{w}_{(m+1)} = \vec{w}_{(m)} - \alpha \nabla J_{SQE}(\vec{w}_{(m)}) \qquad (6.124)$$

In the first iteration ($m=1$) the parameters $\vec{w}_{(1)}$ are randomly chosen. The update in each iteration is in the direction of maximum decrease of $J_{SQE}(\vec{w})$.

Example 6.9 Neural network with error backpropagation.

A neural network is a processing structure consisting of many interconnected processing elements (*neurons*). Each neuron accomplishes a simple task: collecting the outputs from other neurons and producing an output which is simply related to its inputs. The function of the network is distributed over many neurons. This function is not so much determined by the neurons as by the connection structure. In fact, the neurons often share a common functional structure. The connections between neurons determine the total function of the network [Rummelhart et al 1986].

In error backpropagation neural networks, a neuron (figure 6.33a) is a memoryless device. It consists of: a number of input terminals, weight factors, a summator, a transfer function, and an output terminal. The output y is given by:

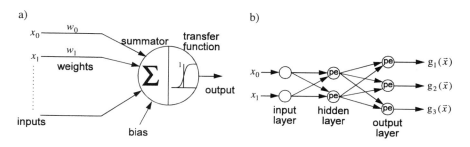

Figure 6.33 Error backpropagation neural network
a) Processing element (pe or neuron) in an error backpropagation neural network
b) Network with one hidden layer

$$y = f\left(bias + \sum_i x_i w_i\right) \qquad (6.125)$$

f() is the transfer function, *bias* is a constant, w_i are weight factors, and x_i are the inputs. The range of the summation is over all inputs connected to the neuron. By using an input vector augmented with a constant "1" the notation can be simplified to $y = f(\vec{w}^t \vec{x})$. Usually, the transfer function is a semilinear sigmoid function:

$$f(x) = \frac{1}{1+\exp(-x)} \qquad (6.126)$$

In a small range around $x = 0$ the function is almost linear. However, seen more globally, the function acts as a Heaviside step-function. Functions of this type are sometimes called *soft threshold-functions*.

The connection structure of an error backpropagation network is strictly feed forward: no closed loops are allowed. The neurons are ordered in layers. The network consists of: an input layer, a number of hidden layers, and an output layer. The input layer merely serves to buffer the input of the network and to distribute these to inputs of the neurons. No real processing is involved here. The neurons in hidden layers only connect to other neurons in the network. The outputs of neurons in the output layer form the output of the network. See figure 6.33b.

As mentioned above, the function of a neural network depends on the choice of the connection structure, and the connections between neurons. In backpropagation networks, the latter is fully determined by the choice of the weight factors between inputs and outputs.

Figure 6.33b shows the connection structure of a neural network suitable to solve the problem of apple classification (example 6.3). The dimension of the input vector is 2. The network has one hidden layer consisting of two neurons. Since we have a 3-class problem the number of output neurons is three. Note that the internal structure of a single neuron resembles the one of a perceptron. In fact, it realises a linear discriminant function. One neuron in the first hidden layer is capable of separating the measurement space into two halfplanes. Two neurons in the hidden layer enable the partitioning of the measurement space into four compartments. A neuron in the output layer is able to select one of these compartments, or to merge some.

Let y_j denote the output of the *j*-th neuron in the hidden layer. Furthermore, let $\vec{w}_j$ denote the weight vector of this neuron. With this notation, the *k*-th output of the network in figure 6.33b is:

$$g_k(\vec{x}) = f\left(w_{k,1} f(\vec{w}_1^t \vec{x}) + w_{k,2} f(\vec{w}_2^t \vec{x}) + w_{k,bias}\right) \qquad (6.127)$$

Apart from the transfer function in the output layer, this function is of the type given in (6.120). Since the transfer function is a monotonically increasing function, classification with or without transfer function in the output layer does not change the performance.

The SQE measure associated with this network is continuous for all weight factors. This implies that we can apply the steepest descent procedure given in (6.124). The functional form of (6.128) is such that the gradient vector $\nabla J_{SQE}(\vec{w})$ can be calculated efficiently by application of the chain rule for differentiation, i.e. error backpropagation.

Figure 6.34a gives an illustration of the training process. The graph shows the SQE measure belonging to the training set in figure 6.6a as a function of the number of epochs that was used to train the network in figure 6.33b. For the sake of visibility the SQE is subsampled with a period of 50 epochs. It can be seen that for

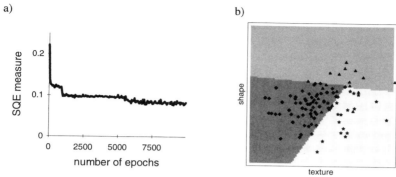

Figure 6.34 Training a neural network
a) Learn curve of the neural network shown in figure 6.33b
b) Decision function of the network trained with the data shown

a large number of epochs the SQE does not seem to decrease much and then it drops down suddenly. Probably, at that point the network has found an important decision boundary. Also, it can be seen that there is no way to find out whether a network has learned enough; i.e. whether a minimum has been reached.

The parameter α is called the *learning rate*. If this parameter is very small, learning will be very slow. In addition, with a small learning rate, the possibility of getting stuck in a local minimum of the SQE measure becomes more likely. On the other hand, choosing a large learning rate speeds up the learning process, and may help to "overlook" a local minimum. A disadvantage may be that the learning process may end up in an oscillation around a minimum.

Another problem of the backpropagation neural network is that it is difficult to find the right connection structure. The choice of the network in figure 6.33b is based on prior knowledge. Beforehand it was known that a topology with three compartments and approximately linear boundaries is optimal. If this knowledge is not available, there is no clue to find the right connection structure.

If the network is too simple, it cannot fulfil the task. On the other hand, if the network is too complex, it gets "overtrained". It means that the network becomes too specialised in the given training set. This is at the cost of the capability to generalise. An example of an overtrained network is given in figure 6.35. The

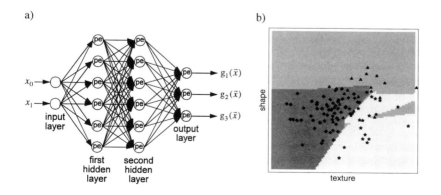

Figure 6.35 Training a network with two hidden layers
a) Network with two hidden layers
b) A decision function of the network given in a

network consists of two hidden layers each with six neurons. Such a network is too complex for the given training set. This is shown in figure 6.35b, which gives a map of the decision function found. Clearly, the network has concentrated on a few irrelevant samples.

6.4 PERFORMANCE MEASURES

The ultimate goal in pattern classification design is to find the "best" classifier. Within a Bayesian framework "best" means the one with minimal risk. Often, the cost of misclassification is difficult to assess, or even fully unknown. In many applications the probability of misclassification - the error rate E - serves as an optimality criterion. For this reason, E is very important to evaluate the performance of a classifier and to compare one classification with another. Unfortunately, the mathematical expressions of the error rate are often difficult to handle.

Therefore, it is common practice to replace the error rate with performance measures that can be expressed more simply thereby allowing a mathematical treatment of the analysis. Often, these measures bound the error rate. As such, the measures are relevant. A disadvantage of these measures is that they often apply in special cases only; for instance, if the conditional densities have a specific functional form.

From the various measures known in literature we discuss two of them. One of them - the interclass/intraclass distance - applies to the multi-class case. It is useful if class information is mainly found in the differences between expectation vectors in the measurement space, while at the same time the scattering of the measurement vectors (due to noise) is class-independent. The second measure - the Chernoff distance - is useful especially in the 2-class case because then it can be used to express bounds on the error rate.

Both measures are used in discriminant analysis (section 6.3.4), feature selection (section 6.5) and feature extraction (section 6.6).

6.4.1 Interclass and intraclass distance

The measure to be discussed in this section is based on the Euclidean distance between pairs of samples in the training set. We assume that the patterns are such that the expectation vectors of the different classes are discriminating. If fluctuations of the measurement vectors around these expectations are due to noise, then these fluctuations will not carry much class information. Therefore, our goal is to arrive at a measure that is a monotonically increasing function of the distance between expectation vectors, and a monotonically decreasing function of the average scattering around the expectations.

As in the previous section, T_S is a (labeled) training set with N_S samples. The classes ω_k are represented by subsets $T_k \subset T_S$; each class having N_k samples ($\Sigma N_k = N_S$). Measurement vectors in T_S - without reference to their class - are denoted by $\underline{x}_n$ (random vectors) or $\bar{x}_n$ (realisations). Measurement vectors in T_k (i.e. vectors coming from class ω_k) are denoted by $\underline{x}_{k,n}$ or $\bar{x}_{k,n}$.

The development starts with the *average squared distance* of pairs of samples in the training set:

$$\overline{\rho^2} = \frac{1}{2N_S^2} \sum_{n=1}^{N_S} \sum_{m=1}^{N_S} (\vec{x}_n - \vec{x}_m)^t (\vec{x}_n - \vec{x}_m) \tag{6.128}$$

This distance increases as the distance between expectation vectors increases. However, it also increases if the average scattering around the expectations increases. This last property makes $\overline{\rho^2}$ useless as a performance measure. To circumvent this, it is needed to separate $\overline{\rho^2}$ into a part describing the average distance between expectation vectors and a part describing distances due to noise scattering. For that purpose, estimations of the conditional expectations ($\vec{\mu}_k = E\{\vec{x}|\omega_k\}$) of the measurement vectors are used, along with an estimate of the unconditional expectation ($\vec{\mu} = E\{\vec{x}\}$). The sample mean of class ω_k is:

$$\hat{\vec{\mu}}_k = \frac{1}{N_k} \sum_{n=1}^{N_k} \vec{x}_{k,n} \tag{6.129}$$

The sample mean of the entire training set is:

$$\hat{\vec{\mu}} = \frac{1}{N_S} \sum_{n=1}^{N_S} \vec{x}_n \tag{6.130}$$

With these definitions, it is easy to show that the average squared distance is:

$$\overline{\rho^2} = \frac{1}{N_S} \sum_{k=1}^{K} \left[\sum_{n=1}^{N_k} \left((\vec{x}_{k,n} - \hat{\vec{\mu}}_k)^t (\vec{x}_{k,n} - \hat{\vec{\mu}}_k) + (\hat{\vec{\mu}} - \hat{\vec{\mu}}_k)^t (\hat{\vec{\mu}} - \hat{\vec{\mu}}_k) \right) \right] \tag{6.131}$$

The first term represents the average squared distance due to scattering of samples around their class-dependent expectation. The second term corresponds to the average squared distance between class-dependent expectations and unconditional expectation.

An alternative way to specify these distances is by means of so called *scatter matrices*. A scatter matrix gives some information about the dispersion of a population of samples around their mean. For instance, the matrix that describes the scattering of vectors from class ω_k is defined as:

$$\mathbf{S}_k = \frac{1}{N_k} \sum_{n=1}^{N_k} \left((\vec{x}_{k,n} - \hat{\vec{\mu}}_k)(\vec{x}_{k,n} - \hat{\vec{\mu}}_k)^t \right) \tag{6.132a}$$

Comparison with equation (6.96) shows that $\mathbf{S}_k$ is close to an unbiased estimate of the class-dependent covariance matrix. In fact, $\mathbf{S}_k$ is the maximum likelihood estimate of $\mathbf{C}_k$. With that, $\mathbf{S}_k$ does not only supply information about the average distance of the scattering, it also supplies information about the eccentricity and orientation of this scattering. This is analogous to the properties of a covariance matrix.

Averaged over all classes the scatter matrix describing the noise is:

$$\mathbf{S}_w = \frac{1}{N_S} \sum_{k=1}^{K} N_k \mathbf{S}_k = \frac{1}{N_S} \sum_{k=1}^{K} \sum_{n=1}^{N_k} (\vec{x}_{k,n} - \hat{\vec{\mu}}_k)(\vec{x}_{k,n} - \hat{\vec{\mu}}_k)^t \tag{6.132b}$$

This matrix is called the *within-scatter matrix* as it describes the average scattering within classes. Complementary to this is the *between-scatter matrix* $\mathbf{S}_b$ which describes the scattering of the class-dependent sample means:

$$\mathbf{S}_b = \frac{1}{N_S} \sum_{k=1}^{K} N_k (\hat{\mu}_k - \hat{\mu})(\hat{\mu}_k - \hat{\mu})^t \qquad (6.132c)$$

Figure 6.36a illustrates the concepts of within-scatter matrices and between-scatter matrices. The training set is taken from figure 6.6a. As with covariance matrices, scatter matrices correspond to ellipses which can be thought of as contours of probability densities associated to these scatter matrices. Of course, strictly speaking the correspondence holds true only if the underlying densities are Gaussian, but even if the densities are not Gaussian, the ellipses give an impression of how the population is scattered. In figure 6.36a the within-scatter is represented by three similar ellipses positioned at the three conditional sample means. The between-scatter is depicted by an ellipse centred at the mixture sample mean.

With the definitions in (6.132) the average square distance in (6.131) is the trace of the matrix $\mathbf{S}_w + \mathbf{S}_b$:

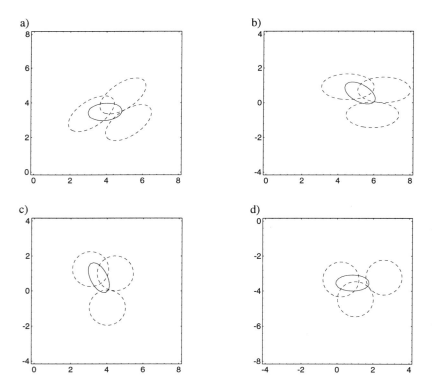

Figure 6.36 Scatter matrices belonging to the samples shown in figure 6.6
a) Within-scatter matrix (dashed) and between scatter matrix depicted by ellipses
b) Rotation of a) such that the within-scattering becomes decorrelated
c) Scaling of the axes of c) such that the within-scattering becomes white
d) Rotation of c) such that the between-scattering becomes decorrelated

$$\overline{\rho^2} = \operatorname{tr}(\mathbf{S}_w + \mathbf{S}_b) = \operatorname{tr}(\mathbf{S}_w) + \operatorname{tr}(\mathbf{S}_b) \tag{6.133}$$

Indeed, this expression shows that the average distance is composed of a contribution due to differences in expectation and a contribution due to noise. The contribution $\operatorname{tr}(\mathbf{S}_w)$ is called *intraclass distance*. The contribution $\operatorname{tr}(\mathbf{S}_b)$ is the *interclass distance*. Equation (6.133) also shows that the average distance is not appropriate as a performance measure, since if $\overline{\rho^2}$ is large this does not imply that the classes are well separated, and vice versa.

A performance measure more suited to express the separability of classes is the ratio between interclass and intraclass distance:

$$\frac{\operatorname{tr}(\mathbf{S}_b)}{\operatorname{tr}(\mathbf{S}_w)} \tag{6.134}$$

The disadvantage of this measure is that it does not take into account the main direction in which the within-class scattering extends seen with respect to the principal direction of the between-class scattering. One way to correct this defect is to transform the measurement space such that the within-scattering becomes *white*. For that purpose we apply a linear operation to all measurement vectors yielding *feature vectors* $\vec{y}_n = \mathbf{A}\vec{x}_n$. In the transformed space the within- and between-scatter matrices become: $\mathbf{AS}_w\mathbf{A}'$ and $\mathbf{AS}_b\mathbf{A}'$, respectively. The matrix $\mathbf{A}$ is chosen such that $\mathbf{AS}_w\mathbf{A}' = \mathbf{I}$.

The matrix can be found by factorisation: $\mathbf{S}_w = \mathbf{V}\Lambda\mathbf{V}'$ where Λ is a diagonal matrix containing the eigenvalues of $\mathbf{S}_w$, and $\mathbf{V}$ a unitary matrix containing corresponding eigenvectors; see appendix B.3.1. With this factorisation it follows that $\mathbf{A} = \Lambda^{-1/2}\mathbf{V}'$. An illustration of the process is depicted in figure 6.36b and 6.36c. The effect of the transform is that the contribution of the noise gets normalised. Therefore, a performance measure that meets all our requirements is:

$$J_{INTER/INTRA} = \operatorname{tr}(\Lambda^{-1/2}\mathbf{V}'\mathbf{S}_b\mathbf{V}\Lambda^{-1/2}) = \operatorname{tr}(\mathbf{V}\Lambda^{-1}\mathbf{V}'\mathbf{S}_b) = \operatorname{tr}(\mathbf{S}_w^{-1}\mathbf{S}_b) \tag{6.134}$$

An application of this performance measure is in linear discriminant analysis. For the sake of simplicity we consider a two-class problem only. In that case the set of linear discriminant functions given in (6.114) can be replaced with a single linear function:

$$g(\vec{x}) = \vec{w}'\vec{x} \tag{6.135}$$

If this function exceeds a suitably chosen threshold, class ω_1 is assumed. Otherwise class ω_2 is assigned.

The linear function $g(\vec{x}) = \vec{w}'\vec{x}$ maps the measurement space into a 1-dimensional space. Application of the mapping to each vector in the training set transforms the within- and between-scatter matrices into scatters $\vec{w}'\mathbf{S}_w\vec{w}$ and $\vec{w}'\mathbf{S}_b\vec{w}$, respectively. Substitution of (6.135) in (6.132) shows that these scatters have the following interpretation:

$$\vec{w}'\mathbf{S}_w\vec{w} = \frac{N_1}{N_1+N_2}\hat{\sigma}_{g_1}^2 + \frac{N_2}{N_1+N_2}\hat{\sigma}_{g_2}^2 \tag{6.136a}$$

$$\vec{w}'\mathbf{S}_b\vec{w} = \frac{N_1 N_2}{(N_1+N_2)^2}\left(\hat{\mu}_{g_1} - \hat{\mu}_{g_2}\right)^2 \tag{6.136b}$$

where $\hat{\sigma}_{g_1}^2$, $\hat{\sigma}_{g_2}^2$, $\hat{\mu}_{g_1}$, and $\hat{\mu}_{g_2}$ are maximum likelihood estimates of the class-dependent variances and expectations of $g(\vec{x})$. With that, the performance measure in the 1-dimensional space of $g(\vec{x})$ is:

$$J_{INTER/INTRA}(\vec{w}) = \frac{\vec{w}^t S_b \vec{w}}{\vec{w}^t S_w \vec{w}} = \frac{\frac{N_1 N_2}{N_1+N_2}\left(\hat{\mu}_{g_1}-\hat{\mu}_{g_2}\right)^2}{N_1 \hat{\sigma}_{g_1}^2 + N_2 \hat{\sigma}_{g_2}^2} \qquad (6.137)$$

Written in this form it is clear that in the transformed domain the performance measure is proportional to the ratio of the difference between sample means and the sample variance.

The goal in discriminant analysis is to find the vector $\vec{w}$ that maximises the performance measure. The obvious way to achieve that is by differentiating the measure with respect to $\vec{w}$, and by equating the result to zero:

$$\frac{\partial J_{INTER/INTRA}(\vec{w})}{\partial \vec{w}} = \vec{0} \qquad (6.138)$$

The solution of this equation is:

$$\vec{w} = S_w^{-1}(\hat{\mu}_1 - \hat{\mu}_2) \qquad (6.139)$$

The derivation of (6.139) is a technical matter. It can be found in Devijver and Kittler [1982] and Schalkoff [1992]. The solution - known as *Fisher's linear discriminant* - is similar to Bayes classification of Gaussian random vectors with class-independent covariance matrices; i.e. the Mahalanobis distance classification. The difference with expression (6.18) is that the true covariance matrix and the expectation vectors have been replaced with estimates based on the training set.

6.4.2 Chernoff-Bhattacharyya distance

The interclass and intraclass distances are based on the Euclidean metric defined in the measurement space. Another possibility is to use a metric based on probability densities. Examples in this category are the Chernoff distance and the Bhattacharyya distance. These distances are useful especially in the 2-class case.

The merit of the Bhattacharyya and Chernoff distance is that an inequality exists with which the distances bound the minimum error rate E_B (i.e. the error rate of Bayes classification with uniform cost function). The inequality is based on the following relationship:

$$\min\{a,b\} \le \sqrt{ab} \qquad (6.140)$$

This inequality holds true for any positive quantities a and b. We will use it in the expression of the minimum error rate. Substitution of (6.13) in (6.14) yields:

$$E_B = \int_{\vec{x}} [1 - \max\{P(\omega_1|\vec{x}=\vec{x}), P(\omega_2|\vec{x}=\vec{x})\}] p_{\vec{x}}(\vec{x}) d\vec{x}$$
$$= \int_{\vec{x}} \min\{p_{\vec{x}|\omega_1}(\vec{x}|\omega_1)P_1, p_{\vec{x}|\omega_2}(\vec{x}|\omega_2)P_2\} d\vec{x} \qquad (6.141a)$$

Together with (6.140) we have the following inequality:

$$E_B \leq \sqrt{P_1 P_2} \int_{\vec{x}} \sqrt{p_{\vec{x}|\omega_1}(\vec{x}|\omega_1) p_{\vec{x}|\omega_2}(\vec{x}|\omega_2)} d\vec{x} \qquad (6.141b)$$

The inequality is called the *Bhattacharyya upper bound*. A more compact notation of it is achieved with the so-called Bhattacharyya distance. This performance measure is defined as:

$$J_{BHAT} = -\ln\left[\int_{\vec{x}} \sqrt{p_{\vec{x}|\omega_1}(\vec{x}|\omega_1) p_{\vec{x}|\omega_2}(\vec{x}|\omega_2)} d\vec{x}\right] \qquad (6.142)$$

With that, the Bhattacharyya upper bound simplifies to:

$$E_B \leq \sqrt{P_1 P_2} \exp(-J_{BHAT}) \qquad (6.143)$$

The bound can be made more tight if the inequality (6.140) is replaced with the more general inequality $\min\{a,b\} \leq a^s b^{1-s}$. This last inequality holds true for any s, a, and b in the interval [0,1]. The inequality leads to the so-called *Chernoff distance* defined as:

$$J_C(s) = -\ln\left[\int_{\vec{x}} p_{\vec{x}|\omega_1}^s(\vec{x}|\omega_1) p_{\vec{x}|\omega_2}^{1-s}(\vec{x}|\omega_2) d\vec{x}\right] \quad \text{with:} \quad 0 \leq s \leq 1 \qquad (6.144)$$

Application of the Chernoff distance in a derivation similar to (6.141) yields:

$$E_B \leq P_1^s P_2^{1-s} \exp(-J_C(s)) \quad \text{for any} \quad s \in [0,1] \qquad (6.145)$$

This so-called *Chernoff bound* encompasses the Bhattacharyya upper bound. In fact, for $s = 0.5$ the Chernoff distance and the Bhattacharyya distance match: $J_{BHAT} = J_C(0.5)$.

A further simplification requires the specification of the conditional probability densities. An important application of the Chernoff and Bhattacharyya distance is with Gaussian densities. Suppose that these densities have class-dependent expectation vectors $\vec{\mu}_i$ and covariance matrices C_i, respectively. Then, it can be shown that the Chernoff distance transforms into:

$$J_C(s) = \tfrac{1}{2} s(1-s)(\vec{\mu}_2 - \vec{\mu}_1)'[(1-s)C_1 + sC_2]^{-1}(\vec{\mu}_2 - \vec{\mu}_1) + \tfrac{1}{2}\ln\left[\frac{|(1-s)C_1 + sC_2|}{|C_1|^{1-s}|C_2|^s}\right] \qquad (6.146)$$

It can be seen that if the covariance matrices do not depend on the classes ($C_1 = C_2$), the second term vanishes, and the Chernoff and the Bhattacharyya distances become proportional to the Mahalanobis distance Δ given in (6.27): $J_{BHAT} = \Delta/8$. Figure 6.37a shows the corresponding Chernoff and Bhattacharyya upper bounds. In this example, the relation between Δ and the minimum error rate is easily obtained using expression (6.28).

Figure 6.37b shows the Chernoff bound in dependency on s. In this particular case, the Chernoff distance is symmetric in s, and the minimum bound is located at $s = 0.5$ (i.e. the Bhattacharyya upper bound). If the covariance matrices are not

a)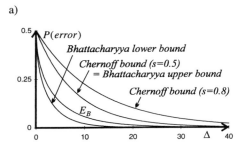
b)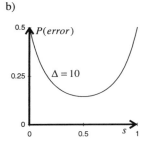

Figure 6.37 Error bounds
a) The minimum error rate with some bounds given by the Chernoff distance. In this example the bound with $s=0.5$ (Bhattacharyya upper bound) is most tight. The figure also shows the Bhattacharyya lower bound

b) The Chernoff bound in dependency on s

equal, the Chernoff distance is not symmetric, and the minimum bound is not guaranteed to be located midway. A numerical optimisation procedure can be applied to find the most tight bound.

There also exists a lower bound based on the Bhattacharyya distance. This bound is expressed as:

$$\tfrac{1}{2}\left[1-\sqrt{1-4P_1P_2\exp(-2J_{BHAT})}\right] \le E_B \qquad (6.147)$$

A proof is found in Devijver and Kittler [1982]. Figure 6.37a is an illustration.

In the Gaussian case, if the expectation vectors are equal ($\bar{\mu}_1 = \bar{\mu}_2$), the first term of (6.146) vanishes, and all class information is represented by the second term. This term corresponds to class information carried by differences in covariance matrices.

6.5 FEATURE SELECTION AND EXTRACTION

The dimension of the measurement vector $\bar{x}$ is N. In practical situations, this dimension may vary from 1 up to 10^4 or more. If pixels from a digital image are used as measurements, the dimension may even be on the order of 10^6. Many elements of $\bar{x}$ can be redundant or even irrelevant with respect to the classification process.

The dimension of the measurement vector cannot be chosen arbitrarily large. For this, two reasons exist. The first reason is that an increase of the dimension can cause a decrease of performance. This topic will be discussed in section 6.5.1.

A second reason is that the computational complexity of a classification becomes too costly if the measurement space is too large. Requirements related to the throughput and the cost price of the classification define the upper limit of the dimension. As an example, consider a classification problem in which the measurement vector is Gaussian. If the covariance matrices are unequal, the classification has a computational complexity which is on the order of KN^2

multiplications and additions; see (6.17). Suppose that the required throughput is about 100 classifications/sec. If a signal-processor with at most 100 Mflops is affordable, then the maximum dimension allowed is on the order of 100.

Two different approaches exist to reduce the dimension of a measurement vector. One is to discard certain elements of the vector and to select the ones that remain. This type of reduction is called *feature selection*. It is discussed in section 6.5.2. The other approach to reduce the dimension is called *feature extraction*. Here, the selection of elements takes place in a transformed space. Section 6.5.3 addresses the problem of how to find suitable transforms.

6.5.1 Performance and dimensionality

As mentioned in section 6.4, the ultimate goal in pattern classification design is often to achieve a low error rate. If the prior probabilities and conditional densities are known, then the Bayes paradigm yields a classification method with minimum error rate E_B. In the multi-class case, this performance measure is given by (6.13) and (6.14):

$$E_B = 1 - \int_{\bar{x}} \max_{\omega \in \Omega} \{ p_{\bar{x}|\omega}(\bar{x}|\omega) P(\omega) \} d\bar{x} \tag{6.148}$$

In the 2-class case, the expression simplifies to (see 6.141):

$$E_B = \int_{\bar{x}} \min \{ p_{\bar{x}|\omega}(\bar{x}|\omega_1) P_1, p_{\bar{x}|\omega}(\bar{x}|\omega_2) P_2 \} d\bar{x} \tag{6.149}$$

Usually, the conditional densities are unknown. This implies that E_B is not available. Instead a decision function $\hat{\omega}(\bar{x})$ derived from a training set is used. Suppose that the error rate of this function is E. Since the decision function is derived from a training set T_S, the error rate must depend on this set. We make this dependency explicit by writing $E(T_S)$ instead of E, and $\hat{\omega}(T_S, \bar{x})$ instead of $\hat{\omega}(\bar{x})$. The fact that the Bayes error rate is minimum implies that:

$$E_B \leq E(T_S) \tag{6.150}$$

With the assumption that the training set is a random selection, the error rate associated with the training set is a random variable. In fact, we must write $\underline{E}(T_S)$.

An interesting question related to performance evaluation of classification methods concerns the estimation of $\underline{E}(T_S)$. We assume that for that purpose a second set of objects (called the *evaluation set* T_{eval}) is available. Similar to T_S, this set is a random selection of objects whose true classes are known. Furthermore, we assume that the evaluation set is statistically independent with respect to the training set. The obvious way to estimate $\underline{E}(T_S)$ is to apply $\hat{\omega}(T_S, \bar{x})$ to each vector in T_{eval} and to determine the relative frequency of misclassifications. This frequency has a binomial distribution with mean and variance:

$$E\{\hat{\underline{E}}(T_S)\} = E(T_S) \qquad \text{Var}\{\hat{\underline{E}}(T_S)\} = \frac{E(T_S)(1 - E(T_S))}{N_{eval}} \tag{6.151}$$

where N_{eval} is the number of objects in the evaluation set. The expression shows that in order to have an accurate estimate it is desirable to have a large evaluation set. As shown in section 6.3 it is also desirable to have a large training set, because in that case the error rate $\underline{E}(T_S)$ will approximate the optimal Bayes error rate E_B.

Since the sample size of objects in the training set and the evaluation set is limited, one has to compromise between classification design on one hand and evaluation on the other hand. A solution would be to use one set for both purposes, i.e. $T_{eval} = T_S$. By doing so, the danger exists that the evaluation is biased. The cause is that the decision function is designed so as to minimise the number of misclassifications of objects in the training set. Therefore, the possibility exists that a decision function becomes specialised for that particular set. This is at the cost of the capability to generalise. By evaluating the decision function with the training set this defect is overlooked. The estimate will be too optimistic.

An illustration of performance estimation based on training and evaluation sets is given in figure 6.38. Here, we have a measurement space the dimension of which varies between 1 and 19. The number of classes is two. The (true) minimum error rate E_B of this example is shown as a function of N. Clearly, this function is non-increasing. Once an element has been added with discriminatory information, the

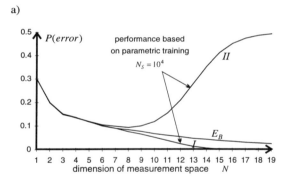

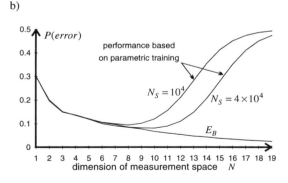

Figure 6.38 Error rates versus dimension of measurement space
a) In curve I the training set and the evaluation set are one. In curve II two different sets have been used. The sizes of the evaluation set and the training set are both 10000
b) Estimates of the error rates based on independent training sets. The curves show the effect of enlargement of the training set

addition of another element cannot destroy this information. Therefore, with growing dimension class information accumulates.

In figure 6.38a the size of a training set is 10^4. With this training set a decision function based on parametric learning (maximum likelihood estimation) has been built. The decision function is evaluated with the same training set (curve I), and also with another set (the evaluation set) resulting in curve II. The interesting thing to notice in figure 6.38a is the behavior of the estimates in relation to the dimension of the measurement space. If the dimension is low, the estimated performances coincide with the optimal error rate E_B. The training set is large enough to use it for both learning and evaluation. If the dimension is high, curve I approaches zero thereby exceeding the performance of the optimal classification. The conclusion is that this evaluation method is much too optimistic. For large dimensions, curve II approaches 0.5 indicating that parametric learning based on the maximum likelihood estimation with this training set is not successful.

In fact, the present example is one with binary measurement data; see example 6.7. The number of states a vector can take is 2^N. If there are no constraints related to the conditional probabilities, then the number of parameters to estimate is on the order of 2^N. As discussed in example 6.7 the number of samples in the training set must be much larger than this. Figure 6.38b shows that with $N_S = 10^4$ the number of useful elements is about $N = 8$. Making the sample size twice as large permits an increment of the dimension by one. Therefore, with $N_S = 4 \times 10^4$ the number of useful elements is about $N = 10$. An illustration of this behavior is given in figure 6.38b.

6.5.2 Feature selection

This section introduces the problem how to determine a subset from the N-dimensional measurement vector such that this subset is most suitable for classification. The problem is formalised as follows. Let $F(N) = \{x_i | i = 0, \cdots, N-1\}$ be the set with elements from the measurement vector $\bar{x}$. Furthermore, let $F_j(D) = \{y_m | m = 0, \cdots, D-1\}$ be a subset of $F(N)$ consisting of $D < N$ elements taken from $\bar{x}$. For each element y_m there exists an element x_i such that $y_m = x_i$. The number of distinguishable subsets for a given D is:

$$q(D) = \binom{N}{D} = \frac{N!}{(N-D)!D!} \tag{6.152}$$

This quantity expresses the number of different combinations that can be made up from N elements, each combination containing D elements and no two combinations containing exactly the same D elements. We will adopt an enumeration of the index j according to: $j = 1, \cdots, q(D)$.

In section 6.4 the concept of a performance measure has been introduced. These measures evaluate statistical properties of a measurement vector according to a certain criterion. Let $J(F_j(D))$ be a performance measure related to the subset $F_j(D)$. The particular choice of $J()$ depends on the problem at hand. For instance, in a multi-class problem, the interclass/intraclass distance $J_{INTER/INTRA}()$ could be used (6.134). This measure assumes the availability of a training set with which the distance can be calculated. In a 2-class problem with Gaussian densities, the Bhattacharyya distance $J_{BHAT}()$ is useful. This distance is expressed in the true parameters of the Gaussian densities; see (6.146). Normally, only a training set is available. In that case, one can estimate the distance by substitution of estimated parameters.

The goal of feature selection is to find a subset $\hat{F}(D)$ with dimension D such that this subsets outperforms all other subsets with dimension D:

$$\hat{F}(D) = F_i(D) \quad \text{with:} \quad J(F_i(D)) \geq J(F_j(D)) \quad \text{for all} \quad j \in \{1, \cdots, q(D)\} \quad (6.153)$$

With an exhaustive search for this particular subset, the problem of feature selection is solved.

Expression (6.153) leaves us with a practical problem. How to accomplish the search process? With an exhaustive search the number of required evaluations of $J(F_j(D))$ is $q(D)$. Even in a simple classification problem this number is enormous. For instance, in the example given in figure 6.38, $N = 19$ and $D \approx 10$. It gives us about 10^5 different subsets. Suppose that the training set consists of $N_S = 4 \times 10^4$ elements, and that the number of operations required to calculate one evaluation is in the order of DN_S. Then, the total number of operations is about 10^{11}. A small increase of the dimension, say to $N = 25$, would require about 10^{13} operations. Needless to say that in problems with moderate complexity exhaustive search is out of the question.

It is obvious that a more efficient search strategy is needed. For this, many options exist, but most of them are suboptimal (in the sense that they cannot guarantee that a subset with best performance will be found). In the remaining part of this section we will consider one search strategy called *branch-and-bound*. This strategy is one of the few that guarantees optimality.

The search process is accomplished systematically by means of a tree structure. The tree consists of $N - D + 1$ levels. Each level is enumerated by a variable l with l varying from D up to N. A level consists of a number of nodes. A node n at level l corresponds to a subset $F_n(l)$. At the highest level $l = N$ there is only one node corresponding to the full set $F(N)$. At the lowest level $l = D$ there are $q(D)$ nodes corresponding to the $q(D)$ subsets amongst which the solution must be found. Levels in between have a number of nodes that is less than or equal to $q(l)$. A node at level l is connected to one node at level $l+1$ (except the node $F(N)$ at level N). In addition, each node at level l is connected to one or more nodes at level $l-1$ (except the nodes at level D). An example of such a tree structure with corresponding subsets is presented in figure 6.39. In this example $N = 6$ and $D = 2$.

A prerequisite of the branch-and-bound strategy is that the performance measure is a monotonically increasing function of the dimension D. The basic assumption behind it is that if we remove one element from a measurement vector, the performance can only become worse. As discussed in the previous subsection, this assumption does not always hold true. Especially, if the performance measure is derived from a finite training set, the assumption is not justified. Therefore, in this sense the branch-and-bound algorithm is not optimal.

The search process takes place from the highest level ($l = N$) by systematically traversing all levels until finally the lowest level ($l = D$) is reached. Suppose that one branch of the tree has been explored up to the lowest level, and suppose that the best performance measure found so far at level $l = D$ is P. Consider a node $F_n(l)$ (at a level $l > D$) which has not been explored as yet. If $J(F_n(l)) \leq P$, it is unnecessary to consider the nodes below $F_n(l)$. The reason is that the performance measure for these nodes can only become less than $J(F_n(l))$ and thus less than P.

The following algorithm traverses the tree structure according to "depth first search with backtrack mechanism". In this algorithm the best performance measure found so far at level $l = D$ is stored in a variable P. Branches whose performance measure is bounded by P are skipped. Therefore, relative to exhaustive search the algorithm is much more computationally efficient. The variable s denotes a subset. Initially this subset is empty. As soon as a combination of D elements has been found that exceeds the current performance, this combination is stored in S.

Algorithm 6.3 Branch-and-bound search

Input: a labeled training set on which a performance measure J() is defined.
1. Initiate: $P = 0$ and $S = \emptyset$;
2. Explore-node($F_1(N)$);

Output: The maximum performance measure stored in P with the associated subset of $F_1(N)$ stored in S.

Procedure: Explore-node($F_n(l)$)
1. If $(J(F_n(l)) \leq P)$ then: return;
2. If $(l = D)$ then:
   2.1. If $(J(F_n(l)) > P)$ then:
      $P = J(F_n(l))$;
      $S = F_n(l)$;
3. For all $(F_m(l-1) \subset F_n(l))$ do: Explore-node($F_m(l-1)$);
4. return;

The algorithm is recursive. The procedure "Explore-node()" explores the node given in its argument list. If the node is not at the lowest level all successive nodes are explored by calling the procedure recursively. The first call is with the full set $F_1(N)$ as argument.

The algorithm listed above does not specify the exact structure of the tree. The structure follows from the specific implementation of the loop in step 3. This loop also controls the order in which branches are explored. Both aspects influence the computational efficiency of the algorithm.

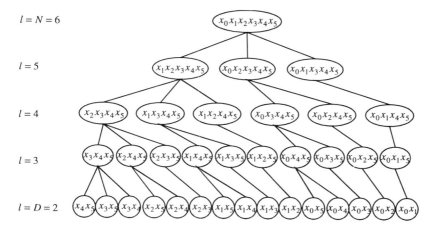

Figure 6.39 Graphical representation of tree structure in behalf of feature selection

Although the branch-and-bound algorithm saves a lot of calculations relative to exhaustive search (especially for large values of $q(D)$), it may still require too much computational effort. In these situations one has to resort to methods which are suboptimal.

6.5.3 Linear feature extraction

Another approach to reduce the dimension of the measurement vector is to use a transformed space instead of the original measurement space. Suppose that $W()$ is a transformation that maps the measurement space $\mathbb{R}^N$ onto a space $\mathbb{R}^D$. Application of the transformation to a measurement vector yields a so-called *feature vector* $\vec{y} \in \mathbb{R}^D$:

$$\vec{y} = W(\vec{x}) \tag{6.154}$$

Classification is based on the feature vector rather than on the measurement vector; see figure 6.40.

The advantage of feature extraction above feature selection is that information from none of the elements of the measurement vector is wasted. Furthermore, in some situations feature extraction is accomplished more easily than feature selection. A disadvantage of feature extraction is that it requires the determination of some suitable transform $W()$. If the transform chosen is too complex, it may miss the ability to generalise. On the other hand, if the transform chosen is too simple, it may constrain the decision boundaries to a form which is inappropriate to discriminate between classes.

This section discusses the design of linear feature extractors. That is the transformation $W()$ is restricted to the class of linear operations. Such operations can be written as a matrix-vector product:

$$\vec{y} = \mathbf{W}\vec{x} \tag{6.155}$$

where $\mathbf{W}$ is a $D \times N$-matrix. The reason for the restriction is threefold. First, a linear feature extraction is computational efficient. Secondly, in many classification problems - though not all - linear features are appropriate. Thirdly, a restriction to linear operations facilitates the mathematical handling of the problem.

An illustration of the computational efficiency of linear feature extraction is the Gaussian case. If covariance matrices are unequal the number of calculations is on the order of KN^2; see (6.16). Classification based on linear features requires about $DN + KD^2$ calculations. If D is very small compared with N, the extraction saves a large number of calculations.

The example of Gaussian densities is also well suited to illustrate the appropriateness of linear features. Clearly, if covariance matrices are equal, then

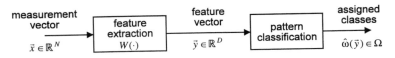

Figure 6.40 Feature extraction

(6.18) shows that linear features are optimal. On the other hand, if expectation vectors are equal and the discriminatory information is in the differences between covariance matrices, linear feature extraction may still be appropriate. This is shown in the example of figure 6.12b where covariance matrices are eccentric, differing only in their orientation. However, in the example shown in figure 6.12a (concentric circles) linear features seem to be inappropriate. In practical situations, covariance matrices will differ both in shape and orientations. Linear feature extraction is likely to lead to some reduction of dimensionality, but this reduction may be less than what is feasible with non-linear feature extraction.

Linear feature extraction may also improve the ability to generalise. In the Gaussian case with unequal covariance matrices the number of parameters to estimate is on the order of KN^2. With linear features this number reduces to $DN + KD^2$. If the number of samples in the training set is also on the order of KN^2, the corresponding classification becomes easily specialised to the training set at hand. But linear feature extraction - provided that $D \ll N$ - retains the ability to generalise.

We assume the availability of a training set and a suitable performance measure $J()$. The design of a feature extraction method boils down to finding the matrix $\mathbf{W}$ that - for the given training set - optimises the performance measure.

The performance measure of a feature vector $\bar{y} = \mathbf{W}\bar{x}$ is denoted by $J(\bar{y})$ or $J(\mathbf{W}\bar{x})$. With this notation, the optimal feature extraction is found as:

$$\mathbf{W} = \underset{\mathbf{W}}{\operatorname{argmax}}\{J(\mathbf{W}\bar{x})\} \tag{6.156}$$

Under the condition that $J(\mathbf{W}\bar{x})$ is continuously differentiable in $\mathbf{W}$ the solution of (6.156) must satisfy:

$$\frac{\partial J(\mathbf{W})}{\partial \mathbf{W}} = 0 \tag{6.157}$$

Finding a solution of either (6.156) or (6.157) gives us the optimal linear feature extraction. The search can be accomplished numerically using the training set. Alternatively, the search can also be done analytically assuming parametrised conditional densities. Substitution of estimated parameters (using the training set) gives the matrix $\mathbf{W}$.

In the remaining part of this section the last approach will be worked out for two particular cases: feature extraction for 2-class problems with Gaussian densities and feature extraction for multi-class problems based on the inter/intra distance measure. The former case will be based on the Bhattacharyya distance.

Bhattacharyya distance

In the 2-class case with Gaussian conditional densities a suitable performance measure is the Bhattacharyya distance. With $J_{BHAT} = J_C(0.5)$ expression (6.146) implicitly gives the Bhattacharyya distance as a function of the parameters of the Gaussian densities of the measurement vector $\bar{x}$. These parameters are the conditional expectations $\bar{\mu}_i$ and covariance matrices $\mathbf{C}_i$. With the substitution $\bar{y} = \mathbf{W}\bar{x}$ the expectation vectors and covariance matrices of the feature vector become $\mathbf{W}\bar{\mu}_i$ and $\mathbf{W}\mathbf{C}_i\mathbf{W}^t$, respectively. For the sake of brevity, let $\bar{m}$ be the difference between expectations of $\bar{x}$:

$$\bar{m} = \bar{\mu}_1 - \bar{\mu}_2 \qquad (6.158)$$

Then, substitution of $\bar{m}$, $\mathbf{W}\bar{\mu}_i$, and $\mathbf{WC}_i\mathbf{W}^t$ in (6.146) gives the Bhattacharyya distance of the feature vector:

$$J_{BHAT}(\mathbf{W}\bar{x}) = \tfrac{1}{4}(\mathbf{W}\bar{m})^t \left[\mathbf{WC}_1\mathbf{W}^t + \mathbf{WC}_2\mathbf{W}^t\right]^{-1}\mathbf{W}\bar{m} + \tfrac{1}{2}\ln\left[\frac{|\mathbf{WC}_1\mathbf{W}^t + \mathbf{WC}_2\mathbf{W}^t|}{2^D\sqrt{|\mathbf{WC}_1\mathbf{W}^t||\mathbf{WC}_2\mathbf{W}^t|}}\right] \qquad (6.159)$$

The first term corresponds to the discriminatory information of the expectation vectors; the second term to discriminatory information of covariance matrices.

Expression (6.159) is in such a form that an analytic solution of (6.157) is not feasible. However, if one of the two terms in (6.159) is dominant, a solution close to the optimal one is attainable. We consider the two extreme situations first.

EQUAL COVARIANCE MATRICES
If the conditional covariance matrices are equal, i.e. $\mathbf{C} = \mathbf{C}_1 = \mathbf{C}_2$, we have already seen that classification based on the Mahalanobis distance (expression 6.18 and 6.25) is optimal (in the sense that it minimises the error rate). In fact, this classification uses a $1 \times N$-dimensional feature extraction matrix given by:

$$\mathbf{W} = \bar{m}^t \mathbf{C}^{-1} \qquad (6.160)$$

We will show that the feature extraction based on maximisation of the Bhattacharyya distance is in line with (6.160).

If covariance matrices are equal, the second term in (6.159) vanishes and the first term simplifies to:

$$J_{BHAT}(\mathbf{W}\bar{x}) = \tfrac{1}{8}(\mathbf{W}\bar{m})^t \left[\mathbf{WCW}^t\right]^{-1}\mathbf{W}\bar{m} \qquad (6.161a)$$

For $\mathbf{W}$ any invertible $N \times N$-matrix, we have:

$$J_{BHAT}(\mathbf{W}\bar{x}) = \tfrac{1}{8}\bar{m}^t\mathbf{C}^{-1}\bar{m} = J_{BHAT}(\bar{x}) \qquad (6.161b)$$

as can easily be shown. However, if we substitute $\mathbf{W} = \bar{m}^t\mathbf{C}^{-1}$ in (6.161a) we have:

$$\begin{aligned} J_{BHAT}(\bar{m}^t\mathbf{C}^{-1}\bar{x}) &= \tfrac{1}{8}(\bar{m}^t\mathbf{C}^{-1}\bar{m})^t\left[\bar{m}^t\mathbf{C}^{-1}\mathbf{CC}^{-1}\bar{m}\right]^{-1}\bar{m}^t\mathbf{C}^{-1}\bar{m} \\ &= \tfrac{1}{8}\bar{m}^t\mathbf{C}^{-1}\bar{m} \\ &= J_{BHAT}(\bar{x}) \end{aligned} \qquad (6.161c)$$

Comparing (6.161b) with (6.161c) shows that it makes no difference whether an arbitrary $N \times N$-matrix or the $1 \times N$-matrix $\mathbf{W} = \bar{m}^t\mathbf{C}^{-1}$ is used. Hence, according to the criterion based on the Bhattacharyya distance, this matrix must be optimal.

EQUAL EXPECTATION VECTORS

If the expectation vectors are equal, $\bar{m}=\bar{0}$, the first term in (6.159) vanishes and the second term simplifies to:

$$J_{BHAT}(\mathbf{W}\vec{x}) = \tfrac{1}{2}\ln\left[\frac{|\mathbf{WC}_1\mathbf{W}^t + \mathbf{WC}_2\mathbf{W}^t|}{2^D\sqrt{|\mathbf{WC}_1\mathbf{W}^t||\mathbf{WC}_2\mathbf{W}^t|}}\right] \qquad (6.162)$$

The $D \times N$-matrix $\mathbf{W}$ that maximises the Bhattacharyya distance can be derived as follows.

The first step is to apply a whitening operation (appendix B.3.1) on $\vec{x}$ with respect to class ω_1. This is accomplished by the linear operation $\Lambda^{-1/2}\mathbf{V}^t\vec{x}$. The matrices $\mathbf{V}$ and Λ follow from factorisation of the covariance matrix: $\mathbf{C}_1 = \mathbf{V}\Lambda\mathbf{V}^t$. $\mathbf{V}$ is an orthogonal matrix consisting of the eigenvectors of $\mathbf{C}_1$. Λ is the diagonal matrix containing the corresponding eigenvalues. The process is illustrated in figure 6.41. The figure shows a 2-dimensional measurement space with samples from two classes. The covariance matrices of both classes are depicted as ellipses. Figure 6.41b shows the result of the operation $\mathbf{V}^t\vec{x}$. The operation corresponds to a rotation of the co-ordinate system such that the ellipse of class ω_1 lines up with the axes. The figure also shows the resulting covariance matrix belonging to class ω_2. The operation $\Lambda^{-1/2}$, shown in figure 6.41c corresponds to a scaling of the axes such that the ellipse of ω_1 degenerates into a circle.

The effect of the operation $\Lambda^{-1/2}\mathbf{V}^t\vec{x}$ is that the covariance matrix associated with ω_1 becomes $\mathbf{I}$ and the covariance matrix associated with ω_2 becomes $\Lambda^{-1/2}\mathbf{V}^t\mathbf{C}_2\mathbf{V}\Lambda^{-1/2}$. The Bhattacharyya distance in the transformed domain is:

$$J_{BHAT}(\Lambda^{-1/2}\mathbf{V}^t\vec{x}) = \tfrac{1}{2}\ln\left[\frac{|\mathbf{I}+\Lambda^{-1/2}\mathbf{V}^t\mathbf{C}_2\mathbf{V}\Lambda^{-1/2}|}{2^N\sqrt{|\Lambda^{-1/2}\mathbf{V}^t\mathbf{C}_2\mathbf{V}\Lambda^{-1/2}|}}\right] \qquad (6.163a)$$

The second step consists of decorrelation with respect to ω_2. Suppose that $\mathbf{U}$ and Γ are matrices containing the eigenvectors and eigenvalues of the covariance matrix $\Lambda^{-1/2}\mathbf{V}^t\mathbf{C}_2\mathbf{V}\Lambda^{-1/2}$. Then, the operation $\mathbf{U}^t\Lambda^{-1/2}\mathbf{V}^t\vec{x}$ decorrelates the covariance matrix with respect to class ω_2. The covariance matrices belonging to the classes transforms into $\mathbf{U}^t\mathbf{U} = \mathbf{I}$ and Γ, respectively. Figure 6.41d illustrates the decorrelation. Note that the covariance matrix (being white) is not affected by the orthonormal operation $\mathbf{U}^t$.

The matrix Γ is a diagonal matrix. The diagonal elements are denoted by $\gamma_i = \Gamma_{i,i}$. In the transformed domain $\mathbf{U}^t\Lambda^{-1/2}\mathbf{V}^t\vec{x}$, the Bhattacharyya distance is:

$$J_{BHAT}(\mathbf{U}^t\Lambda^{-1/2}\mathbf{V}^t\vec{x}) = \tfrac{1}{2}\ln\left[\frac{|\mathbf{I}+\Gamma|}{2^N\sqrt{|\Gamma|}}\right] = \tfrac{1}{2}\sum_{i=0}^{N-1}\ln\tfrac{1}{2}\left(\sqrt{\gamma_i}+\frac{1}{\sqrt{\gamma_i}}\right) \qquad (6.163b)$$

The merit of the expression is that it shows that we have found a transformed domain in which the contributions to the Bhattacharyya distance of any two elements are independent. The contribution of the i-th element is:

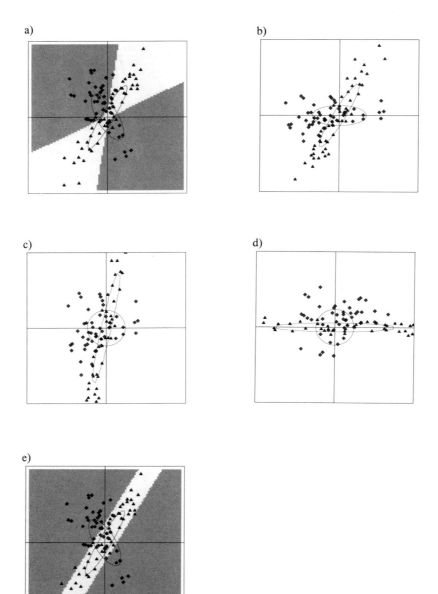

Figure 6.41 Linear feature extraction with equal expectation vectors
a) Covariance matrices with decision function
b) Decorrelation of ω_1-samples
c) Whitening of ω_1-samples
d) Decorrelation of ω_2-samples
e) Decision function based on linear feature extraction using only one feature from the space depicted in d)

$$\tfrac{1}{2}\ln\tfrac{1}{2}\left(\sqrt{\gamma_i}+\frac{1}{\sqrt{\gamma_i}}\right) \qquad (6.163c)$$

Therefore, if in the transformed domain D elements have to be selected, the selection with maximum Bhattacharyya distance is found as the set with the largest contributions. If the elements are sorted according to:

$$\sqrt{\gamma_0}+\frac{1}{\sqrt{\gamma_0}} \geq \sqrt{\gamma_1}+\frac{1}{\sqrt{\gamma_1}} \geq \cdots \geq \sqrt{\gamma_{N-1}}+\frac{1}{\sqrt{\gamma_{N-1}}} \qquad (6.164)$$

then the first D elements are the ones with optimal Bhattacharyya distance. Let $\mathbf{U}_D$ be an $N \times D$-submatrix of $\mathbf{U}$ containing the D eigenvectors of $\Lambda^{-1/2}\mathbf{V}^t\mathbf{C}_2\mathbf{V}\Lambda^{-1/2}$. The optimal linear feature extractor is:

$$\mathbf{W} = \mathbf{U}_D^t \Lambda^{-1/2} \mathbf{V}^t \qquad (6.165)$$

and the corresponding Bhattacharyya distance is:

$$J_{BHAT}(\mathbf{W}\vec{x}) = \tfrac{1}{2}\sum_{i=0}^{D-1}\ln\tfrac{1}{2}\left(\sqrt{\gamma_i}+\frac{1}{\sqrt{\gamma_i}}\right) \qquad (6.166)$$

Figure 6.41d shows the decision function following from linear feature extraction backprojected in the 2-dimensional measurement space. The linear feature extraction reduces the measurement space to a 1-dimensional feature space. Application of Bayes classification in this space is equivalent to a decision function in the measurement space defined by two linear, parallel decision boundaries. In fact, the feature extraction is conceived as a projection of the samples on a line orthogonal to these decision boundaries.

THE GENERAL CASE

If both the expectation vectors and the covariance matrices depend on the classes, an analytic solution of the optimal linear extraction problem is not feasible. A suboptimal method, i.e. a method that is likely to yield reasonable solutions without a guarantee that the optimal solution will be found, is the one that seeks features in the subspace defined by the differences in covariance matrices.

An example is given in figure 6.42. Figure 6.42a shows the covariance matrices of figure 6.41a, but this time one covariance matrix has been translated. This translation corresponds to a situation with unequal expectation vectors. The figure also shows the optimal decision function. One way to obtain a linear feature extraction is to apply the same simultaneous decorrelation technique as in figure 6.41. Application of such a technique gives a Bhattacharyya distance according to:

$$J_{BHAT}(\mathbf{U}^t\Lambda^{-1/2}\mathbf{V}^t\vec{x}) = \tfrac{1}{4}\sum_{i=0}^{N-1}\frac{d_i^2}{1+\gamma_i}+\tfrac{1}{2}\sum_{i=0}^{N-1}\ln\tfrac{1}{2}\left(\sqrt{\gamma_i}+\frac{1}{\sqrt{\gamma_i}}\right) \qquad (6.167)$$

where d_i are the elements of the transformed difference of expectation, i.e. $\vec{d} = \mathbf{U}^t\Lambda^{-1/2}\mathbf{V}^t\vec{m}$.

a) 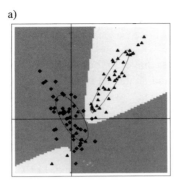 b)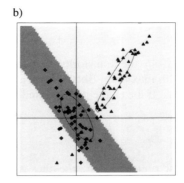

Figure 6.42 Linear feature extraction
Covariance matrices and expectation are both class-dependent.
a) Optimal decision function
b) Decision function corresponding to a linear feature extraction

Expression (6.167) shows that in the transformed space the optimal features are the ones with largest contributions, i.e. with largest

$$\frac{1}{4}\frac{d_i^2}{1+\gamma_i}+\frac{1}{2}\ln\frac{1}{2}\left(\sqrt{\gamma_i}+\frac{1}{\sqrt{\gamma_i}}\right) \tag{6.168}$$

This extraction method is suboptimal. It is based on the assumption that features that are appropriate for differences in covariance matrices are also appropriate for differences in expectation vectors. Figure 6.42 is a situation in which this assumption holds. Figure 6.42b shows the decision boundaries after application of linear feature extraction. Note that these boundaries are orthogonal to the ones in figure 6.41e.

Interclass and intraclass distance

In the multi-class case, the interclass/intraclass distance discussed in section 6.4.1 is a performance measure that may yield suitable feature extractors. The condition for this to happen is that most class information is in the differences between expectation vectors, and that the noise is more or less Gaussian distributed.

Starting point of the derivation is the expression of the performance measure given in the space defined by $\bar{y}=\Lambda^{-1/2}\mathbf{V}^t\bar{x}$. As in section 6.4.1 Λ is a diagonal matrix containing the eigenvalues of $\mathbf{S}_w$, and $\mathbf{V}$ a unitary matrix containing corresponding eigenvectors. In the transformed domain the performance measure is expressed as (6.134):

$$J_{INTER/INTRA}=\mathrm{tr}(\Lambda^{-1/2}\mathbf{V}^t\mathbf{S}_b\mathbf{V}\Lambda^{-1/2}) \tag{6.169}$$

A further simplification of the expression occurs when a second unitary transform is applied. The purpose of this transform is to "decorrelate" the between-scatter matrix. Suppose that Γ is a diagonal matrix whose diagonal elements $\gamma_i=\Gamma_{i,i}$ are the eigenvalues of the between-scatter matrix $\Lambda^{-1/2}\mathbf{V}^t\mathbf{S}_b\mathbf{V}\Lambda^{-1/2}$. Let $\mathbf{U}$ be an unitary matrix containing the eigenvectors corresponding to Γ. Then, in the domain defined by the operation:

$$\bar{y} = \mathbf{U}^t \Lambda^{-1/2} \mathbf{V}^t \bar{x} \tag{6.170}$$

the performance measure becomes:

$$J_{INTER/INTRA} = \text{tr}(\Gamma) = \sum_{i=0}^{N-1} \gamma_i \tag{6.171}$$

The performance measure itself is not affected by the operation $\bar{y} = \mathbf{U}^t \Lambda^{-1/2} \mathbf{V}^t \bar{x}$. In other words: $\text{tr}(\mathbf{S}_w^{-1} \mathbf{S}_b) = \text{tr}(\Gamma)$.

Figure 6.36d illustrates the operation given in (6.170). The operation $\mathbf{U}^t$ corresponds to a rotation of the co-ordinate system such that the between-scatter matrix lines up with the axes.

The merit of expression (6.171) is that the contribution of elements adds up independently. Therefore, in the space defined by $\bar{y} = \mathbf{U}^t \Lambda^{-1/2} \mathbf{V}^t \bar{x}$ it is easy to select the best combination of D elements. It suffices to determine the D elements from $\bar{y}$ whose eigenvalues γ_i are largest. Suppose that the eigenvalues are sorted according to $\gamma_i \geq \gamma_{i+1}$, and that the eigenvectors corresponding to the D largest eigenvalues are collected in $\mathbf{U}_D$ being a $N \times D$-submatrix of $\mathbf{U}$. Then, the linear feature extraction:

$$\mathbf{W} = \mathbf{U}_D^t \Lambda^{-1/2} \mathbf{V}^t \tag{6.172}$$

is optimal according to the interclass/intraclass distance. The performance measure associated with this feature space is:

$$J_{INTER/INTRA}(\mathbf{W}\bar{x}) = \sum_{i=0}^{D-1} \gamma_i \tag{6.173}$$

The feature space defined by $\bar{y} = \mathbf{W}\bar{x}$ can be thought of as a linear subspace of the measurement space. This subspace is spanned by the D row vectors in $\mathbf{W}$. Figure 6.43 gives two examples. In figure 6.43a the 2-dimensional measurement space from figure 6.36a is depicted. Application of (6.172) with $D=1$ yields a linear subspace shown as a thick line. Strictly speaking this subspace should pass through

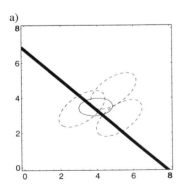
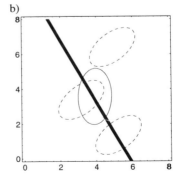

Figure 6.43 Feature extraction based on the interclass/intraclass distance
The feature space is a linear subspace (thick line) of the measurement space
a) A configuration for which linear feature extraction significantly decreases the performance
b) A configuration for which linear feature extraction is successful

the origin, but for the sake of visibility it has been moved towards the centre of the densities. The feature extraction itself can be conceived as an orthogonal projection of samples on this subspace. Therefore, decision boundaries defined in the feature space correspond to hyperplanes orthogonal to the linear subspace, i.e. planes satisfying equations of the type $\mathbf{W}\bar{x} = constant$.

Another characteristic of feature extraction based on $J_{INTER/INTRA}$ is that the dimension of any reasonable feature space does not exceed $K-1$. This follows from expression (6.132c) which shows that $\mathbf{S}_b$ is the sum of K outer products of vectors (of which one vector linearly depends on the others). Therefore, the rank of $\mathbf{S}_b$ cannot exceed $K-1$.

In the example of figure 6.43a the eigenvalues are $\gamma_0 = 0.8$ and $\gamma_1 = 0.2$. Here, feature extraction involves a significant loss of performance. This is due to the specific spatial configuration of class centres which does not allow a reduction of dimension. A counterexample is in figure 6.43b where the eigenvalues are $\gamma_0 = 3.2$ and $\gamma_1 = 0.5$, respectively. Reducing the dimension from $N=2$ to $D=1$ leads to only a small decrease of performance.

REFERENCES

Devijver, P.A. and Kittler, J.: *Pattern Recognition - A Statistical Approach*, Prentice Hall International, Englewood Cliffs NJ, 1982.

Gelb, A. (ed): *Applied Optimal Estimation*, MIT Press, Massachusetts, 1974.

Kalayeh, H.M. and Landgrebe, D.A.: *Predicting the Required Number of Training Samples*, IEEE Tr. PAMI, Vol. PAMI-5, No. 6, 664-667, November 1983.

Kreyszig, E.: *Introductory Mathematical Statistics - Principles and methods*, J. Wiley & Sons Inc, New York, 1970.

Patrick, E.A.: *Fundamentals of Pattern Recognition*, Prentice-Hall Inc, Englewood Cliffs NJ, 1972.

Rummelhart, D.E. and McClelland, J.L., *Parallel Distributed Processing: Explorations of the Microstructure of Cognition*, MIT Press, Massachusetts, 1986.

Schalkhoff, R.J.: *Pattern Recognition - Statistical, Structural and Neural Approaches*, J. Wiley & Sons Inc, New York, 1992.

Sorenson, H.W.: *Parameter Estimation: Principles and Problems*, M. Dekker, New York, 1980.

7
IMAGE ANALYSIS

Statistical pattern classification and parameter estimation are basic disciplines in measurement science. Typical in the approach is that measurements from a sensory system are organised as vectors. The connection between the object of interest on one hand and the measurement vector on the other hand is given by means of probability densities.

An image, being a collection of a number of pixels, can be looked upon as a vector. Therefore, in theory the classification of an object (or estimation of parameters of the object) could be accomplished by direct application of the theory from Chapter 6 to the image of the object. Unfortunately, there are a number of limitations:

- The Bayesian approach to classification and estimation is not appropriate in dealing with objects that are described with "non-vector-like" data structures.
- Usually, the relation between object and image is too complex to be described with conditional probability densities.
- The number of pixels in an image normally ranges between 32×32 and 1024×1024. With that, the dimension of a measurement vector would be between 10^3 and 10^6. Direct processing of such a vector - with techniques discussed in Chapter 6 - is impractical.

One way to overcome the limitations of the Bayesian approach is with *structural pattern recognition* [Pavlidis 1977]. Characteristic to this approach is the decomposition of patterns into *primitives*.

An example of structural pattern recognition is in automatic interpretation of engineering diagrams, for instance, the diagram of an electric circuit shown in the image of figure 7.1a. Such a diagram exists of a number of symbolic representations of components (resistors, capacitors, diodes, etc.) together with a network that symbolises the wiring of the corresponding electric circuit. In fact, the information conveyed by the line drawing is that of an *attributed graph*. A formal representation of such a graph - together with a visualisation - is shown in figure 7.1b and c.

A graph consists of a triple "$V, A, \psi()$" where V is a set of *vertices* or *nodes*. A is a set of *arcs* (also called *edges, link* or *branches*). The function $\psi()$ associates a pair of (unordered) vertices with each arc of the graph. In figure 7.1 arcs correspond to

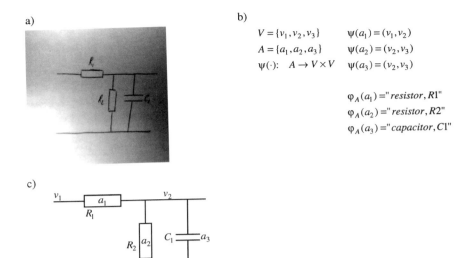

Figure 7.1 Description of an object with structured data
a) Image of a line drawing of an electric circuit
b) Graph representing the electric circuit
c) Visualisation of the graph

electrical symbols of the network. The connections between the symbols are the nodes. Together, the arcs and the nodes are the primitives with which the structure is formed. The function $\psi()$ defines the relations between nodes and arcs.

The representation "$V, A, \psi()$" is not fully complete. In fact, the structure should be extended such that it becomes clear that a_1 and a_2 are resistors and that a_3 is a capacitor. Furthermore, in many occasions, components are specified symbolically (e.g. R1) or metrically (e.g. 10Ω). The obvious way to extend the graph is with a function $\varphi_A()$ which gives the desired information. In the example, the function would be such that: $\varphi_A(a_1) =$"$resistor, \text{ R1, } 10\Omega$".

Sometimes it is desirable to add a list of properties (symbolic attributes or perhaps numerical information) to the vertices of the graph. In electric circuits the vertices are accompanied by symbols representing the voltage. We can introduce a function $\varphi_V()$, defined on V, which represents this list. As a whole, the data structure $(V, A, \psi())$, together with $\varphi_A()$ and $\varphi_V()$, is called an *attributed graph*. Figure 7.2 gives a visual representation. The vertices are shown as dots, the arcs as line segments. The information carried by the functions $\varphi_A()$ and $\varphi_V()$ is shown on labels attached to the vertices (node lists) and arcs (arc lists).

Another extension to ordinary graphs is the assignment of directions to arcs. In the example of electric diagrams this is useful if components like diodes are allowed. A *directed graph* (or *digraph*) is a graph in which the range of the function $\psi()$ is an ordered pair of vertices instead of an unordered pair. Graphs that are not directed are *unidirected graphs*.

In figure 7.1 the components of the electric diagram are considered as primitives. The structure becomes *hierarchical* if a primitive itself is structured. In an electric diagram, most symbols of the components can be built with line

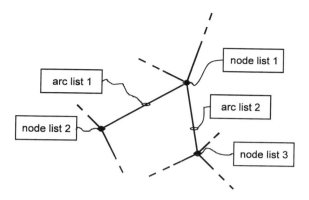

Figure 7.2 Representation of structured data with an attributed graph

segments that are either straight or smoothly curved. Corner points or T-junctions (bifurcations) form the connection between line segments. Therefore, the symbol of a component can be regarded as a graph. Corner points and T-junctions are the vertices, line segments are the arcs. Figure 7.3 is a graph representation of the symbol of a resistor. The structure consists of a few straight line segments together with two terminals. In addition, labels attached to the arcs of the structure indicate whether the length of the line segment should be long or not. A further stipulation of this description would be to add labels to the vertices giving the required angles between the line segments incident to the vertex.

Structural pattern recognition is based on the assumption that objects can efficiently be modeled with hierarchically structured data of the type discussed above. Moreover, it assumes that this structure will also be reflected in the image of the objects. With this assumption, the complex problem of the recognition of an object can be broken down into a number of subproblems defined at various hierarchical levels. It is hoped for that tackling the subproblems - one at a time - is much easier than tackling the entire problem at one go.

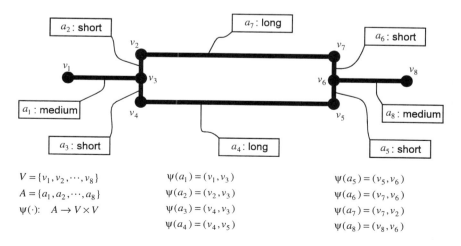

Figure 7.3 Representation of a graph describing the symbol of a resistor

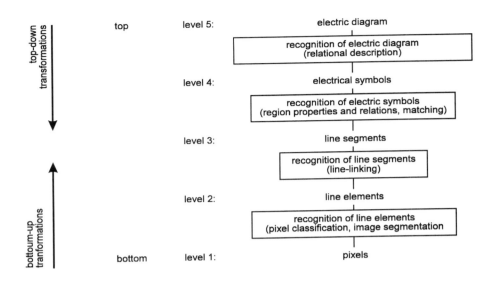

Figure 7.4 Hierarchical representation of a structural pattern recognition system for the automatic interpretation of line drawings of electric circuits

An example of a structural method to recognise diagrams of electric circuits is shown in figure 7.4. At the lowest level of the hierarchy we have the pixels of the image of the line drawing. The second level consists of a subset of these pixels called *line elements*. These elements are the primitives with which line segments are built (the third level). As discussed above, line segments are the primitives of symbols, which form the primitives of the whole diagram.

Certain actions can be undertaken in order to transform one level into another. This can be done in top-down direction like in "computer graphics" where the ultimate goal is to create a line drawing of the circuit given a symbolic network description. In measurement systems the situation is opposite. Here, assertions concerning the circuit must be established given a line drawing. Thus, transformations in bottom-up direction are required.

The recognition of a pattern at a certain level can take place with various techniques. We will briefly discuss some approaches.

Statistical pattern classification
For each primitive to recognise a set of measurements are acquired and organised as a vector. A decision function maps each vector into a class which is assigned to the primitive. Examples from figure 7.4 are:

- The transition from level 1 to level 2 consists of classifications of the pixels to classes "line element" or "no line element". It takes place by means of a measurement vector comprising the grey levels in the neighborhood of the pixel.
- At level 3, the classification of a line segment into classes "short", "medium", or "long" takes place based on a measurement of the length of the line segment.

Syntactic pattern recognition

The composition of a pattern from a set of primitives and a set of relations between primitives must satisfy certain rules. For one class of patterns, the set of rules often differ from the rules of other classes. Therefore, certain relations encountered between the primitives of an observed pattern indicate that the pattern belongs to a certain class. For instance, the syntactic rules of the symbol of a resistor are that the symbol consists of a rectangle with terminals attached to the short sides (figure 7.3). The syntactic rules of a capacitor is different: two parallel line segments with terminals perpendicular to both segments. Given a configuration of a number of line segments one can check whether one of these two syntaxes conform with the observed configuration.

On one hand the syntactic rules must be stringent enough in order to enable discrimination between different classes. On the other hand, if the rules are too stringent, the classification may suffer from a lack of robustness.

There are several ways to implement a syntactic pattern classification system. One way is to use concepts from *formal languages* (or *grammars*). In that case, classification is based on *parsing* (similar to syntax checkers in computer language compilers). Another way is to use "if-then-else" rules. For instance:

> *if* (there are two long, parallel line segments)
> *then*:   *if* (there are two terminals midway the two line segments)
>   *and* (the line segments are not connected)
>   *then*: the configuration is a capacitor
>   *else-if*: (there are two short line segments connecting the endpoints of
>   the long line segments)
>   *and* (each short line segments has one terminal)
>   *then*: the configuration is a resistor.

Semantics

Besides syntactic rules, a pattern must also satisfy the rules brought forth by the semantics of the pattern. Exploitation of this type of rules contributes to the robustness of a recognition system. For instance, in electric diagrams some configurations of components are more likely to occur than others. Some semantic rules are:

- A shortcut of a component is very unlikely.
- Two current sources are not allowed to be connected serially.
- Two voltage sources are not allowed to be connected in parallel.

Probably, the robustness of human interpretation of images is partly due to its (semantic) knowledge about the application domain.

Overview of the chapter

This chapter consists of a number of sections that describe operations with which the transformation from one level to another can take place. The levels on which these operations operate are: pixels (image segmentation), regions (region properties), and relational structures (relational description).

The last two sections discuss techniques to arrive at a final description of the object. Sad as it is, there is (still) no unifying theory that arrives at such a description. In fact, methods and methodology encountered in computer vision literature are very divergent. Probably, the reason is that the application domains and (especially) the nature of the vision tasks diverge so much. With this in mind, it becomes clear that the techniques described in this chapter are exemplary.

7.1 IMAGE SEGMENTATION

The goal of image segmentation is to process the data of an acquired image so as to arrive at a meaningful partitioning of the image plane. The compartments resulting from this process are called *regions* or *segments*. The partitioning is meaningful if a correspondence is known to exist between segments on one hand and parts of the object being imaged on the other hand. Often, the assumption behind image segmentation is that homogeneity of a certain local property of the image data (e.g. irradiance) refers to a homogeneous part of the surface of the object. See Chapter 3.

There are several methods to represent a segmented image. One possibility is to use a map of the (discretised) image plane. Each pixel position (n,m) carries a label indicating to which segment the pixel belongs. Suppose that the partitioning ends up in K different segments. The k-th segment is represented by a symbol ω_k. Then, a "region map" representation of a segmented would be: $f_{n,m} \in \{\omega_k | k = 1, \cdots, K\}$. In software implementations it is convenient to replace the symbol ω_k with the number k. With that, the representation of a segmented image is close to the representation of a grey level image. In fact, the visualisation of segmented image is often done by displaying k as a grey level.

Another representation of segmented images is one based on boundaries between segments. Here, the "edges" in an "edge map" represent the boundaries; i.e. all locations between segments. Other representations of segmented images will be introduced in subsequent sections.

Broadly speaking, the methodology of image segmentation can be divided in three categories: pixel classification, region growing and edge based methods. These categories are the topics in the next subsections.

7.1.1 Pixel classification

As the name suggests, pixel classification is an application of statistical pattern classification. Suppose that an image of an object is denoted by $f_{n,m}$. A position of a pixel is given by the indices (n,m). Our task is to assign a class to each pixel (n,m). The set of possible classes is $\Omega = \{\omega_k | k = 1, \cdots, K\}$. The pixels all from the same class form a segment.

The assignment of a class to a pixel (n,m) is guided by a measurement vector $\vec{x}_{n,m}$. The formation of such a vector can take place in different domains, see figure 7.5. In the sequel, we assume that, whatever the domain is from which measurement vectors are formed, the method to arrive at a vector is space invariant. That is, the method does not depend on the position.

If $f_{n,m}$ is a grey level image, the easiest way to obtain a measurement vector is to use $f_{n,m}$ as a 1-dimensional measurement vector $\vec{x}_{n,m} = [f_{n,m}]$. The advantage of such a measurement vector is that the assignment can be kept very simple: a monadic multi-threshold operation (table 5.1) suffices. However, with a 1-dimensional measurement vector the error rate can easily become unacceptable

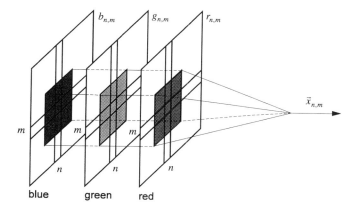

Figure 7.5 Formation of a measurement vector in the spatial and/or spectral domain

high. Extension of the measurement vector could improve the quality of the classification. One way to extend the measurement vector is to consider the spatial domain. For instance, by collecting grey levels in the neighborhood of (n,m) in the vector $\vec{x}_{n,m}$. Alternatively, one can add measurements from the spectral domain by using multiple spectral bands. Of course, the image sensor must be equipped for this.

Another domain which could be exploited is the temporal domain (motion based segmentation). This is useful if the object, the image formation system, or the illumination system is time-varying. The usage of multiple images of a static object seen from different points of view (i.e. stereoscopic vision) is closely related to the usage of time-varying image sequences. The book of Jaehne [1991] contains an introduction to image based motion analysis.

Once a measurement vector has been made available, the further design of the pixel classificator is as discussed in Chapter 6. Often, it is needed to limit the computational complexity of a classificator. The reason is that classification is performed on each pixel. Therefore, the number of calculations is proportional to the number of pixels in the image. Since the latter may be very large (e.g. on the order of 10^6), the classification itself must be kept simple. One way to reduce the complexity of the classification is by exploiting feature extraction (section 6.5.2). With that, pixel classification has the structure given in figure 7.6.

Grey level thresholding

In this subsection we consider the situation in which the image is achromatic and the measurement vector is defined in the spatial domain. Starting point is the assumption that the measurement vector is 1-dimensional: $\vec{x}_{n,m} = [f_{n,m}]$. As an example we take the segmentation of the image of a blood cell (figure 7.7a).

Figure 7.6 Computational structure of pixel classification

According to the model discussed in example 3.1 the image plane is partitioned into three regions: background, cytoplasm and nucleus. Corresponding to that, segmentation involves classification with a set of three classes. We have to determine a decision function that maps the vector $\underline{\bar{x}}_{n,m}$ onto this set.

If the conditional probabilities of the measurement vector are not available, a training set must be used. Such a set can be created interactively using so-called *regions of interest* (ROIs). In an interactive image analysis system, a user can manually define a ROI by control of a mouse whose movement is coupled to the position of a cursor which overlays the display of the image. The definition of three ROIs, each ROI completely belonging to one segment, implicitly yields a labeled training set. Figure 7.7c shows Parzen estimates of the conditional probability densities. These estimates are based on such a training set.

Once the conditional probabilities are available, the design of a classification is straightforward. Figure 7.7c shows a decision function which assigns the class with maximum likelihood. If prior probabilities are assumed to be equal, this decision function has minimum error rate. Application of the decision function to $\underline{\bar{x}}_{n,m}$ is equivalent to the application of a multi-threshold operation to the grey level $f_{n,m}$. The segmented image resulting from this operation is shown in figure 7.7e.

The overlapping zones in the conditional densities in figure 7.7c indicate that the error rate of the classification will be high. Extension of the measurement vector in the spatial domain might improve the classification. In order to explore this, assume that a $(2K+1) \times (2L+1)$-neighborhood is available for each pixel of the input image. The neighborhood of pixel (n,m) is defined in equation (5.18). The i-th element of the vector $\underline{\bar{x}}_{n,m}$ becomes:

$$\bar{x}_{n,m_i} = f_{n+k,m+l} \quad \text{where:} \quad i = (k+K)(2K+1) + (l+L) \quad \text{and} \quad \begin{cases} k = -K, \cdots, K \\ l = -L, \cdots, L \end{cases} \tag{7.1}$$

With that, the dimension of the measurement vector is $N = (2K+1)(2L+1)$.

From section 6.1.3 it is known that a very simple classification scheme results when the measurement vector is assumed to be Gaussian with class-dependent expectation vector and class-independent white noise:

$$\underline{\bar{x}}_{n,m} = \bar{\mu}_k + \underline{\bar{n}} \quad \text{with:} \quad E\{\underline{\bar{n}}\} = \bar{0} \quad \text{and} \quad E\{\underline{\bar{n}}\,\underline{\bar{n}}^t\} = \sigma_n^2 \mathbf{I} \tag{7.2}$$

In this model, $\bar{\mu}_k$ are the expectation vectors of grey levels coming from a neighborhood that entirely fits in the segment associated with ω_k. With a flat model of grey levels (see section 3.1) we assume that these expectations are equal. Hence, within the class ω_k elements of $\bar{\mu}_k$ are constant:

$$\bar{\mu}_k = c_k [1 \ 1 \ \cdots \ 1]^t \tag{7.3}$$

Substitution of (7.3) in (6.18) shows that in order to calculate the posterior probability of ω_k it suffices to calculate the variable:

$$[1 \ 1 \ \cdots \ 1] \underline{\bar{x}}_{n,m} \tag{7.4}$$

This operation is equivalent to convolution of the image $f_{n,m}$ with a PSF-matrix consisting entirely of "1". If the result of the operation in (7.4) is denoted by $y_{n,m}$, then:

a)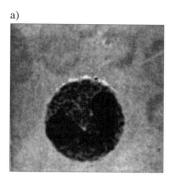

b)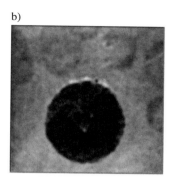

c)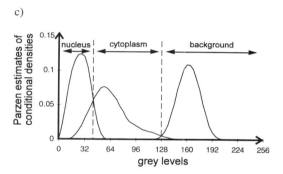

e)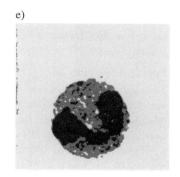

d)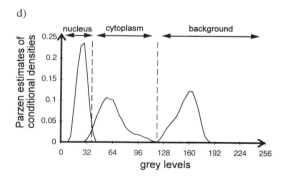

f)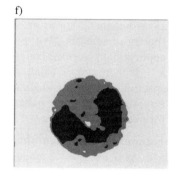

Figure 7.7 Pixel classification with multi-thresholding
a) Original image of a blood cell
b) Low-pass filtered image (5×5-averaging)
c) Parzen estimation of conditional probability densities of grey levels of original together with corresponding Bayes decision function for pixel classification
d) Ditto of low-pass filtered image
e) Original image segmented with the decision function given in c
f) Low-pass filtered image segmented with the decision function given in d

$$y_{n,m} = [1 \quad 1 \quad \cdots \quad 1]\bar{x}_{n,m} = \sum_{k=-K}^{K} \sum_{l=-L}^{L} f_{n-k,n-l} \qquad (7.5)$$

The conclusion is that with a flat model of the grey levels the averaging operator defined in (5.25a) is suitable as feature extractor. Figure 7.7b shows the original image averaged in a 5×5 neighborhood. Estimates of the conditional probability densities of $y_{n,m}$ are given in figure 7.7d. It can be seen that the overlapping zones are much smaller than in figure 7.7c. The corresponding segmented image, given in figure 7.7f, contains less classification errors than the one obtained without averaging.

The flat model, expressed in (7.2) and (7.3), is incorrect for neighborhoods which are on the boundary of a segment. As a result, the localisation of the boundary will be affected by application of the averaging operator. In fact, the precision with which the boundary is located becomes less as the size of the neighborhood increases. We can improve the ratio between classification and localisation a little by selecting a low-pass filter with a PSF that gradually decays to zero. For this reason a Gaussian low-pass filter (5.26) is often preferred above averaging.

Spectral features

This subsection discusses the use of multi-spectral images in the context of pixel classification. The surface of an image sensor receives radiant energy with wavelengths distributed according to a density function (generally referred to as *spectrum*), see sections 2.1.1, 4.1 and 5.1.2. If the sensor is achromatic its output is a single band proportional to the radiant energy weighted by the spectral responsivity function of the sensor. In color image sensors the output consists of three bands with three different responsivity functions. Usually, the three bands correspond to red, green and blue.

There is no physical reason to fix the number of bands to three. Neither it is required to restrict the responsivity function to the visible part of the electromagnetic spectrum. In some applications (satellite imagery) the number of spectral bands is five or more, including infrared and near-infrared. Other applications have two bands (dual mode X-ray imagery).

If there are N bands available, then for each pixel (n,m) we can arrange the spectral data in an N-dimensional measurement vector $\bar{x}_{n,m}$. Application of the Bayesian approach to pattern classification (Chapter 6) gives the "optimal" method for pixel classification.

Example 7.1 Segmentation of the image of a printed circuit board

As an example we consider the RGB-image of a printed circuit board shown on the color plate as figure I. The three individual bands are also shown (figure III). Suppose that our task is to find those segments that correspond to:
- electric components (i.e. capacitors and resistors)
- integrated circuits (ICs)
- metallic parts
- printed board

With that, the segmentation involves a pattern classification problem with four classes ($K = 4$) and a measurement vector with dimension $N = 3$.

Pixel classification based on an achromatic image (with spectral responsivity almost uniform over the visible part of the spectrum) is unsatisfactory. This follows

a)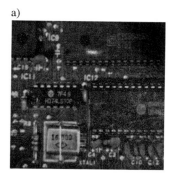

b)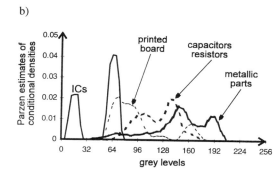

c)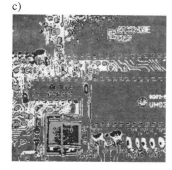

Figure 7.8 Grey level thresholding
a) Achromatic image of printed circuit board (PCB) with electronic parts mounted
b) Parzen estimates of conditional probability densities of grey levels
c) Segmented image

from figure 7.8. Estimates of the conditional probability densities of grey levels from figure 7.8a are given in figure 7.8b. The corresponding segmented image is shown in figure 7.8c. Due to overlapping zones in the conditional densities the error rate is unacceptable large. The inter/intra distance $J_{INTER/INTRA}$ (equation 6.134) is about 1.3, also indicating that the performance is poor.

An improvement is achieved by using two spectral bands. A possibility is to use the red and green channels from figure I; see figure II. If the three spectral bands in figure I are denoted by $r_{n,m}$, $g_{n,m}$ and $b_{n,m}$, the measurement vector becomes $\vec{x}_{n,m} = [r_{n,m} \; g_{n,m}]^t$. By interactive definition of ROIs we can create a training set consisting of measurement vectors whose true classes are known. A scatter diagram of this training set is depicted in figure 7.9a. The inter/intra distance $J_{INTER/INTRA}$ of the set is about 5.1 indicating that a substantial improvement is obtained. With parametric training (assumption of Gaussian densities) a decision function becomes available. Figure 7.9a shows this function. The corresponding segmented image is depicted in figure 7.9b. It can be seen that the error rate is considerably less than the one based on a single achromatic band.

A direct implementation of the decision function based on Gaussian densities requires on the order of $4K$ multiplications and additions per pixel. Although in the realisation such a computational complexity is not a difficult problem, there exists an alternative that is often computationally more efficient, and also more versatile. This implementation uses a 2-dimensional look-up table (LUT). If each spectral band is quantised to 256 levels (8 bits), the number of entries of the LUT needed to store all possible events of the measurement vector is $256 \times 256 \approx 65 \times 10^3$.

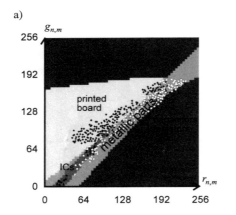

Figure 7.9 Pixel classification based on feature selection. The blue channel is left out. The red and green channels are retained. See also figure II
a) Scatter diagram and decision function
b) Segmented image

Therefore, preloading the LUT with the decision function requires about 65×10^3 classifications. After that, future classifications can be accomplished by a simple table look-up operation. A LUT can also be used as an efficient implementation of other classification schemes, including nearest neighbor rules and neural networks.

With three spectral bands (each having 256 quantisation levels) the number of entries (about 16×10^6) of a LUT is impractical. A solution would be to reduce the number of quantisation levels. With 6 bits per channel a true-color image would require a LUT with about 250×10^3 entries. An alternative is to apply feature extraction to the measurement vector in order to arrive at a dimension $D=1$ or $D=2$.

With three spectral bands the measurement vector becomes $\vec{x}_{n,m} = [r_{n,m} \; g_{n,m} \; b_{n,m}]^t$. Application of a linear feature extraction method in which the within-scatter matrix is whitened and the between-scatter matrix is decorrelated (sections 6.41. and 6.5.2) yields a transformation:

$$\vec{y}_{n,m} = \mathbf{W}\vec{x}_{n,m} \quad \text{with:} \quad \mathbf{W} = \begin{bmatrix} -0.25 & 0.15 & 0.09 \\ -0.04 & 0.18 & -0.03 \\ -0.03 & 0.18 & -0.22 \end{bmatrix}$$

The eigenvalues of the associated between-scatter matrix are $\gamma_0 = 4.5$, $\gamma_1 = 1.1$ and $\gamma_2 = 0.8$. According to (6.173) a reduction of the dimension from $N=3$ to $D=1$ or $D=2$ yields an inter/intra distance $J_{INTER/INTRA}$ around 4.5 and 5.6, respectively. Classification without feature extraction (i.e. $D=3$) would give $J_{INTER/INTRA} \approx 6.4$.

The inter/intra distances associated with the different features discussed so far are summarised in table 7.1. Feature extraction with $D=1$ gives a performance that is much better than the achromatic case. Feature selection using only the red and green channel is slightly better than feature extraction with $D=1$. Feature extraction with $D=2$ gives a further improvement. The best performance is achieved with the full measurement vector.

The concept of feature extraction is illustrated in figure IV, V and 7.10. Application of the transform $\vec{y}_{n,m} = \mathbf{W}\vec{x}_{n,m}$ gives pixels represented by 3-dimensional feature vectors. The three elements of the vectors can be mapped onto the red- green and blue-channels of a display system. Figure IV shows the result.

Table 7.1 Inter/intra distance depending on features used

Features	$J_{INTER/INTRA}$
achromatic	1.3
feature extraction from $N=3$ to $D=1$	4.5
feature selection: red and green	5.1
feature extraction from $N=3$ to $D=2$	5.6
red, green and blue ($D=3$)	6.4

Feature extraction can be regarded as feature selection in the feature space. Since the feature space is designed such that elements appear in decreasing importance, optimal feature selection is accomplished by discarding the last element(s).

When $D=2$, the third element in $\bar{y}_{n,m}$ should be discarded. A colored representation of this feature vector is depicted in figure V. Here, the two elements from $\bar{y}_{n,m}$ are mapped on the red and green channel of the display system. The scatter diagram together with a decision function are given in figure 7.10. The corresponding segmented image is also shown.

If $D=1$ is chosen, we have to discard the last two elements from $\bar{y}_{n,m}$. The resulting image is shown in figure 7.10c. Parzen estimates of the conditional probability densities together with the corresponding segmented image are depicted in figure 7.10d and e.

It can be seen that the quality of the segmented images in figure 7.8, 7.9 and 7.10 is roughly in accordance with the distance measures given in table 7.1. There is a small discrepancy in the transition from the red-green space to the 2-dimensional feature space. Table 7.1 predicts an improvement of the $J_{INTER/INTRA}$ from 5.1 to 5.6. In the segmented images this improvement can hardly be perceived. Reasons for this small disagreement may be: a) quantisation effects, b) non-Gaussian-like probability densities, c) the training set is not fully representative.

Texture features

It is difficult to give a stipulative definition of the word *texture*. A loose description would be: "a surface that consists of a large number of - more or less - similar primitives which are arranged - more or less - in a regular pattern. Examples of textured surfaces are (figure 7.11):

- brick wall
- grass
- textile
- a wire screen
- sand

The common factor in these objects is that they are made of a large number of particles. Some textures have particles that are quite similar (e.g. bricks). Other textures are made of irregular particles (e.g. grass blades). The arrangement of particles can be regular (as in a wall or a wire screen) or irregular (grass, sand). In any case the arrangement satisfies a certain syntax. The syntax is either deterministic or stochastic.

Often, the arrangement of primitives of a textured surface can also be seen in the image of that surface. With a perspective projection in which the image plane is parallel to the surface of the object a stationary second order statistics is often appropriate as a model describing the image of that texture. Let $f_{n,m}$ be the image. Then, the second order stationary statistics is given by the probability density:

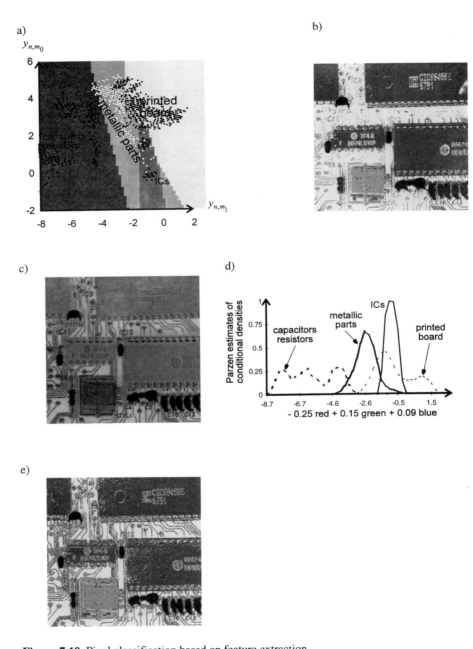

Figure 7.10 Pixel classification based on feature extraction
a) Scatter diagram and decision function of a feature vector with $D=2$. See also figure V
b) Segmented image based on feature vector with $D=2$
c) Image resulting from feature extraction with $D=1$
d) Parzen estimates of conditional densities of grey levels in c)
e) Segmented image based on feature vector with $D=1$

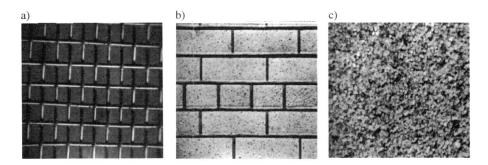

Figure 7.11. Images of textured surfaces (reproduced with permission from P. Brodatz, published by Dover Publications Inc, New York)
a) Woven aluminium wire.   b) Brick wall.   c) Beach sand.

$$p_{\underline{f},\underline{f}}(f_1,f_2;a,b) \tag{7.6}$$

If the surface is not parallel to the image plane, the perspective projection of the surface can be described with a transformation $\bar{x} = M\bar{X}$, where $\bar{X}$ and $\bar{x}$ are the homogeneous co-ordinates of the surface and the image plane, respectively. M is a concatenation of rotations, translation and perspective projection; see expression (2.7). Figure 7.12 gives an example of a textured surface (a rectangular grid of points) that has undergone such an operation. Note that although the texture on the surface is periodic, its image is not. Consequently, the second order statistics is not stationary. If the parameters of the perspective projection are known, we can apply a geometrical image transformation that brings on stationarity in the resulting image. In fact, the perspective projection of a textured surface can be used to estimate the orientation of the surface [Korsten 1989].

Statistical pattern classification is applicable in methods for segmentation of images of textured surfaces. As an example, we consider the case in which the grey levels are stationary, and in whcih the probability density is Gaussian. Furthermore, we consider only two types of textures. Image segmentation can be accomplished by means of a measurement vector $\bar{x}_{n,m}$ found in the local neighborhood of the pixel (n,m), see equation (7.1). With that, we have a 2-case classification problem with conditionally Gaussian measurement vector. Solutions to this problem have been discussed in section 6.5.3.

Figure 7.12. Perspective projection of an orthogonal grid of points

Suppose that the expectation vector and covariance matrix of $\underline{\tilde{x}}_{n,m}$ associated with class ω_k ($k=1,2$) are denoted by $\bar{\mu}_k$ and $\mathbf{C}_k$, respectively. As mentioned above, a statistical description of textures involves the second order statistics. Therefore, we must assume that differences between $\mathbf{C}_1$ and $\mathbf{C}_2$ are important. In fact, one way to describe the randomness in a textured image is by means of covariance matrices.

The answer to the question whether the expectation vectors are important depends on the type of textures being imaged. For the moment we assume that the differences between $\bar{\mu}_1$ and $\bar{\mu}_2$ are zero. This brings us to the linear feature extractor discussed in section 6.5.3:

$$\underline{\bar{y}}_{n,m} = \mathbf{W}\underline{\tilde{x}}_{n,m} \tag{7.7}$$

with the $D \times N$-matrix $\mathbf{W}$ given in expression (6.165). The resulting feature vector $\underline{\bar{y}}_{n,m}$ has conditional covariance matrices $\mathbf{I}$ and Γ, respectively. Γ is a diagonal matrix with elements γ_i. The expectation vector of $\underline{\bar{y}}_{n,m}$ is likely to be zero, because $\mathbf{W}$ suppresses components in $\underline{\tilde{x}}_{n,m}$ which do not contribute to classification performance. The log-likelihood ratio of the feature vector is (see equation 6.24):

$$\Lambda(\bar{y}_{n,m}) = \tfrac{1}{2}\left(\log|\Gamma| + \bar{y}^t_{n,m}(\Gamma^{-1} - \mathbf{I})\bar{y}_{n,m}\right) =$$
$$= constant + \frac{1}{2}\sum_{i=0}^{D-1}\left(\frac{1}{\gamma_i^2} - 1\right) y^2_{n,m_i} \tag{7.8}$$

In terms of image operations the feature extraction and log-likelihood ratio in (7.7) and (7.8) has the following interpretation. Let $\bar{w}^t_i$ be the i-th row vector in the matrix $\mathbf{W}$:

$$\mathbf{W} = \begin{bmatrix} \bar{w}^t_0 \\ \bar{w}^t_1 \\ \vdots \\ \bar{w}^t_{D-1} \end{bmatrix} \tag{7.9}$$

With that, the i-th element of the feature vector $\underline{\bar{y}}_{n,m}$ is found as:

$$y_{n,m_i} = \bar{w}^t_i \bar{x}_{n,m} \tag{7.10}$$

Comparison with expression (5.21) shows that $\bar{w}^t_i \bar{x}_{n,m}$ is equivalent to convolution, the PSF-matrix of which is:

$$\mathbf{w}_i = \begin{bmatrix} \bar{w}_{i_0} & \bar{w}_{i_{(2L+1)}} & \cdots & \bar{w}_{i_{(2L+1)2K}} \\ \bar{w}_{i_1} & & & \vdots \\ \vdots & & & \vdots \\ \bar{w}_{i_{2L}} & \cdots & \cdots & \bar{w}_{i_{(2L+1)2K+2L}} \end{bmatrix} \tag{7.11}$$

Hence, the linear feature extraction $\bar{y}_{n,m} = \mathbf{W}\bar{x}_{n,m}$ is equivalent to a parallel bank of D image filters with PSF-matrix w_i. According to (7.8) the resulting D images are squared, scaled by factors:

$$c_i = \frac{1}{2}\left(\frac{1}{\gamma_i^2} - 1\right) \tag{7.12}$$

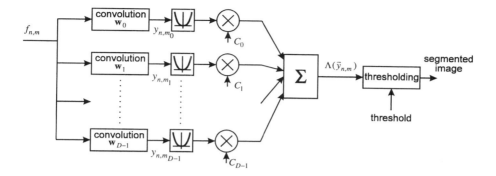

Figure 7.13 Computational structure for texture segmentation based on covariance models

and summed together to yield the log-likelihood ratio $\Lambda(\bar{y}_{n,m})$. Class ω_1 is assigned to all pixels whose log-likelihood ratio is above a certain threshold. Figure 7.13 gives an overview of all image operations involved.

The constants C_i in (7.12) depend on the eigenvalues γ_i associated with the covariance matrix belonging to ω_2 (the eigenvalues associated with ω_1 are normalised). Eigenvalues for which $\gamma_i < 1$ correspond to components of the measurement vector whose energy will be relatively large if the underlying class is ω_1. Consequently, the constant C_i is positive and the contribution of y_{n,m_i} to $\Lambda(\bar{y}_{n,m})$ is excitatory with respect to ω_1. On the other hand, if $\gamma_i > 1$, class ω_2 is dominant and C_i is negative. In that case, the contribution of y_{n,m_i} to $\Lambda(\bar{y}_{n,m})$ is inhibitory with respect to ω_1.

Example 7.2 Segmentation of images of barbed tape

Figure 7.14a is the image of two pieces of barbed tape. The left part has a regular texture. Due to weaving faults the right part is irregular. Suppose that the goal is to find the regions with irregular texture. Simply thresholding the image (figure 7.14b) does not give satisfactory results. The reason is that the expectation of the grey levels in both parts are equal.

Furthermore, application of morphological operations to the thresholded image - with the goal to improve the discrimination of the two regions - is difficult. It is not so easy to find a suitable structuring element.

a)
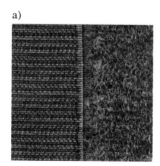

b)
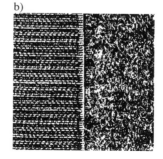

Figure 7.14 Segmentation of barbed tape
a) Image of two pieces of barbed tape
b) Segmentation by grey level thresholding

An alternative is to create a training set consisting of neighborhoods taken from both regions and to use the training set in order to arrive at a suitable feature extraction. In this example a 9×9-neighborhood suffices. Such a neighborhood is large enough to contain the primitives from which the texture is built. On the other hand, 9×9 is small enough to allow parametric training. With a 9×9-neighbourhood the dimension of the measurement vector is $N = 81$. In order to accurately estimate the covariance matrices the number of samples in the training set should be on the order of 10^3. Such a number is not prohibitive.

We can use the training set to calculate the Bhattacharyya distance J_{BHAT} as a function of the dimension D of the feature vector, see equation (6.166). Figure 7.15a shows J_{BHAT} when the features are sorted according to decreasing importance. The figure shows that with $D = 16$ the Bhattacharyya distance is greater than three indicating that with equal prior probabilities the error rate will be below 2.5%, see (6.143). Thus, if we assume that such a performance is satisfactory, the dimension of the feature vector would be $D = 16$.

Application of the design method discussed in section 6.5.2 yields a 16×81-dimensional feature matrix $\mathbf{W}$. The row vectors $\vec{w}_i^t$ of $\mathbf{W}$ are depicted in figure 7.15b as PSF-matrices $\mathbf{w}_i$. The figure also gives an impression of the constants C_i defined in (7.12). The constants are shown as either black bars (the corresponding C_i is positive), or white bars (the corresponding C_i is negative).

The feature extractor consists of two filters which are excitatory with respect to the regular texture. The PSF-matrices of these filters reflect the periodic nature (horizontal lines whose period is about 4Δ). The filters are band-pass with maximum transfer at frequencies of about $v \approx 1/4\Delta$. The phase shift between both filters is about $\pi/2$ radians. The need for two filters stems from the fact that the squared output of single band-pass filter tends to be periodic with doubled frequency. The addition of a second filter with $\pi/2$ radians phase-shift equalises the output; i.e. $\cos^2(v_1 y) + \sin^2(v_1 y) = 1$.

The remaining 14 filters shown in figure 7.15b are all excitatory with respect to the irregular texture. The PSF-matrices reflect the most important eigenvectors of the irregularities.

a)

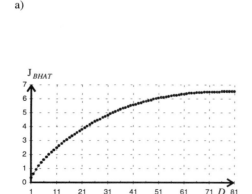

b)

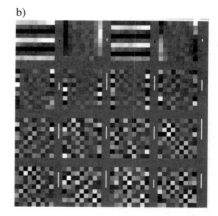

Figure 7.15 Design of a feature extraction based on covariance models
a) The found Bhattacharyya distance in dependency on the dimension of the feature space
b) The found PSF-matrices and corresponding constants C_i. The PSF-matrices are visualised by a grey level ranging from dark (negative) to light (positive). The constants are shown as vertical bars right from the PSF-matrices. Black bars indicate positive constants, white bars negative constants

With $D=2$ the Bhattacharyya distance is about 0.7 corresponding to an upper bound of the error rate of 25%. The output of the first filter in figure 7.15b (top-left) is shown in figure 7.16a. It can clearly be seen that the filter emphasises the horizontal line structures in the left part of the image. The second filter emphasises the randomness in the right part (figure 7.16b). The log-likelihood ratio corresponding to the feature extractor with $D=2$ is shown in figure 7.16c. Thresholding this image results in a segmented image (figure 7.16d). The figure shows that the second band-pass filter cannot be missed since leaving out this filter gives rise to erroneous horizontal line structures in the left part with frequency $v = 1/2\Delta$ as predicted above.

With $D=16$ the log-likelihood ratio and the corresponding thresholded image is given in figure 7.16e and 7.16f, respectively. The error rate, which is below 2.5% as predicted, can be reduced further by application of some simple morphological operations (e.g. opening and closing).

Spot detection

As stated in section 3.3, some objects in the scene may be so small that their sizes fall down the resolution of the imaging device. The image of such an object - called spot - is governed by the PSF $h(x, y)$ of the imaging device.

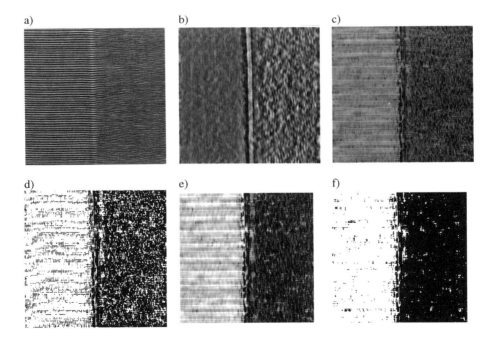

Figure 7.16 Segmentation of the image with barbed tape (figure 7.14a)
a) Output of first feature extraction filter
b) Output of second feature extraction filter
c) Log-likelihood ratio based on a) and b)
d) Segmented image based on first two feature extraction filters
e) Log-likelihood ratio based on 16 feature extraction filters
f) Segmented image based on 16 feature extraction filters

a)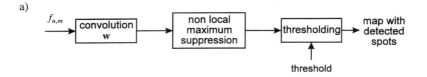

b)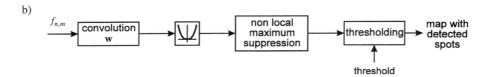

Figure 7.17 Computational structure for spot detection
a) For spots having positive contrasts only
b) For spots with positive and spots with negative contrast

The simplest model to describe the presence of spots is with expression (3.20):

$$f(x,y) = C_0 + \sum_i C_i h(x-x_i, y-y_i) + \underline{n}(x,y) \qquad (7.13)$$

C_0 is a background level and C_i are the amplitudes of the spots. Usually, both C_0 and C_i are unknown. The goal of spot detection is to determine the positions (x_i, y_i) of the spots given a discrete representation $f_{n,m}$ of the image $f(x,y)$.

The obvious way to find these positions is with the correlation techniques discussed in section 5.3. The strategy is to calculate the cross correlation between the observed image and a suitably chosen template. After application of non-local maximum suppression (page 105), positions for which the cross correlation exceeds a threshold are marked as a spot. The process can be looked upon as feature extraction followed by classification. See figure 7.17a.

Since the background level is unknown, the template should be chosen such that the influence of the background level is neutralised. Suppose that $h_{n,m} = h(n\Delta, m\Delta)$ is the discretised PSF defined on a $(2K+1)\times(2L+1)$ orthogonal grid. The following measure of match (see eq. 5.71) is appropriate:

$$g_{n,m} = \sum_{k=-K}^{K} \sum_{l=-L}^{L} f_{n+k,m+l}(h_{k,l} - \bar{h}) \qquad (7.14a)$$

where $\bar{h}$ is the average of all samples $h_{n,m}$, see equation (5.70). The operation in (7.14a) can be regarded as a convolution between $f_{n,m}$ and a PSF-matrix **w**:

$$g_{n,m} = f_{n,m} * w_{n,m} \qquad (7.14b)$$

The elements of **w** are: $w_{n,m} = h_{-n,-m} - \bar{h}$. In case of 3×3-neighborhoods, the following templates are often used:

$$\mathbf{w} = \begin{bmatrix} 0 & -1 & 0 \\ -1 & 4 & -1 \\ 0 & -1 & 0 \end{bmatrix} \quad \text{and:} \quad \mathbf{w} = \begin{bmatrix} -1 & -1 & -1 \\ -1 & 8 & -1 \\ -1 & -1 & -1 \end{bmatrix} \qquad (7.15a)$$

In effect, convolution with these kind of PSF-matrices accomplishes a discrete approximation of the Laplacian of the image, see equation (5.60). A disadvantage is the noise sensitivity of this operation. The influence of the noise can be reduced a little by using a larger neighborhood, e.g.:

$$\mathbf{w} = \begin{bmatrix} -1 & -1 & -1 & -1 & -1 \\ -1 & -1 & -1 & -1 & -1 \\ -1 & -1 & 24 & -1 & -1 \\ -1 & -1 & -1 & -1 & -1 \\ -1 & -1 & -1 & -1 & -1 \end{bmatrix} \quad \text{or} \quad \mathbf{w} = \begin{bmatrix} -1 & -1 & -1 & -1 & -1 \\ -1 & 1 & 2 & 1 & -1 \\ -1 & 2 & 4 & 2 & -1 \\ -1 & 1 & 2 & 1 & -1 \\ -1 & -1 & -1 & -1 & -1 \end{bmatrix} \qquad (7.15b)$$

The second PSF-matrix is appropriate for spots having a certain spatial extension. A noise analysis (left as an exercise to the reader) reveals that further enlargement of the neighbourhood does not significantly reduce the noise sensitivity.

In most applications the model stated in expression (7.13) is too simple. Often, in addition to the objects of interest the scene also contains objects that are not of particular interest. Instead they disturb the model. Interference takes place whenever a disturbing object is within a distance of $K\Delta$ or $L\Delta$ of the object of interest. The probability on an interference increases with the size of the neighborhood. Therefore, the size should not be chosen too large. In practical situations a 5×5-neighbourhood is often a good balance between noise reduction on one hand, and reduction of object interference on the other hand.

Another aspect in spot detection is the situation in which some spots have positive contrast ($C_i > 0$) and other spots negative contrast ($C_i < 0$). If this is the case, the feature extraction should be extended with a "square law" function or "rectifier" in order to have both types of spots been detected, see figure 7.17b. Unfortunately, the addition of such a function impairs the detection process further, because the side lob of the response to a negative spot may be mixed up with the main lob coming from a spot with positive contrast.

An illustration of spot detection is given in figure 7.18. Convolution of the image in figure 7.8a with the PSF-matrix $\mathbf{w}_2$ in (7.15b) yields the image shown in figure 7.18a. Non-local maximum suppression and thresholding lead to a map of detected spots shown in figure 7.18b. In order to compare this map with the original, the map is also shown as an overlay (figure 7.18c). In this example, many small parts of objects that one would intuitively mark as a "spot" (e.g. the leads of ICs) are detected correctly. However, the map also shows many falsely detected spots. These so-called *false alarms* stem from the image of objects that are not included in the model of expression (7.13).

Line detection

The previous subsection dealt with the detection and localisation of objects that are small with respect to the resolution of the imaging device. This subsection discusses another class of "small" objects: objects whose images have length, but no width. This occurs in situations where the width of an object falls below the spatial resolution of the imaging device, see section 3.3. Expression (3.18) is a simple model of the image describing the grey levels in the vicinity of a line element at a position (x_1, y_1) in the image plane:

$$f(x,y) = C_0 + C_1 h_{lsf}(x - x_1, y - y_1, \varphi) + \underline{n}(x,y) \qquad (7.16)$$

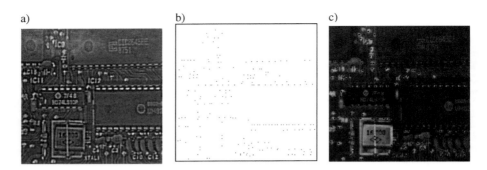

Figure 7.18 Detection of spots with positive contrast
a) Spot enhancement applied to the image shown in figure 7.8a
b) Detected spots
c) Figure 7.10c overlaid by the detected spots

C_0 is the background level. $h_{lsf}()$ is the LSF of the imaging device measured at an angle φ with respect to the x-axis. C_1 is the amplitude (magnitude). In contrast with example 3.2 on page 46, we assume that the background level is unknown. As a consequence, simply thresholding a low-pass filtered version of the input image does not solve the problem. For example, consider the image shown in figure 7.1a. The background level slowly varies due to an inhomogeneous illumination of the object. Therefore, it is not possible to find a global threshold function which detects all line elements, see figure 7.19a.

The model expressed in (7.16) does not specify rules concerning the emplacement of line elements. In contrast with spots, line elements are not isolated. In a digital representation in which the continuous image plane is replaced with the vertices of a rectangular grid, configurations of line elements are restricted by two properties: *thinness* and *continuation*. Thinness requires an absence of side neighbors across a line segment (a line has no width) and a certain amount of connectedness between line elements along the line. Loosely speaking one could state that all line elements must have two neighboring line elements, except for a few branch points (junctions, bifurcations) and a few end points (only one neighbouring line element). Continuation requires consistency in the direction of a set of neighboring line elements. The last requirement is based on the observation that the raggedness of a line is limited.

The obvious way to find the positions of all line elements seems to be with the correlation techniques from section 5.3. However, in the generalisation from spots to lines, one has to face two problems. The first complication is that the appearance of a line element is controlled by a nuisance parameter, namely its orientation φ. The second complication is the constraints enforced by the properties of thinness and continuation mentioned above.

The classical way to detect line elements is to correlate the image with a number of templates. Each template aims at the detection of a line element with a specific orientation. The templates are chosen such that - taken together - all orientations are covered. Usually the orientation is quantised to four directions: horizontal, vertical and diagonal. Known templates for these directions are:

$$\mathbf{w}_0 = \begin{bmatrix} -1 & -1 & -1 \\ 2 & 2 & 2 \\ -1 & -1 & -1 \end{bmatrix} \quad \mathbf{w}_1 = \begin{bmatrix} -1 & 2 & -1 \\ -1 & 2 & -1 \\ -1 & 2 & -1 \end{bmatrix} \quad \mathbf{w}_2 = \begin{bmatrix} -1 & -1 & 2 \\ -1 & 2 & -1 \\ 2 & -1 & -1 \end{bmatrix} \quad \mathbf{w}_3 = \begin{bmatrix} 2 & -1 & -1 \\ -1 & 2 & -1 \\ -1 & -1 & 2 \end{bmatrix} \quad (7.17)$$

For each image position, the cross correlations between the image data and the four templates are determined. The maximum of the four correlations represents the degree of match at that image position. Comparing this maximum with some well chosen threshold is the most simple method to complete the detection process. Unfortunately, such a process yields configurations of line elements which violates the constraint of thinness. A more sophisticated method is to add non-local maximum suppression schemes, applied to each of the four correlated images. In order to prevent the connectedness between line elements from being broken the suppression must be directional, i.e. matched to the orientation of the template. For instance, a template for horizontal line elements must be followed by a process that suppresses all pixels which - seen in vertical direction - are non-local maximums.

Figure 7.19b shows the result of the procedure applied to the image in figure 7.1a. The performance of the method is poor. In spite of non-local maximum suppression the figure shows "thick" line segments. These so-called *multiple responses* are the responses coming from two or more different templates that produce maximums at neighboring positions.

An improvement of the performance is achieved when the four templates are replaced with one rotationally invariant template. In fact, templates for spot detection are also suitable to enhance line elements. Of course, the non-local maximum suppression for spots should be adapted for line elements. Otherwise it would affect the connectedness between neighbouring line elements. A possibility is to use a combination of two directional suppression techniques, e.g. to suppress pixels that have a maximum neither in the horizontal nor in the vertical direction.

As an example, consider the image operation $g_{n,m} = f_{n,m} * w_{n,m}$ given in (7.14) with w one of the PSF-matrices given in (7.15). The operation enhances line elements regardless of their orientations. A non-local maximum suppression scheme suitable for line elements is to mark the pixels (n,m) for which:

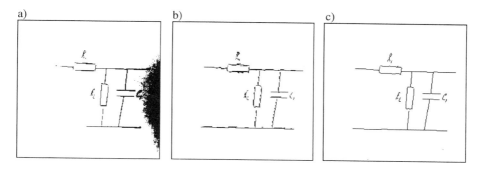

Figure 7.19 Line detection
a) Thresholding
b) Using four line templates
c) Using a rotationally invariant template

$$((g_{n,m-1} \leq g_{n,m}) \text{ and } (g_{n,m+1} < g_{n,m}))$$
$$\text{or} \qquad (7.18)$$
$$((g_{n-1,m} \leq g_{n,m}) \text{ and } (g_{n+1,m} < g_{n,m}))$$

All marked pixels for which $g_{n,m}$ is larger than a threshold are considered as line elements. Application of the procedure to the image in figure 7.1a yields the map shown in figure 7.19c.

Instead of thresholding the enhanced image $g_{n,m}$ one can equally well compare the original image $f_{n,m}$ with a space-variant threshold that depends on the average of the neighborhood of $f_{n,m}$. Such an image operation is called *adaptive thresholding*.

The defects of the operations discussed so far are essentially the same as the one of spot detection: noise sensitive, not robust against the presence of "non-line-like" disturbing objects, and multiple responses (especially if line elements with positive and negative contrasts are to be detected).

The sensitivity to noise can be reduced by increasing the size of the neighborhood of the template, see example 3.2 on page 46.

The image operation $g_{n,m} = f_{n,m} * w_{n,m}$ can be regarded as a feature extraction which transforms a measurement vector $\bar{x}_{n,m}$ into a single feature $g_{n,m}$. The robustness against disturbing objects can be improved by extracting supplementary features from the image data. Suppose that the extraction is such that some features enhances line-like structures, while other features enhance all kinds of non-line-like features. As an example, figure 7.20a gives a graphical representation of five templates (PSF-matrices) with dimension 7×7. The first two templates correspond to first order directional derivatives with diagonal orientations (see section 5.2.3). As such these templates extract the gradient of the image. The components of the gradient vector are non-line-like. The last three templates correspond to the Laplacian operator and two rotated versions of $\partial^2/\partial x \partial y$ operators. The second order derivative operators enhance line-like structures.

The robustness is achieved by subtraction of the non-line-like features from the line-like features. A computational structure suitable for that purpose is given in figure 7.21. Except for the non-local maximum suppression, the structure is identical to the one given for texture segmentation in figure 7.13. In fact, the model

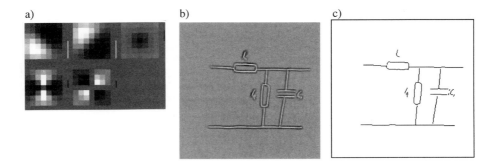

Figure 7.20 Line detection based on covariance models
a) Templates
b) Log-likelihood ratio $\Lambda_{n,m}$
c) Map with detected line elements

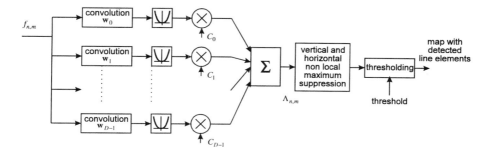

Figure 7.21. Computational structure for line detection

which was used to derive the templates in figure 7.20a was quite similar to the covariance model used to describe textures (see page 222). Van der Heijden [1991] gives a stipulation of the model.

The weight constants c_i in figure 7.21 are depicted as vertical bars in figure 7.20a. White bars denote negative constants. Note that the templates and weight constants are such that - as a whole - the operation is rotationally invariant.

Application of the computational structure to the image in figure 7.1a with the templates shown yields a "log-likelihood ratio image" as depicted in figure 7.20b. Figure 7.20c shows the corresponding map with detected line elements.

7.1.2 Region based segmentation

The region based approach to segmentation focuses its attention on areas of the image plane in which a property of the image data is homogeneous. As such, some pixel classification techniques discussed in the previous section can be regarded as region based. However, these techniques were conducted under supervision: in the design phase ROIs were interactively defined with which a labeled training set was created. With that, prior to the actual segmentation, estimations of the conditional probabilities of the image data were available.

Supervised training is only valuable in applications where similar vision tasks must be accomplished repeatedly. For instance, in the visual quality control of a production process, we may expect that successive products are similar, and that the imaging conditions do not vary much from scene to scene and from object to object.

The methods to be discussed in this section are devoid of labeled training sets. We have to find a partitioning of the image plane with only rough prior knowledge concerning the conditional probability densities of the image data. The advantage of not using too much prior knowledge is that the methods become more versatile. In complex vision tasks, the objects to be analysed vary from scene to scene, and the imaging conditions may be unpredictable. Therefore, it is hard to create a representative labeled training set.

The lack of such a labeled set transforms the segmentation problem into an unsupervised classification problem in which the image data can be regarded as an unlabeled training set. Cluster analysis applied to the training set gives a solution. As an example, consider the image in figure 7.22a that shows an object with its background. Clearly, the histogram of the grey levels in this image is bimodal (figure 7.22b). An analysis applied to this histogram could reveal that there are two clusters separated by the minimum found between the two modes of the histogram. Thresholding this image at this particular grey level yields a segmented image as shown in figure 7.22c. It can be seen that the resulting quality is low due to the overlap of grey levels of the two classes.

The lack of a training set hampers the design of a segmentation method. Fortunately, some of the missing prior knowledge concerning the *radiometry* (irradiance, color, etc.) of a region can be compensated by prior knowledge of the *geometry* (shape, size, compactness) of that region. In other words, we may facilitate the (unsupervised) cluster analysis in the radiometric domain by adding information from the spatial domain.

One way to take advantage of geometric information is by means of *region aggregation*. Here, first some pixels are selected for which enough evidence exists that they belong to a particular class. These so-called *seed pixels* are used to assign classes to neighboring pixels whose classes are not so clear. The process repeats until all pixels have been classified. In figure 7.22d, pixels with grey levels in a range in the middle between the two maximums of the histogram are shaded. The classes of the pixels in this area are ambiguous. The other pixels (either black or white) are considered as initial seed pixels the classes of which are unambiguous.

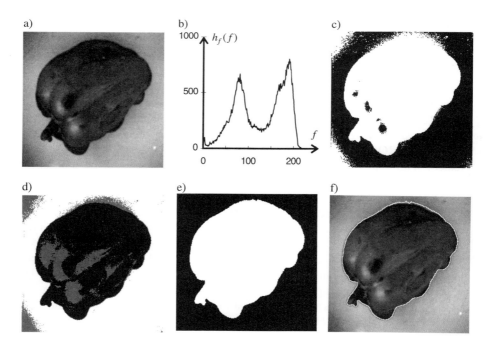

Figure 7.22 Region aggregation
a) Original
b) Histogram
c) Thresholded image
d) Initial segmentation. Black and white are seed pixels; shaded pixels are ambiguous
e) Final segmentation
f) Boundary of segmented image shown as an overlay

Repeated dilations of the seed pixels into the shaded area gives the segmented image in figure 7.22e.

Region aggregation is an example of *region growing*. This term refers to an iterative segmentation process in which neighboring regions (or parts of regions) are merged provided that these regions share some common property. The iteration proceeds until an appropriate segmentation has been found.

Another example in which region growing is applied is a segmentation technique called "*split-and-merge*". The operations in "split-and-merge" are conducted according to a data structure called *quartic picture tree* (quadtree, QPT). This structure is applicable only when the image is square ($N = M$) and N is a power of 2, i.e. $N = 2^L$. The root of the QPT corresponds to the complete image plane, the leaves to single pixels. Each node - except the leaves - has four children. The number of levels in the tree is $L+1$, see figure 7.23.

A node in the QPT represents a particular region of the image plane. For instance at level 1 we have four nodes corresponding to four square areas with $N/2 \times N/2$ pixels. At level 2 we have 16 nodes corresponding to 16 subareas; and so on. With the association of nodes with regions the QPT defines a resolution pyramid as shown in figure 7.23b. The grey level at one node is the average of the grey levels of its four children. In fact, many image operations can efficiently be implemented by exploiting this pyramidal data structure [Jolion and Rosenfeld 1994].

In its initial phase the "*split-and-merge*" segmentation process consists of split and merge operations applied to the square regions associated to the QPT. A split operation is started up whenever it is found that the homogeneity of the image data is insufficient. In that case the region is always split into the four subregions in accordance with the four children of the corresponding node in the QPT. Likewise,

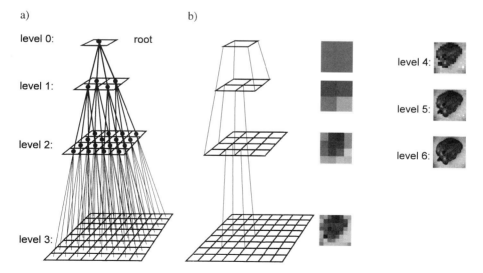

Figure 7.23 The quartic picture tree of an image consisting of 8×8 pixels
a) Each node of the QPT corresponds to a rectangular area in the image plane
b) The resolution pyramid of figure 7.22a shown up to level 6

four adjacent regions that share the same parent (a *quadruplet*) may be merged together if the homogeneity is sufficient.

A recursion of split and merge operations ends up when all regions found are homogeneous, and each quadruplet - taken together - would be inhomogeneous. At this stage, the segmented image corresponds to a *cutset* of the QPT. A cutset is a minimal set of nodes which separates the root from the leaves. The last phase of the segmentation is a grouping process in which all pairs of adjacent regions are recursively merged together whenever allowed according to the homogeneity criterion.

In order to test the homogeneity of a region a so-called *uniformity predicate* UP(A) must be available. This is a binary function that maps a region A onto the values "true" and "false". The mapping depends on the image data within that region. Strictly speaking, the function must be such that if for some region A for which $UP(A) = true$, then for any subset $B \subset A$ we have $UP(B) = true$.

If $f_{n,m}$ is a grey level image used as input image of the segmentation process, then a simple example of a uniformity predicate is :

$$UP_1(A) = true \quad \text{iff} \quad \max_{(n,m) \in A} \{f_{n,m}\} - \min_{(n,m) \in A} \{f_{n,m}\} \leq threshold \quad (7.19)$$

Other uniformity predicates will be discussed briefly at the end of this section.

Split-and-merge segmentation is an algorithm consisting of at least four steps. Detailed implementation issues of the following algorithm are given by Pavlidis [1977]:

Algorithm 7.1 Split-and-merge
  Input:
  - image data $f_{n,m}$
  - uniformity predicate
  - the initial segmentation level l
  1. *Initiation:* Choose an initial segmentation by selecting the nodes at level l in the QPT.
  2. *Tree-merge:* Evaluate the uniformity predicate for each quadruplet in the current segmentation. On assertion: merge the quadruplet by replacing the quadruplet with the corresponding parent node. Repeat recursively until no quadruplet can be found for which the uniformity predicate holds.
  3. *Split:* Visit the remaining nodes at level l. Evaluate its uniformity predicates. On rejection: split the node into four regions by replacing the node with its four children. Repeat recursively until all nodes are found uniform or until the bottom of the tree has been reached.
  4. *Adjacent region merge:* For each region in the current segmentation: find an adjacent region, evaluate the uniformity predicate for the union of the two regions, on assertion: merge the two regions. Repeat recursively until no pair of adjacent regions can be found for which the uniformity predicate holds.
  Output: a segmented image.

Figure 7.24 is an illustration of the algorithm. The figure shows the regions of the grey level image (figure 7.24a) found after each successive step. The initial level of the segmentation is $l = 6$.

The segmented image in figure 7.24e is *over-segmented*. There are more regions than actually desired: some regions are erroneously split up. One could try to circumvent over-segmentation by relaxing the uniformity predicate (i.e. to select less stringent thresholds). However, the danger exists that by doing so other regions become *under-segmented*: regions are wrongly merged.

A remedy for over-segmentation might be to repeat the last step of the split-and-merge algorithms a few times with increasing thresholds of the uniformity predicate. Small regions can be eliminated by using loose uniformity predicates for regions whose area are less than some bound. Figure 7.24f gives an example.

The merit of split-and-merge segmentation is twofold. In the first place, the use of the QPT increases the computational efficiency of the algorithm. The effectiveness can be demonstrated with the uniformity predicate given in equation (7.19). Here, the predicate of a region depends on only two parameters which are global with respect to the region: the minimum and the maximum grey level within the region. The globalness of the parameters implies that merging two regions is a cheap operation. For instance, if two adjacent regions A_1 and A_2 with minimums min_1, min_2 and maximums max_1, max_2, respectively, must be merged, then the parameters of the new region $A_1 \cup A_2$ becomes: $max_{1 \cup 2} = \max\{max_1, max_2\}$ and $min_{1 \cup 2} = \min\{min_1, min_2\}$. Thus, the uniformity predicate in (7.19) can be evaluated without scanning the image data. The only occasions that scanning is needed is during initiation (in order to calculate the parameters of each initial region), and after a split operation (in order to calculate the parameters of the four children).

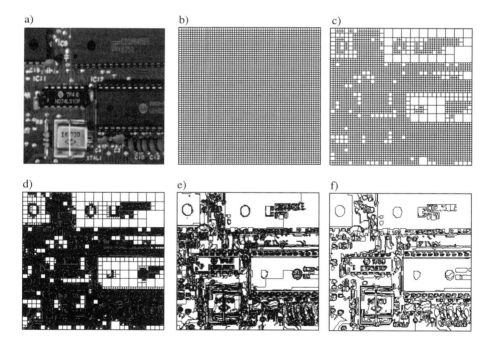

Figure 7.24 Split-and-merge segmentation
a) Input image
b) Initial segmentation consisting of 4096 square regions
c) Tree-merge reduces the number of regions to 2983
d) Split increases the number of regions to 37354
e) Adjacent region merge: 7805 regions
f) Further improvement obtained by slowly weakening the uniformity predicate: 1245 regions

Another merit of split-and-merge is its versatility. Since the homogeneity criterion depends directly on the uniformity predicate used, other criterions are implemented simply be replacing the predicate with another. With a modular implementation of the algorithm, the replacement of a predicate requires only the rewriting of a few lines of the code.

An interesting class of predicates follows from the image models given in section 3.1. The deterministic part of the image within a segment is given in a parametric form, the parameters of which depend on the particular segment. For instance, in a flat model with additive noise the image data is (see equations 3.1 and 3.3):

$$c_k + \underline{n}_{n,m} \tag{7.20}$$

where c_k is an unknown constant and $\underline{n}_{n,m}$ white Gaussian noise with variance σ_n^2. More general, let $h(j)_{n,m}$, $j = 0, \cdots, J-1$ be a set of J linearly independent functions, then an extension of the flat model is:

$$\sum_{j=0}^{J-1} c_{k,j} h(j)_{n,m} + \underline{n}_{n,m} \tag{7.21}$$

In a flat model, $J = 1$ and $h(0)_{n,m} = 1$. In a sloped model $J = 3$, $h(0)_{n,m} = 1$, $h(1)_{n,m} = n$ and $h(2)_{n,m} = m$. The predicate UP(A) is true, if a norm of the residuals:

$$\|\varepsilon_{n,m}\| = \left\| f_{n,m} - \sum_{j=0}^{J-1} c_{k,j} h(j)_{n,m} + \underline{n}_{n,m} \right\| \tag{7.22}$$

is less than a threshold. The parameters $c_{k,j}$ are chosen so as to minimise the norm. Often, the Euclidean norm is used, implying that the "optimal" parameters follow from least squares estimation. Since the model is linear, the LSE-solution given in equation (6.68) is valid.

Another way to define a uniformity predicate based on the linear models given in (7.21) is with *hypothesis testing* [Kreyszig 1970]. As an example, consider the flat model in (7.20). Suppose that two adjacent regions A_1 and A_2 are found. The sample mean $\hat{\mu}_1$ and sample variance $\hat{\sigma}_1^2$ are defined as:

$$\hat{\mu}_1 = \frac{1}{N_1} \sum_{n,m \in A_1} f_{n,m}$$

$$\hat{\sigma}_1^2 = \frac{1}{N_1 - 1} \sum_{n,m \in A_1} (f_{n,m} - \hat{\mu}_1)^2 \tag{7.23a}$$

where N_1 is the number of pixels in region A_1. Likewise definitions hold for region A_2.

If the variance σ_n^2 of the noise is known, the hypothesis that the two regions share the same expectation c_k can be tested by checking whether:

$$|\hat{\mu}_1 - \hat{\mu}_2| \leq threshold \times \sigma_n \sqrt{\frac{1}{N_1} + \frac{1}{N_2}} \tag{7.23b}$$

Note that if the two regions share the same c_k, the test variable $\hat{\mu}_1 - \hat{\mu}_2$ is Gaussian with zero expectation. The variable *threshold* defines the significance level of the test. If the variance σ_n^2 is unknown, the test in (7.23c) must be replaced with Student's t test:

$$-threshold \leq \frac{\hat{\mu}_1 - \hat{\mu}_2}{\sqrt{\left(\frac{N_1+N_2}{N_1 N_2 (N_1+N_2-2)}\right)\left((N_1-1)\hat{\sigma}_1^2 + (N_2-1)\hat{\sigma}_2^2\right)}} \leq threshold \qquad (7.23c)$$

The test variable in this expression has a t-distribution with $N_1 + N_2 - 2$ degrees of freedom.

The uniformity predicate is true if (and only if) the hypothesis that the two regions share the same c_k is accepted. In that case, the two regions are allowed to merge. Again, the update of the parameters can be done without scanning the image data:

$$\hat{\mu}_{1 \cup 2} = \frac{1}{N_1+N_2}(N_1 \hat{\mu}_1 + N_2 \hat{\mu}_2)$$
$$\hat{\sigma}_{1 \cup 2}^2 = \frac{1}{N_1+N_2-1}\left((N_1-1)\hat{\sigma}_1^2 + N_1 \hat{\mu}_1^2 + (N_2-1)\hat{\sigma}_2^2 + N_2 \hat{\mu}_2^2 - (N_1+N_2)\hat{\mu}_{1 \cup 2}^2\right) \qquad (7.23d)$$

The split-and-merge algorithm is also suitable as a tool to segment images of textured objects. In these cases the uniformity predicate must be defined such that it tests the homogeneity of parameters characterising the texture within the region. Some of these so-called *texture measures* will be discussed in section 7.2.1.

If we drop the restriction on a uniformity predicate that it should only depend on the grey levels of the region, we can also include shape features as a directive to conditionally split and/or merge regions. Figure 7.25a is an image containing nuts. Some of them are physically connected. An initial segmentation based on grey levels gives rise to some under-segmented regions (two nuts are seen as one region) and over-segmented regions (two regions correspond to one nut), see figure 7.25b. The segmentation used in this example is based on the predicate defined by (7.23c). The over-segmentation is due to the shading of the surfaces of the nuts.

In the example from figure 7.25b, region properties like: "small", "thin", "elongated", "not-compact" indicate that a region is due to over-segmentation. The *compactness* of a region is often defined as the ratio of area to the square of the perimeter. Under-segmented regions are characterised by a *constriction* of the boundary. Such a constriction can be defined as the pair of boundary points whose minimum arc length is larger than a threshold, and whose width is smallest amongst all other pair of points, see figure 7.25c.

Shape based segmentation can be implemented with a set of rules that are applied iteratively. For instance, in figure 7.25 the set of rules could be:

- *if* (a region contains a constriction and the width is smaller than a threshold)
  then: split the region at the line passing through its constriction points

- *if* (the area of a region is smaller than a threshold)
      and (the compactness of the region is smaller than a threshold)
  then: merge the region to its "closest" adjacent region

Application of these rules to figure 7.25b yields the segmented image shown in figure 7.5d.

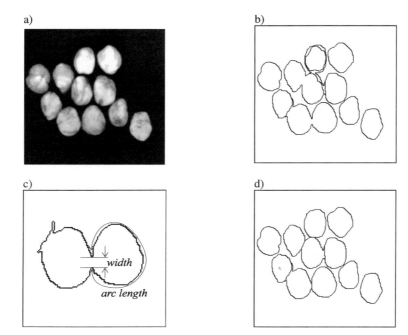

Figure 7.25 Shape based segmentation
a) Original
b) Segmentation based on grey levels
d) Detection of a constriction
c) Further improvement by exploiting shape

In a further development of a *rule based* segmentation method there is no reason to restrict ourselves to shape features only. Other region properties, as well as relations between regions, can be exploited. These properties and relations will be discussed in next sections.

7.1.3 Edge based segmentation

Another approach to image segmentation is to locate points in the image plane where the image data is discontinuous. As such this approach is supplementary to region based methods. Edge based methods are grounded on the assumption that physical 3-dimensional edges in the scene, such as object boundaries, shadow boundaries, abrupt changes in surface orientation or material properties, are clues for the characterisation of the scene. The 3-dimensional edges are often seen in the image data as discontinuities of certain local image properties: irradiance, chromaticity values, texture, etc..

In a broad sense the term *edge detection* refers to the detection and localisation of discontinuities of these image properties. In a more restrictive sense, it only refers to locations of significant change of irradiance. Points of these locations are called *edges* or *edge elements*.

As an example, consider the image shown in figure 3.6. Here, the object boundaries and edges of the cube are seen as step functions in the image data

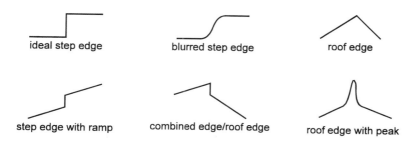

Figure 7.26 Examples of different types of image discontinuities

(blurred by the PSF of the imaging device and contaminated by noise). The step function is not the only type of discontinuity which may appear. Other types are the so-called *roof edge*, the *step edge with ramp*, and combinations, see figure 7.26.

The physical connotations of the different types of discontinuities are diverse. Step edges are often caused by object boundaries and occlusion. A roof edge, eventually combined with a step edge, often corresponds to a sudden change of orientation of the surface of an object. Specular reflection may cause a peak added to the roof edge. Shadow boundaries in the scene are often seen as step edges with a ramp.

Restricting the discussion to step-like image discontinuities, edges are points in the image plane which correspond to physical 3-dimensional edges, the image of which is characterised by a change in irradiance given by the ESF of the imaging device (see equation 3.17).

The classical approach to find edges is essentially the same as in the enhancement/threshold technique in spot and line detection. Table 7.2 presents some known operators to extract edge features in the image. Each operator consists of two or more templates with which the input image is convolved. For each pixel position the resulting outputs can be arranged in feature vectors $\vec{y}_{n,m}$ with elements:

$$y_{n,m_i} = f_{n,m} * w_{n,m_i} \qquad i = 0,1,\cdots \tag{7.24}$$

Thresholding the norm $\|\vec{y}_{n,m}\|$ gives the final map of detected edges. Usually, the Euclidean l_2-norm is used. Sometimes the l_1-norm or the l_∞-norm (see appendix A.1) is preferred in order to save computational effort.

The performance of the classical operators is rather poor when the image is degraded by some noise. The classification of one pixel is done independently from the classification of neighboring pixels. As a result, noise will fragment the found configuration of edge elements, making it difficult to recover the boundaries of the imaged objects. One way to decrease the noise sensitivity is to enlarge the neighborhood size of the templates in order to suppress high-frequency components. For instance, a generalisation of the Prewitt templates to a 5×5-neighborhood would give:

$$\begin{bmatrix} -1 & -1 & 0 & 1 & 1 \\ -1 & -1 & 0 & 1 & 1 \\ -1 & -1 & 0 & 1 & 1 \\ -1 & -1 & 0 & 1 & 1 \\ -1 & -1 & 0 & 1 & 1 \end{bmatrix} \quad \text{and} \quad \begin{bmatrix} 1 & 1 & 1 & 1 & 1 \\ 1 & 1 & 1 & 1 & 1 \\ 0 & 0 & 0 & 0 & 0 \\ -1 & -1 & -1 & -1 & -1 \\ -1 & -1 & -1 & -1 & -1 \end{bmatrix}$$

Unfortunately, low-pass filtering will blur the image. As a result the edge configurations tend to be several pixels wide. This makes it more difficult to localise the true boundary and to resolve detailed structures.

Canny operators
The edge enhancement operators discussed so far have been designed without explicitly taking into account the various attributes that qualifies the process. These attributes are as follows:

- *Good detection*: the operation should have a low probability of missing an edge. At the same time it should have a low probability of falsely marking non-edges.
- *Good localisation*: the operation should locate an edge at the correct position.
- *Immunity to interfering objects*: The detection of an edge should not be affected by the presence of edge structures coming from neighbouring objects.

One of the first edge operators with a methodic design are the *Canny operators*. The design is based on a 1-dimensional model of the signal. Canny [1986] formulates the detection problem as follows: An observed signal $f(x)$ has the form:

$$f(x) = \varepsilon \cdot U(x - x_0) + \underline{n}(x) \tag{7.25}$$

where $U(x)$ is a unit step function; x_0 is the position of the step; ε is a discrete variable taking values of either 0 or 1; and $\underline{n}(x)$ is white Gaussian noise with power spectrum $S_{nn}(u) = 1$, see figure 7.27a. The variable ε models the presence ($\varepsilon = 1$) or the absence ($\varepsilon = 0$) of the edge. In order to detect and to estimate the position of the step the observed signal is convolved with an impulse response function $w(x)$. The positions of local maximums of the convolved signal are marked as edges whenever these maximums exceed a threshold, see figure 7.27b.

Table 7.2 Templates for edge enhancement

Operator	Templates			
Roberts	$\begin{bmatrix} 0 & -1 \\ 1 & 0 \end{bmatrix}$	$\begin{bmatrix} -1 & 0 \\ 0 & 1 \end{bmatrix}$		
Prewitt	$\begin{bmatrix} 1 & 0 & -1 \\ 1 & 0 & -1 \\ 1 & 0 & -1 \end{bmatrix}$	$\begin{bmatrix} 1 & 1 & 1 \\ 0 & 0 & 0 \\ -1 & -1 & -1 \end{bmatrix}$		
Sobel	$\begin{bmatrix} 1 & 0 & -1 \\ 2 & 0 & -2 \\ 1 & 0 & -1 \end{bmatrix}$	$\begin{bmatrix} 1 & 2 & 1 \\ 0 & 0 & 0 \\ -1 & -2 & -1 \end{bmatrix}$		
Prewitt compass templates	$\begin{bmatrix} 1 & 1 & -1 \\ 1 & -2 & -1 \\ 1 & 1 & -1 \end{bmatrix}$ $\begin{bmatrix} -1 & 1 & 1 \\ -1 & -2 & 1 \\ -1 & 1 & 1 \end{bmatrix}$	$\begin{bmatrix} 1 & 1 & 1 \\ 1 & -2 & -1 \\ 1 & -1 & -1 \end{bmatrix}$ $\begin{bmatrix} -1 & -1 & 1 \\ -1 & -2 & 1 \\ 1 & 1 & 1 \end{bmatrix}$	$\begin{bmatrix} 1 & 1 & 1 \\ 1 & -2 & 1 \\ -1 & -1 & -1 \end{bmatrix}$ $\begin{bmatrix} -1 & -1 & -1 \\ 1 & -2 & 1 \\ 1 & 1 & 1 \end{bmatrix}$	$\begin{bmatrix} 1 & 1 & 1 \\ -1 & -2 & 1 \\ -1 & -1 & 1 \end{bmatrix}$ $\begin{bmatrix} 1 & -1 & -1 \\ 1 & -2 & -1 \\ 1 & 1 & 1 \end{bmatrix}$

a)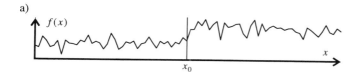

b)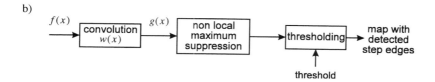

c) d)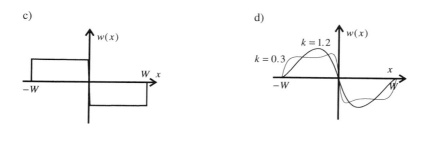

e)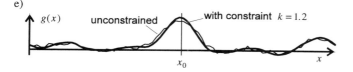

Figure 7.27 Canny operator for the detection of step functions in 1-dimensional signals
a) Observed signal
b) Computational structure of the operator
c) Optimal impulse response function without multiple response constraint
d) Some optimal response with multiple response constraint
e) Resulting output of filter

Within the given computational structure the problem is to find the "optimal" impulse response $w(x)$. The third attribute "immunity to interfering objects" is taken into account (to a certain extent) by bounding the spatial support (width) of the impulse response to an interval $-W < x < W$: the convolution has a so-called *finite impulse response* (FIR).

Good detection is expressed mathematically in terms of the signal-to-noise ratio SNR. This ratio is defined as the quotient between the response on the step (measured at $x = x_0$) and the standard deviation of the noise (see also example 3.2):

$$SNR = \frac{\int_{x=-\infty}^{0} w(x)dx}{\sqrt{\int_{x=-W}^{W} w^2(x)dx}} \qquad (7.26a)$$

The probabilities of missing an edge and having a false alarm are both monotonically decreasing functions of SNR. Hence, maximisation of SNR improves the quality of the detection.

Good localisation is defined with the standard deviation $\sigma_{\Delta x}$ of the localisation error Δx. Edges are marked at the maximum of the filter output. Equivalently, one could mark the zero crossings of the first derivative of the output. In a first order Taylor series approximation, the shift of a zero crossing due to noise is found as the noise amplitude divided by the slope of the first derivative. Therefore:

$$\sigma_{\Delta x} \approx \frac{\sqrt{\int_{x=-\infty}^{\infty} w_x^2(x)dx}}{|w_x(0)|} \qquad (7.26b)$$

where $w_x(x)$ is the first derivative of $w(x)$. Localisation is defined as the reciprocal of $\sigma_{\Delta x}$:

$$LOC = 1/\sigma_{\Delta x} \qquad (7.26c)$$

In order to have both good detection and good localisation $w(x)$ must be selected such that SNR and LOC are simultaneously maximised. Using Schwarz inequality for integrals Canny manages to prove that the filter with optimal detection and localisation ability has an impulse response as shown in figure 7.27c (a so-called *boxcar function*). Note that - seen in one direction - the Prewitt templates in table 7.2 are also boxcar functions.

Figure 7.27e gives the input signal convolved with the boxcar. The figure shows that due to noise the number of local maximums nearby the true edge position is not restricted to one. The extra local maximums lead to spuriously detected edges nearby the true edge. These *multiple responses* are eliminated if the convolution is designed in such a way that the mean distance $\bar{x}_{max}$ between neighboring local maximums is larger than a minimum interval. The mean distance can be expressed as:

$$\bar{x}_{max} = 2\pi \sqrt{\frac{\int_{x=-W}^{W} w_x^2(x)dx}{\int_{x=-W}^{W} w_{xx}^2(x)dx}} \qquad (7.26d)$$

Canny uses this expression by stating that the mean distance between local maximums must be set to some fraction k of the operator width:

$$\bar{x}_{max} = kW \qquad (7.26e)$$

The impulse response $w(x)$, Canny is looking for, simultaneously optimises SNR and LOC, but is constrained to have $\bar{x}_{max}$ set to kW. Figure 7.27d shows the solution for two particular choices of k.

Canny has pointed out that a reasonable choice for k is about 1.2. For this parameter setting the impulse response is reasonably approximated by the first derivative of a Gaussian. In many implementations this approximation is used.

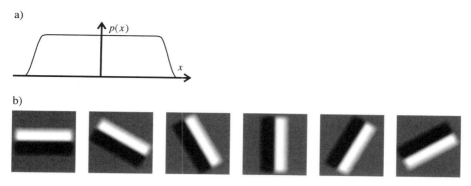

Figure 7.28 Canny operator in two dimensions
a) Projection function
b) Impulse responses corresponding to six directional operators for step edges

In the generalisation from one to two dimensions Canny uses an edge model similar to equation (3.17), i.e. the image data in the surrounding of an edge element is given by the ESF with the orientation φ aligned along the direction of the boundary. With fixed orientation, an edge can be detected and localised by application of a convolution, the PSF-matrix of which is such that:

- parallel to the edge direction image data is smoothed in order to reduce the noise
- orthogonal to the edge direction the 1-dimensional detection filter $w(x)$ is applied.

If at a certain position the output of the convolved image - scanned orthogonal to the edge direction - is a local maximum, and if the output at that position exceeds a threshold, then the pixel is marked as an edge.

The smoothing along the edge is accomplished by a FIR filter, the response of which is called the *projection function*. This function must be flat in the middle part and must smoothly decay to zero at each end of the smoothing interval. Figure 7.28a gives an example of a projection function. Graphical representations of the PSF-matrix of the directional operator are depicted in figure 7.28b.

In practice, the orientation of a hypothesised edge is not known beforehand. One way to overcome this problem is to use a number of directional operators. Each operator covers a part of the interval $[0, 2\pi]$ of edge orientations. The complete set of operators should span the whole range. At each position the operator whose output is maximum amongst all others is selected. The output of that operator is non-maximum suppressed in the direction normal to its main direction. Thresholding completes the process.

The variance of the noise at the output of an operator is inversely proportional to the width of the projection function. From that point of view a broad projection function is desirable. On the other hand, if the projection function is chosen too broad, the neighborhood of the operator exceeds the area in which the edge model is valid. Moreover, choosing a large neighborhood also increases the probability of interference of neighboring boundaries and irregularities of the boundary (e.g. corners, branches). It is obvious that the optimal size of the neighborhood depends on local properties of the image. The Canny edge detector, in its most advanced form, utilises multi-resolutions. The scale on which the operator works is locally adapted to the image data.

In the simplest form of a Canny's edge detector the projection function is a 1-dimensional Gaussian (with standard deviation σ_p) and the detection filter $w(x)$ is approximated by the first derivative of a 1-dimensional Gaussian (also with standard deviation σ_p). In that case, the resulting operator is equivalent to a first derivative operator combined with 2-dimensional Gaussian low-pass filtering, see equation (5.65). For instance, the PSFs of the operator for edges in vertical and in horizontal directions become:

$$\frac{\partial}{\partial x}\left(\frac{1}{2\pi\sigma_p^2}\exp\left(-\frac{x^2+y^2}{2\sigma_p^2}\right)\right) \quad \text{and} \quad \frac{\partial}{\partial y}\left(\frac{1}{2\pi\sigma_p^2}\exp\left(-\frac{x^2+y^2}{2\sigma_p^2}\right)\right) \tag{7.27}$$

respectively. The standard deviation σ_p determines the scale on which the operator is working.

It is interesting to note that with a symmetric Gaussian low-pass filter the whole operation can be made rotationally invariant with only two directional operators. The first derivative in any direction can be determined exactly from a linear combination of the derivatives in two directions, see equation (5.53). The first order directional derivative in the direction α takes its maximum when α is the direction of the gradient.

$$\operatorname*{argmax}_{\alpha}\{\dot{f}_\alpha(x,y)\} = \angle(\vec{\nabla}f(x,y)) \quad \text{and} \quad \max_{\alpha}\{\dot{f}_\alpha(x,y)\} = \|\vec{\nabla}f(x,y)\| \tag{7.28}$$

Hence, non-local maximum suppression of the first order directional derivative must take place in the direction of the gradient. The suppression can be implemented equally well by zero crossing detection of the second directional derivative also taken in the direction of the gradient. From differential geometry it is known that these zero crossings coincide with the locations for which:

$$f_x^2(x,y)f_{xx}(x,y) + 2f_{xy}(x,y)f_x(x,y)f_y(x,y) + f_{yy}(x,y)f_y^2(x,y) = 0 \tag{7.29}$$

The convolution masks given in equation (5.65) are the obvious means to calculate the first and higher order derivatives. Figure 7.29 gives a graphical representation of an implementation of a Canny operator based on equation (7.29). In digital form, the continuous PSFs are replaced with sampled versions. Listing 7.1 gives a possible realisation of zero crossing detection.

In its simplest form, the Canny operator is similar to the Marr-Hildreth operator discussed in section 5.2. Both operators use derivatives that are estimated using 2-dimensional Gaussians. The main difference is in the definition of the zero crossings. Marr and Hildreth prefer to use the Laplacian:

$$f_{xx}(x,y) + f_{yy}(x,y) = 0 \tag{7.30}$$

The most important reason for this preference is that it saves some computational effort. Unfortunately, a noise analysis (left as an exercise to the reader) reveals that the savings is at the cost of the localisation ability. Furthermore, the behavior of the zero crossings of the Laplacian near corners and junctions differs from that of the directional derivative. For instance, the former swing more widely around corners.

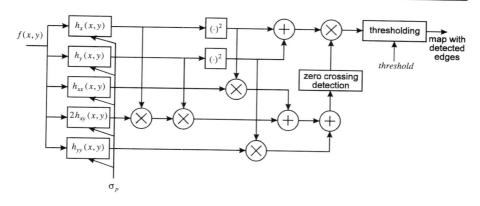

Figure 7.29 Computational structure of Canny's edge detector in two dimensions
The projection function and the edge detection filter are approximated by a 1-dimensional Gaussian and its first derivative, respectively

Example 7.3 Quality control in food industry: inspection of nuts

Figure 7.30a is the image of some samples taken from the raw input stream of a roastinghouse. Some of these nuts are impure (figure 7.30b). In order to assure the quality of the end product almost the first stage of the production process consists of the detection and removal of impurities. The purity of a nut depends on many physical features: size, mass density, gross color, shape, spotlessness, etc. Some features (e.g. mass density) are most easily measured with mechanical systems. Vision techniques can be exploited to measure most other features.

For that purpose a carefully designed image acquisition system is needed. The step following image acquisition is segmentation. Figure 7.30c, d and e are results of the Canny operation based on Gaussian derivatives working at various scales. Most striking in these edge maps is that boundaries of the segments are fragmented. The map does not define regions unless the threshold is selected such that all zero crossings are marked as edge. However, lowering the threshold increases the

Listing 7.1 Zero crossing detection

```
zero_x(f,g,N,M)               /* zero crossings in f are marked in g as "1";
                                 remaining pixels as "0".                      */
double **f;                   /* pointer to input image f, declared as a
                                 2-dimensional array of doubles.               */
int **g;                      /* pointer to output image g, declared as a
                                 2-dimensional array of integers.              */
int N,M;                      /* number of rows and columns                    */
{
  int i,j;                    /* counters                                      */

  for (i=0;i<N;i++)
  for (j=0;j<M;j++) g[n][m]=0;              /* reset output image              */
  for (i=0;i<N;i++)
  for (j=1;j<M-1;j++)                       /* scan row by row:                */
  {  if ((f[n][m]==0) && (f[n][m-1]*f[n][m+1]<0)) g[n][m]=1;
     if (f[n][m]*f[n][m-1]<0)
       if (fabs(f[n][m])<fabs(f[n][m-1])) g[n][m]=1;
       else g[n][m-1]=1;
  }
  for (i=1;i<N-1;i++)
  for (j=0;j<M;j++)                         /* scan column by column:          */
  {  if ((f[n][m]==0) && (f[n-1][m]*f[n+1][m]<0)) g[n][m]=1;
     if (f[n][m]*f[n-1][m]<0)
       if (fabs(f[n][m])<fabs(f[n-1][m])) g[n][m]=1;
       else g[n-1][m]=1;
  }
}
```

number of false alarms. Closed boundaries are only guaranteed at the cost of many spurious regions (figure 7.30f).

With increasing scale the number of spurious regions becomes less. However, the homotopy of the non-spurious regions is also effected. Furthermore, at a large scale the gross shape of regions is retained, but details are lost.

Figure 7.30h is a map of detected edges resulting from the Marr-Hildreth operation. The zero crossings are given in figure 7.30g. The map is to be compared with figure 7.30d (both maps are on the same scale). It appears that with respect to localisation the Canny operator is slightly better. There is no appreciable difference between the detection qualities.

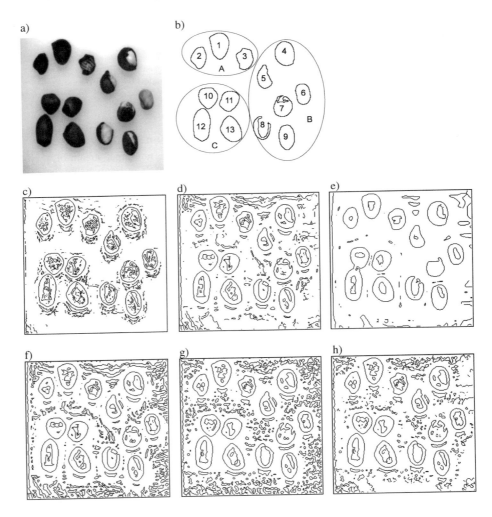

Figure 7.30 Food inspection
a) Nuts
b) Some types of impurities. A: empty, fragmented shells; B: burst shells; C: healthy nuts.
c-e) Results of Canny operation on different scales ($\sigma_p = \Delta$, $\sigma_p = 2\Delta$ and $\sigma_p = 4\Delta$).
f) Zero crossings of the directional derivative at scale $\sigma_p = 2\Delta$.
g) Zero crossings of the Laplacian at scale $\sigma_p = 2\Delta$
h) Result of Marr-Hildreth operator ($\sigma_p = 2\Delta$).

Edge linking

As noted above the process of edge detection often yields fragmented boundaries. As such, the edge map does not fully specify the geometry of the regions. Edge linking is the process that aims to complete the boundary, and to delete spuriously detected edge elements from the edge map.

A common factor in the various approaches to edge linking is the assumption that configurations of edge elements must possess the properties of thinness and continuation. Gaps are filled so as to find configurations that comply with these properties.

MORPHOLOGICAL OPERATIONS

If the fragmentation of the boundary is not too severe, some of the morphological operations from section 5.4 are appropriate to fill small gaps and to delete spurious edges and tiny spurs. As an example, consider the map of line elements shown in figure 7.19c. Here, the connectedness near some junctions is broken. Closing the set of found edge elements by a disk-like structuring element fills small gaps (figure 7.31a). Thinning the resulting set with the structuring elements given in figure 5.36c produces an 8-connected skeleton (figure 7.31b). Finally, application of the pruning operation removes tiny spurs (figure 7.31c).

EXTRAPOLATION

Gaps with a width of a few pixels may be bridged by extrapolation of edge segments. For that purpose it is needed to estimate the parameters of the tangent of a non-closed segment at one of its end points. Edge linking is done by continuing the segment along the tangent until an element from another edge segment has been found. In order to decrease the sensitivity to estimation errors of the parameters of the tangent it is advantageous to search in a sector surrounding the tangent instead of restricting the search to the tangent solely.

In some implementations the direction of the tangent is quantised to eight angles. The advantage is that on an orthogonal grid the definition of two sectors suffices. The remaining sectors follow from mirroring and reversal. Figure 7.31a is an edge map with two edge segments. The first segment is roughly vertically oriented. The search area associated to the end point of the segment is shaded.

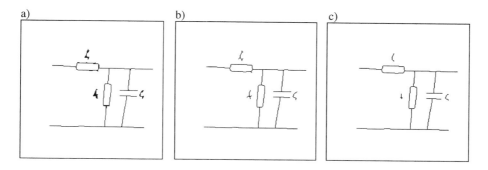

Figure 7.31 Edge linking with morphological operations
a) Closing the detected line elements
b) Skeltonising
c) Pruning

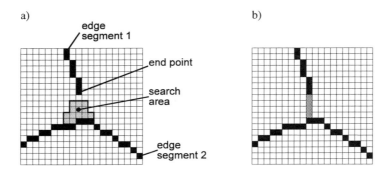

Figure 7.31 Extrapolation of an edge segment
a) The search area of an end point of a vertical oriented edge segment
b) Result of edge linking

Linking occurs by locating the edge element nearest to the end point (figure 7.31b). Zhou et al [1989] describe an implementation based on extrapolation of linear edge segments.

GRAPH SEARCHING

The goal of graph searching is to find a minimum cost path between nodes in a graph. In the present context, graph searching is useful if the image data can be mapped onto a graph. One way to accomplish this is to regard pixels as nodes. Arcs are defined between a pixel and its direct neighbors. Figure 7.32a shows the graph of an image in which 8-connectivity is applied. A search is initiated for each end point of an edge segment encountered in the edge map. The search goes on until a path has been found connecting the end point with another edge segment (preferable the end point of that segment). With properly selected search criterion graph searching can bridge wide gaps.

The search process takes place by setting up a tree structure. The root of the tree is the starting node of the search. In successive steps the tree grows by expanding the leaves of the tree with neighboring nodes, see figure 7.32a. In the expansion of a node, the number of successors is often limited to three so as to keep the expansion "in front of" the path, see figure 7.32b.

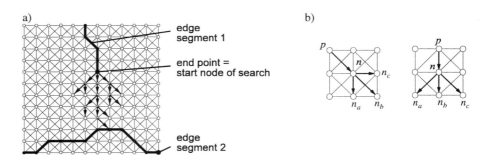

Figure 7.32 Graph searching
a) The search graph
b) Definition of successors of a node n with predecessor p

In order to guide the search a cost is assigned to each path. Often, the cost of the nodes in a path are defined recursively. For instance, suppose that a path consists of M nodes: $\{n_i | i = 1, \cdots, M\}$, and that the cost of this path is denoted by $C(n_1, \cdots, n_M)$. Then, the cost of the path when expanded with a successor n_{M+1} is:

$$C(n_1, \cdots, n_{M+1}) = C(n_1, \cdots, n_M) + F(n_{M+1} | n_{M-L}, \cdots, n_M) \tag{7.31}$$

That is, the growth of the cost due to the expansion with n_{M+1} only depends on n_{M+1} and the $L+1$ predecessors of n_{M+1}. The cost function $F(\cdot)$ could be defined such that it covers the following aspects:

- *Adaptation to edge operation output.* If the gradient $\vec{\nabla} f$ of the grey levels is used, compliance with the edge strength and orientation could be expressed as:

$$\begin{aligned} F_1(n_{M+1}) &= Constant - |\vec{\nabla} f| \\ F_2(n_{M+1} | n_M) &= |\angle \vec{\nabla} f - \angle(n_{M+1}, n_M)| \end{aligned} \tag{7.32}$$

$\vec{\nabla} f$ is the gradient vector estimated at the position corresponding to node n_{M+1}. The angle $\angle(n_{M+1}, n_M))$ denotes the direction between n_M and n_{M+1}.

- *Smoothness of the boundary.* If smooth boundaries are desirable, the curvature measured near the expansion should be low. The curvature can be estimated using, for instance, the last seven nodes. Two lines are defined: one passing nodes n_{M+1} and n_{M-2} and one passing n_{M-2} and n_{M-5}. If the angle between the two lines is denoted $\angle(n_{M-5}, n_{M-2}, n_{M+1})$, then the cost of "raggedness" could be expressed as:

$$F_3(n_{M+1} | n_{M-5}, n_M) = |\angle(n_{M-5}, n_{M-2}, n_{M+1})| \tag{7.33}$$

Another method to quantify the cost of raggedness is to use the difference between consecutive angles. With that, the cost is only zero if the boundary deviates from a circle or a line.

- *Distance from the goal.* The ultimate goal is to link edge segments. We may prefer short bridges above long bridges. This preference can be expressed in a cost function by use of the distance transform of all edge segments (excluding the segment that contains the start node) to the present node:

$$F_4(n_{M+1}) = d_{\text{edge segments}}(n_{M+1}) \tag{7.34}$$

This cost function accomplishes an attraction towards edge segments that are nearby.

The general structure of a graph search algorithm is as follows:

Algorithm 7.2 Heuristic search

    Input: start node s.
    1. Create an empty list - called OPEN - in which nodes can be stored.
    2. Move s to OPEN.
    3. Remove the top entry n from the list OPEN.
    4. If n is not a goal node, expand it and append the successors to OPEN. Create a pointer associated with each successor and direct it to n.
    5. If n is a goal node, go to 8.
    6. Reorder the entries in the list OPEN.

7. Go to 3.
8. Trace back through the pointers from n to s.
Output: the path from the start node to a goal node.

The criterion used in step 6 to reorder the list of paths is crucial. If the criterion is such that the node n with minimal cost $C(s,\cdots,n)$ is put on the top, then the algorithm always finds a minimum cost path. Unfortunately, the algorithm is *breadth first* then, implying that many short paths must be expanded in order to guarantee minimum cost. Using estimated costs in the reordering of the list makes the algorithm faster (the so-called *A algorithm*). However, this may be at the cost of optimality. Details of implementations are given by Martelli [1972].

CURVE FITTING AND CORRELATION TECHNIQUES: HOUGH TRANSFORM

The assumption behind curve fitting and related techniques is that a boundary consists of a number of *boundary segments*. Each segment is described by a curve of a given functional form. The parameters of the curves define the boundary. For instance, suppose that a boundary is modeled by a piecewise linear contour. Then, each curve is a straight line segment which can be defined with four parameters (e.g. the co-ordinates of the two end points).

Curve fitting is the estimation of the parameters of a curve given a set of detected edge elements. Correlation involves the determination of a measure of match between detected edge elements and hypothesised curves.

An elegant method which combines the detection of curves with the estimation of the corresponding parameters is provided by the so-called *Hough transform*. Although in principle the Hough transform applies to any class of parametrised curves we confine ourselves to a particular class: straight lines in the xy-plane. Each straight line can be given in the following parametric form:

$$x \sin \varphi - y \cos \varphi + r = 0 \quad \text{with:} \quad \varphi \in [0, \pi], \ r \in \mathbb{R} \tag{7.35}$$

where φ is the direction of the line with respect to the x-axis, and $|r|$ is the shortest distance to the origin.

Equation (7.35) defines a one-to-one correspondence between a set of lines and points in the $r\varphi$-domain. For instance, the three lines in figure 7.33a correspond to the three points in figure 7.33b. On the other hand, if we consider a point (x_0, y_0) in the xy-plane, then the set of all lines passing through that point with arbitrary orientation establishes a relation between r and φ:

$$r(\varphi) = y_0 \cos \varphi - x_0 \sin \varphi \tag{7.36}$$

That is, a one-to-one correspondence exists between points in the xy-domain and sinusoids in the $r\varphi$-domain.

Suppose that a number of edges has been detected. Each edge corresponds to a point in the image plane. As such it defines also a sinusoid in the $r\varphi$-domain. If a number of edges are collinear, then the corresponding sinusoids cross at a single point. For instance, in figure 7.33c the seven points in the xy-plane form two subsets containing four collinear points. In figure 7.33d these seven points correspond to seven sinusoids in the $r\varphi$-domain. The sinusoids of the two subsets cross at two single points (r_0, φ_0) and (r_1, φ_1), respectively. Each intersection in the $r\varphi$-domain corresponds to a line passing through the collinear points.

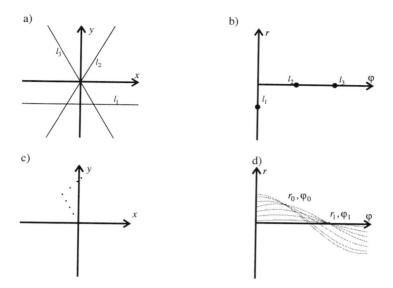

Figure 7.33 The Hough transform
a-b) Lines in the xy-plane correspond to points in the $r\varphi$-domain
c-d) Points in the xy-plane correspond to sinusoids in the $r\varphi$-domain

The Hough transform is a mapping in which each detected edge element is converted into a sinusoid in the $r\varphi$-domain. For that purpose the $r\varphi$-domain is quantised into cells. If a sinusoid falls in a cell, that particular cell is incremented by one count. For an image plane whose area is limited to $N\Delta \times M\Delta$ the range of the parameter space $(r-\varphi)$ is limited to $\Delta\sqrt{N^2+M^2} \times \pi$. Therefore, the cells needed to cover all possible sinusoids can be regarded as a finite 2-dimensional array. This permits a digital implementation of the Hough transform:

Algorithm 7.3 Hough transform for lines
    input:
- map with detected edges: $f_{n,m}$
- sampling periods Δr and $\Delta\varphi$
- 2-dimensional accumulator array $accu_{i,j}$
1. For all i,j: set $accu_{i,j} = 0$
2. For each (n,m) for which $f_{n,m} = 1$:
    2.1. For all j:
        2.1.1. $\varphi = j\Delta\varphi$
        2.1.2. $x = n\Delta$
        2.1.3. $y = m\Delta$
        2.1.4. $r = y\cos\varphi - x\sin\varphi$
        2.1.5. $i = r/\Delta r$
        2.1.6. $accu_{i,j} = accu_{i,j} + 1$
    output: Hough transform stored in $accu_{i,j}$

As an example, figure 7.34a shows an edge map derived from an image of a cube. Figure 7.34b is the corresponding Hough transform. An edge segment of the cube coincides with a cluster in the Hough transform. Therefore, application of a cluster analysis to the Hough transform gives us the r–φ parameters of the nine edge

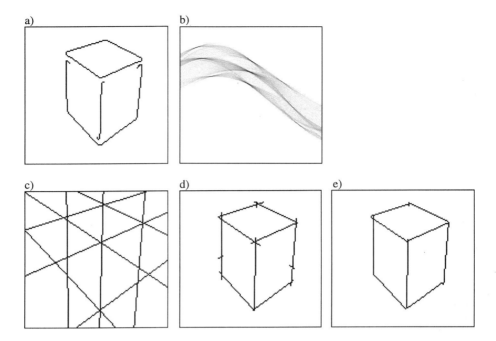

Figure 7.34 Edge linking with the Hough transform
a) Map with detected edges
b) Hough transform for lines
c) Detected lines
d) Detected lines masked by a dilated version of the detected edges
e) Pruned

segments. Equation (7.35) associates a line in the image plane with each pair of parameters, see figure 7.34c. In order to arrive at linked edges it is necessary to mask the lines with a dilated version of the original edge map (figure 7.34d). Finally, tiny spurs can be removed with a pruning operation (figure 7.34e).

The only step in the edge linking process that is non-trivial is the cluster analysis. The simplest method is to find the positions of all local maximums in the Hough transform and to retain those positions for which the maximums exceed a given threshold. Unfortunately, this method may lead to false detections whenever the contributions of two edge segments to the Hough transform interfere. A method, superior to local maximum detection, is the following algorithm:

Algorithm 7.4 Cluster seeking in Hough transform
  Input:
  • Hough transform (i.e. accumulator array)
  • *threshold*
  1. Find the position (r, φ) of the global maximum in Hough transform.
  2. If the global maximum is less than *threshold*, then exit.
  3. For all edges in the edge map that are within 1 pixel period reach of the line associated with (r, φ):
      Remove the components of the Hough transform which are due to these edge elements.
  4. Go to 1.

A refinement of the Hough transform can be obtained if additional information from the edge detection is used. Most edge detectors have a provision to estimate the orientation of the edge. With that, the necessity to increment the accumulator for all orientations φ vanishes. A small interval around the estimated orientation suffices. The advantage of doing so is that it saves much computational effort. Moreover, it facilitates the detection of clusters.

The generalisation of the Hough transform to parametric curves other than straight lines is easy. The equation of a circle:

$$(x-a)^2 + (y-b)^2 - r^2 = 0 \tag{7.37}$$

shows that in order to detect and to localise circles with unknown radius and position a 3-dimensional parameter space is needed. Detection of line segments would require a 4-dimensional space. Ellipses are described with five parameters. Unfortunately, the computational complexity of the Hough transform increases exponentially as the number of parameters. Therefore, the technique is impractical for curves with a large number of parameters.

7.2 REGION PROPERTIES

The step following after image segmentation is the characterisation of the regions. The description of a region may include many aspects. For instance:

- geometry (shape, size, position, orientation)
- radiometry (irradiance, tristimulus values)
- texture properties.
- relations with other regions.

Often, it is desirable that a parameter describing one aspect is unaffected by other aspects. For instance, shape parameters should be invariant to position, scale, and orientation. Texture parameters should not depend too much on shape. And so on.

Representation

The obvious way to represent regions is with the region map introduced in Chapter 3. The map $r_{n,m}$ is an assignment of labels $\omega_1, \omega_2, \cdots$ (or more briefly $1, 2, \cdots$) to pixels. Alternatively, we may represent each region by its own bitmap. Here, a single pixel denotes whether it belongs to a particular region, or not. The conversion from region map representation to bitmap representation is easy. For instance, the bitmap $b_{n,m}$ associated to a region ω_k is found as:

$$b_{n,m} = \begin{cases} 1 & \text{if } r_{n,m} = k \\ 0 & \text{elsewhere} \end{cases} \tag{7.38}$$

Other representations are also useful, but they will not be considered here, except for representations based on boundaries (section 7.2.2).

Often, a region consists of a number of connected components (section 5.4.1). In order to extract features from the individual components it is necessary to have an algorithm that identifies each component. For instance, figure 7.35a is a bitmap consisting of 13 connected components. The purpose of *component labeling* is to assign unique labels - often numbers - to all components, see figure 7.35b.

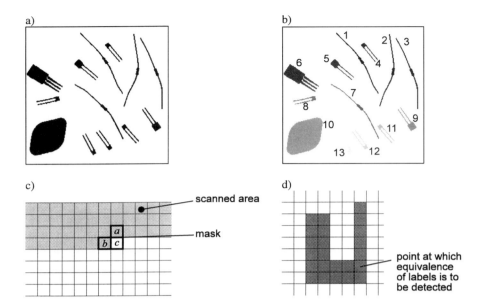

Figure 7.35 Component labeling
a) A bitmap containing 13 connected components
b) Result of component labeling
c) The mask which scans the bitmap row-by-row
d) A component may be given two different labels. In a second pass, these equivalent labels must be replaced with a unique label

The classical way to handle this is to scan the region map row-by-row by a mask as depicted in figure 7.35c. If at a certain position the region map indicates that pixel c belongs neither to pixel a nor to pixel b, a new connected component may be encountered. Therefore, in a first pass this pixel is given a new label that is stored in an output map. If the region map indicates that c belongs to either a or b, it is given the label of the corresponding pixel. If the region map indicates that c belongs both to a and to b, and that at the same time the labels of a and b in the output map differ, then c is given the label of a, but it is also noted that the label of b is equivalent to the one of a. (Figure 7.35d shows a connected component in which such an equivalence occurs). In a second pass, all equivalent labels are replaced with unique labels. Listing 7.2 is a realisation of the algorithm. Haralick and Shapiro [1992] describes some variants of the algorithm that are less expensive with respect to computational effort and storage requirements.

7.2.1 Region parameters

The parameters which are the subject in this section are region-oriented. In order to measure the parameters it is necessary to scan the whole area of the region. This is in contrast with parameters that are boundary-oriented.

Listing 7.2 Component labeling.

```
int component_labeling(in,out,N,M)       /* labels the 4-connected components.        */
int **in,**out;                          /* region map and output map, respectively.  */
int N,M;                                 /* number of rows and columns.               */
{
        #define MAXEQUI 20000                    /* length of equivalence table         */

        int equivalences[MAXEQUI+1];             /* notebook to store equivalences      */
        int label,i,j,a,b,c,y,z,v,w;

        for(i=0;i<=MAXEQUI;i++) equivalences[i] = i;
        label=1;
        out[0][0] = label++;                     /* first label to spent                */
                                                 /* init first pixel                    */
        for (j=1; j<M; j++)                      /* init first row                      */
        {       c = in[0][j];
                b = in[0][j-1];
                if (c!=b) out[0][j] = label++;
                else out[0][j] = out[0][j-1];
        };

        for (i=1; i<N; i++)                      /* scan rows                           */
        {       c = in[i][0];                    /* init first column                   */
                a = in[i-1][0];
                if (c!=a) out[i][0] = label++;
                else out[i][0] = out[i-1][0];
                for (j=1;j<M; j++)               /* scan columns                        */
                {       c = in[i][j];
                        b = in[i][j-1];
                        a = in[i-1][j];
                        y = (b==c);
                        z = (a==c);
                        if (!z && !y) out[i][j] = label++;
                        else if (!z && y) out[i][j] = out[i][j-1];
                        else if (z && !y) out[i][j] = out[i-1][j];
                        else if (z && y)
                        {       out[i][j] = out[i-1][j];
                                w = equivalences[out[i][j-1]];
                                v = equivalences[out[i-1][j]];
                                if (v != w) update(equivalences,label,v,w);
                        }
                }
        }
        z=1;
        for (i=1;i<label;i++)                    /* change labels into consecutive numbers */
        {       j = equivalences[i];
                if (j == i) equivalences[i] = z++;
                else equivalences[i] = equivalences[j];
        };
        for(i=0;i<N;i++)                         /* change equivalent labels to unique labels */
        for(j=0;j<M;j++) out[i][j] = equivalences[out[i][j]];
        return(z-1);                             /* return number of found components */
}
static int update(equivalences,label,v,w)
int v,w,label,equivalences[];
{       int i,mmax,mmin;                         /* change max(v,w) into min(v,w) */
        mmax = (v>w) ? v : w;
        mmin = (v<w) ? v : w;
        for (i=1;i<label;i++) if (equivalences[i] == mmax) equivalences[i] = mmin;
}
```

Geometric parameters

The geometry of a planar region concerns aspects like: size, position, orientation, and shape. Many of these aspects are covered by a family of parameters called *moments*. In probability theory, moments are used to characterise probability density functions, e.g. expectation (first order moment), variance, covariance (second order central moments). In the present context, we use the same definitions, but replace the density function with a binary, 2-dimensional function with level "1" within the region, and level "0" elsewhere, i.e. a bitmap. The moments of order $p+q$ of a region represented by the bitmap $b_{n,m}$ are:

$$M_{p,q} = \sum_{n=0}^{N-1}\sum_{m=0}^{M-1} n^p m^q b_{n,m} = \sum_{n,m \in \text{region}} n^p m^q \tag{7.39}$$

A definition in which $b_{n,m}$ is replaced with the grey level image $f_{n,m}$ also exists. Then, however, the corresponding moments are no longer geometric parameters. The zero order moment of a density function is the volume covered by that function. In our bitmap definition, the altitude of the function is either one or zero, implying that the volume coincides with the area of the region. In fact, $M_{0,0}$ is the number of pixels inside the region, so that the area is given in units of pixel area (Δ^2).

The first order moments $M_{0,1}$ and $M_{1,0}$ are related to the balance point $(\bar{x},\bar{y})$ of the region:

$$\bar{x} = M_{1,0}/M_{0,0} \quad \text{and} \quad \bar{y} = M_{0,1}/M_{0,0} \tag{7.40}$$

As such these parameters are appropriate to determine the position of the region given in units of pixel period (Δ). The point $(\bar{x},\bar{y})$ is called the *centre of gravity*, or *centroid*.

In order to make the description independent on position, moments can be calculated with respect to the centroid. The results are the so-called *central moments*:

$$\mu_{p,q} = \sum_{n=0}^{N-1}\sum_{m=0}^{M-1}(n-\bar{x})^p(m-\bar{y})^q b_{n,m} = \sum_{n,m \in \text{region}}(n-\bar{x})^p(m-\bar{y})^q \tag{7.41}$$

If the ordinary moments are known, it is less expensive to derive the central moments from the ordinary moments than to evaluate expression (7.41) directly. For instance:

$$\begin{aligned}\mu_{0,0} &= M_{0,0} \\ \mu_{0,1} &= \mu_{1,0} = 0 \\ \mu_{0,2} &= M_{2,0} - \bar{x}M_{1,0} \\ \mu_{1,1} &= M_{1,1} - \bar{x}M_{0,1}\end{aligned} \tag{7.42}$$

etc.

The second order central moments exhibit a number of properties that are comparable with covariance matrices in probability theory and the moments of inertia associated with rotating bodies in mechanics. The *principal axes* of a region are spanned by the eigenvectors of the matrix:

$$\begin{bmatrix} \mu_{2,0} & \mu_{1,1} \\ \mu_{1,1} & \mu_{0,2} \end{bmatrix}$$

(see also appendix B.2 and B.3.1). The *principal moments* are the corresponding eigenvalues.

$$\begin{aligned}\lambda_{\max} &= \tfrac{1}{2}(\mu_{2,0}+\mu_{0,2})+\tfrac{1}{2}\sqrt{\mu_{2,0}^2+\mu_{0,2}^2-2\mu_{0,2}\mu_{2,0}+4\mu_{1,1}^2} \\ \lambda_{\min} &= \tfrac{1}{2}(\mu_{2,0}+\mu_{0,2})-\tfrac{1}{2}\sqrt{\mu_{2,0}^2+\mu_{0,2}^2-2\mu_{0,2}\mu_{2,0}+4\mu_{1,1}^2}\end{aligned} \tag{7.43a}$$

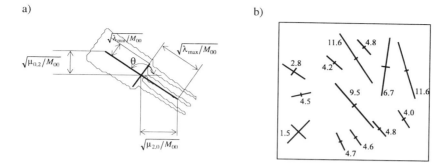

Figure 7.36 Features derived from normalised second order moments
a) Principal components analysis
b) Principal axes, moments and eccentricities of the connected components from figure 7.35a

The direction of the largest principal moment is:

$$\theta = \tan^{-1}\left(\frac{\lambda_{max} - \mu_{2,0}}{\mu_{1,1}}\right) \quad (7.43b)$$

θ is often used to specify the orientation of a region.
The *eccentricity* of a region can be defined as the ratio between square roots of the two principal moments:

$$eccentricity = \sqrt{\frac{\lambda_{max}}{\lambda_{min}}} \quad (7.44)$$

The parameter depends solely on the shape, not on size and orientation. Figure 7.36 illustrates the use of principal component analysis. For each connected component in figure 7.35 $\sqrt{\lambda_{max}/M_{00}}$ and $\sqrt{\lambda_{min}/M_{00}}$ are depicted as two orthogonal axes oriented at the angle θ. The eccentricity is added as a number.
Since $\mu_{0,0} = M_{0,0}$ is the area of the region, it can be used as a measure of size. As such it is useful in order to normalise the moments so as to arrive at a size independent description. The *normalised central moments* are:

$$\eta_{p,q} = \frac{\mu_{p,q}}{\mu_{0,0}^{\alpha}} \quad \text{with:} \quad \alpha = \frac{p+q}{2}+1 \quad (7.45)$$

From the normalised moments up to order three, seven parameters can be extracted called *moment invariants*:

$$\begin{aligned}
h_1 &= \eta_{2,0} + \eta_{0,2} \\
h_2 &= (\eta_{2,0} - \eta_{0,2})^2 + 4\eta_{1,1}^2 \\
h_3 &= (\eta_{3,0} - 3\eta_{1,2})^2 + (3\eta_{2,1} - \eta_{0,3})^2 \\
h_4 &= (\eta_{3,0} + \eta_{1,2})^2 + (\eta_{0,3} + \eta_{2,1})^2 \\
h_5 &= (\eta_{3,0} - 3\eta_{1,2})(\eta_{3,0} + \eta_{1,2})\big((\eta_{3,0} + \eta_{1,2})^2 - 3(\eta_{0,3} + \eta_{2,1})^2\big) + \\
&\quad (3\eta_{2,1} - \eta_{0,3})(\eta_{0,3} + \eta_{2,1})\big(3(\eta_{3,0} + \eta_{1,2})^2 - (\eta_{0,3} + \eta_{2,1})^2\big)
\end{aligned} \quad (7.46)$$

$$h_6 = (\eta_{2,0} - \eta_{0,2})\left((\eta_{3,0} + \eta_{1,2})^2 - (\eta_{0,3} + \eta_{2,1})^2\right) + 4\eta_{1,1}(\eta_{3,0} + \eta_{1,2})(\eta_{0,3} + \eta_{2,1})$$

$$h_7 = (3\eta_{2,1} - \eta_{0,3})(\eta_{3,0} + \eta_{1,2})\left((\eta_{3,0} + \eta_{1,2})^2 - 3(\eta_{0,3} + \eta_{2,1})^2\right) - $$
$$(\eta_{3,0} - 3\eta_{1,2})(\eta_{0,3} + \eta_{2,1})\left(3(\eta_{3,0} + \eta_{1,2})^2 - (\eta_{0,3} + \eta_{2,1})^2\right)$$

These parameters depend neither on position, nor size, nor orientation. The invariance of these shape descriptors has been established by Hu [1962]. Figure 7.38 show some resulting features extracted from the components in figure 7.35a. The components can be divided into three object classes: capacitors (C), resistors (R), and transistors (T).

A disadvantage of the moment invariants is that it is difficult to give them a geometric interpretation. Other region parameters, not particularly related to moments, do have such an interpretation. These parameters include:

Thickness: minimum number of erosions needed to let the region vanish.
Maximum chord: straight line segment joining the two points farthest from each other.
Minimum chord: line segment perpendicular to the maximum chord, and of minimum length such that a rectangular box bounding the region can be formed.
Diameter: length of the maximum chord.
Elongation: ratio of the lengths of maximum chord and minimum chord.
Compactness: ratio of the area to the square of the perimeter.

Thickness depends on the structuring element (usually a 4-n or 8-n disk) used in the erosion. The parameter is approximately invariant to translation and rotation, but it depends linearly on the scale of the region. It can be made invariant to scale change somewhat by normalising it with respect to the square root of the area. The thickness of a skeleton is small; the one of a disk is large.

Compactness differs from thickness in the sense that a disk-like region with a rippled boundary is low whereas the thickness is high.

Radiometric and textural parameters

The radiometry of a region concerns measurements related to physical quantities defined within the region. First and second order statistical descriptions of the irradiance have already been given in Chapter 3, see tables 3.1 and 3.3. Estimation of these parameters based on observed data in the regions is straightforward with the techniques discussed in section 6.3.1.

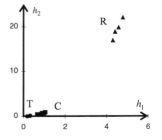
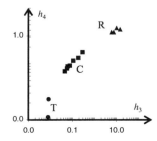

Figure 7.37 Scatter diagrams of some moment invariants of the components in figure 7.35b

As mentioned in section 7.1.1 the frontal view of a homogeneously textured surface corresponds to a region with stationary statistical properties. Often (not always) the second order statistics suffices to characterise the texture. Therefore, the description of a textured region involves the estimation of the joint probability density $p_{f,f}(f_1,f_2;a,b)$ of the irradiance $f(x,y)$ and $f(x+a,y+b)$.
In digital images, both the irradiance and the spatial co-ordinates are discrete: $f_{n,m}$ replaces $f(x,y)$. The so-called *co-occurrence matrix* $\mathbf{P}(a,b)$ is an estimate of $p_{f,f}(f_1,f_2;a,b)$. Suppose that $f_{n,m}$ has K quantisation levels. Then, $\mathbf{P}(a,b)$ is a $K \times K$-dimensional matrix. The (i,j)-th element $\mathbf{P}_{i,j}(a,b)$ of the matrix $\mathbf{P}(a,b)$ is proportional to the number of pixel pairs the first pixel of which is separated from the second pixel by a displacement (a,b) and of which the first and second pixel have grey levels i and j, respectively. An algorithm to calculate $\mathbf{P}(a,b)$ is as follows:

Algorithm 7.5 Grey level co-occurrence
input:
- grey level image $f_{n,m}$
- region A
- displacement (a,b)
1. Initiate: set all elements of $\mathbf{P}(a,b)$ to zero; set $N_{region} = 0$.
2. For all pixels $(n,m) \in A$:
    2.1. For all pixels $(n+a, m+b) \in A$:
        2.1.1. Increment $\mathbf{P}_{f_{n,m}, f_{n+a,m+b}}(a,b)$ and $\mathbf{P}_{f_{n+a,m+b}, f_{n,m}}(a,b)$.
        2.1.2. Increment N_{region}.
3. For all elements: $\mathbf{P}_{i,j}(a,b) = \mathbf{P}_{i,j}(a,b)/2N_{region}$.
output: co-occurrence matrix $\mathbf{P}(a,b)$.

Note that this procedure can be regarded as the estimation of the parameters of a multinomial distribution (example 6.7, equation 6.98). For a digital image with 256 grey levels the number of elements in $\mathbf{P}(a,b)$ is 65384. Often, the number of pixels in a region is less than that. As a result, the variance of the estimation is prohibitory large. In order to reduce the variance it is common practice to requantise the grey level scale to, for instance, 16 levels. With that, the number of elements of $\mathbf{P}(a,b)$ becomes 256.

Another concern is the choices of the displacement vector (a,b). The joint probability density $p_{f,f}(f_1,f_2;a,b)$ is defined for any real vector (a,b). In order to limit the computational effort, the choices of (a,b) are usually restricted to a few integer samples in horizontal, vertical and diagonal directions: $\{(0,0), (0,1), (0,2), \cdots\}$, $\{(1,0), (2,0), \cdots\}$, and $\{(1,1), (2,2), \cdots\}$. Suppose that the number of samples in each direction is five, then the total number of co-occurrence matrices to calculate is 17. With $K = 16$ this makes the number of descriptors of the texture 4352; still a large number.

One way to reduce the number of descriptors is to derive a few parameters from each co-occurence matrix. Examples are given in table 7.3. The parameters μ and σ^2 are the sample mean and variance, respectively. Another way to characterise textures is by measuring the energy in certain frequency bands. For instance, the covariance in table 7.3 can be evaluated without using the co-occurrence matrix:

$$\hat{C}_{nn}(a,b) = \frac{1}{N_{a,b}} \sum_{\substack{(n,m) \in region \\ \text{and} \\ (n+a,m+b) \in region}} (f_{n,m} - \mu)(f_{n+a,m+b} - \mu) \qquad (7.47)$$

where $N_{a,b}$ is a normalising constant equal to the number of pixel pairs which fits the region.

The Fourier transform $\hat{S}_{nn}(u,v) = F\{\hat{C}_{nn}(a,b)\}$ is an estimate of the power spectrum related to the textured region. The texture energy within a frequency range is calculated by windowing the power spectrum with energy transfer functions, say $H_k(u,v)$, and integrating the resulting power:

$$E_k = \iint_{u\,v} H_k(u,v)\hat{S}_{nn}(u,v)dudv \tag{7.48}$$

Radial symmetric band-pass filters and sector-like filters are often used as transfer functions (figure 7.38). Similar results are achieved by replacing the transfer function in the Fourier domain with appropriate convolutions applied to the original data. This brings us close to the texture based pixel classificator shown in figure 7.13.

7.2.2 Contours

The boundary of a region is an unambiguous representation of it. Therefore, parameters describing the boundary implicitly characterise the region.

Contour tracing

With a morphological operation it is easy to convert a bitmap of a region into a set consisting of its boundary points, see section 5.4. Unfortunately, the representation of such a set is not very useful. The boundary becomes more descriptive if the representation is that of a closed curve. The *contour* of a region refers to an ordered

Table 7.3 Texture features derived from co-occurrence matrix

Feature	Definition
correlation	$\dfrac{1}{\sigma^2}\sum_{i=0}^{K-1}\sum_{j=0}^{K-1}(i-\mu)(j-\mu)\mathbf{P}_{i,j}(a,b)$
covariance	$\sum_{i=0}^{K-1}\sum_{j=0}^{K-1}(i-\mu)(j-\mu)\mathbf{P}_{i,j}(a,b)$
energy	$\sum_{i=0}^{K-1}\sum_{j=0}^{K-1}\mathbf{P}_{i,j}^2(a,b)$
entropy	$\sum_{i=0}^{K-1}\sum_{j=0}^{K-1}\mathbf{P}_{i,j}(a,b)\ln \mathbf{P}_{i,j}(a,b)$
inertia	$\sum_{i=0}^{K-1}\sum_{j=0}^{K-1}(i-j)^2 \mathbf{P}_{i,j}(a,b)$
local homogeneity	$\sum_{i=0}^{K-1}\sum_{j=0}^{K-1}\dfrac{1}{1+(i-j)^2}\mathbf{P}_{i,j}(a,b)$
maximum probability	$\max_{i,j} \mathbf{P}_{i,j}(a,b)$

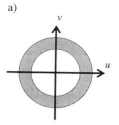
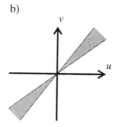

Figure 7.38 Energy transfer functions for texture features
a) Radial symmetric   b) Sector-like

set of boundary points (figure 7.39). Suppose that (x_0, y_0) is a point of the boundary of a region. The perimeter is denoted by P. Let s be the running arc length of the boundary of the region traversed in counter clockwise direction starting from the point (x_0, y_0). The traversal ends in (x_0, y_0) when all boundary points have been visited. This occurs when $s = P$.

In digital form, the contour of a region is a path consisting of the boundary points of the regions. There are many representations of such paths. One of them is the so-called *Freeman chain code*, see figure 7.39. Another representation is that of an ordered list of co-ordinates: a *boundary list*.

The following algorithm determines the boundary list of a region given one of the boundary points (i.e. the start point). The algorithm uses 8-connectivity.

Algorithm 7.6 Contour tracing
Input:
- region
- start point $p_{init} = (n_0, m_0)$
1. Create an empty boundary list and put p_{init} on top of it.
2. Set $p = p_{init}$.
3. From the eight neighbors of p_{init} choose a point, say q, that is not part of the region. That is, $q \in N_8(p_{init})$ and $q \notin region$.
4. Set $q_{init} = q$.
5. Determine the eight neighbors r_i (where $i = 0, \cdots, 7$) of p such that:
   - $r_i \in N_8(p_{init})$.
   - $r_0 = q$.
   - $r_0, r_1, \cdots, r_7$ is a counter clockwise oriented closed contour (see figure 7.39e).
6. Select j such that:
   - $r_0, \cdots, r_{j-1} \notin region$.
   - $r_j \in region$.
7. Set $p = r_j$ and $q = r_{j-1}$.
8. Append p to boundary list.
9. If $(p = p_{init})$ and $(q_{init} \in \{r_0, \cdots, r_{j-1}\})$ exit.
10. Go to 5.
Output: an ordered boundary list.

Fourier descriptors

The boundary list of a region corresponds to a parametric curve $\{x(s), y(s)\}$ in the continuous domain. As noted before, s is the running arc length relative to the start point $\{x(0), y(0)\}$. If P is the perimeter of the region, the co-ordinates $\{x(s), y(s)\}$ can

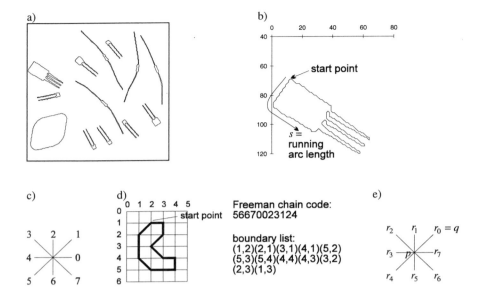

Figure 7.39 Contours
a) Boundaries of regions
b) Contour tracing
c) Definition of Freeman's chain code
d) Two representations of a contour
e) Example of a mask used in contour tracing

be regarded as two periodic functions with period P. Since the contour is closed, it has no gaps. Thus, both functions are continuous.

Another useful representation of the contour is obtained by considering the image plane as a complex plane consisting of a real axis and an imaginary axis: $z = x + jy$ with $j = \sqrt{-1}$. In that case, the contour is a complex, periodic and continuous function:

$$z(s) = z(s + P) \qquad (7.49)$$

Fourier series expansion of a periodic function involves the determination of the complex amplitudes Z_k of harmonic functions with frequencies that are multiples of $1/P$:

$$z(s) = \sum_{k=-\infty}^{\infty} Z_k \exp\left(\frac{2\pi jks}{P}\right)$$

$$Z_k = \frac{1}{P} \int_{s=0}^{P} z(s) \exp\left(\frac{-2\pi jks}{P}\right) ds \qquad \text{with:} \qquad k = \cdots, -2, -1, 0, 1, \cdots \qquad (7.50)$$

The complex harmonics are the *Fourier descriptors* (FDs) of the contour. Some properties are:

1. *Area* and *perimeter*:

$$\text{Area} = \pi \sum_{k=-\infty}^{\infty} k|Z_k|^2 \qquad P^2 = 4\pi^2 \sum_{k=-\infty}^{\infty} k^2 |Z_k|^2 \qquad (7.51)$$

Assuming that the region behaves regular (e.g. non-fractal like) the perimeter must be finite. For contours with piecewise continuous tangent angles $Z_k = O(1/k^2)$ as $|k| \to \infty$. A finite number of terms in (7.51) can establish an approximation of the contour with arbitrary accuracy.

2. *Position* and *translation*:
The position of a contour is reflected in the zero-th harmonic only. In fact, z_0 is the centre of gravity of the contour (not of the region). A change of z_0 is equivalent to a translation of the contour.

3. *Scaling* and *size*:
If the contour is blown up by a factor α, then all FDs are scaled by a factor α. The Fourier expansion of the first harmonic alone yields a circle with radius $|Z_1|$. This radius can be used as a measure of size.

4. *Rotation* and *change of start point*:
If a contour is rotated by an angle α in counter clockwise direction, the FDs are multiplied by $\exp(j\alpha)$. If the start point is shifted by an angle $\beta = 2\pi s_0/P$ (s_0 is the distance along the arc by which the start point is shifted), then the FDs are multiplied by $\exp(jk\beta)$.

5. *Symmetries*:
Consider two contours which are mirror images of one another about the x-axis. Then the FDs of one contour are the complex conjugate of the other. Furthermore, if a contour has M-rotational symmetry, then only the harmonics $1 \pm nM$ (with $n = 0, 1, \cdots$) can be non-zero.

In the conversion from a (discrete) boundary list to a set of Fourier descriptors we have to face some difficulties that are due to quantisation. The boundary list forms a non-uniformly sampled representation of a continuous contour. In some intervals the sampling period is $\sqrt{2}\Delta$ whereas in other intervals it is Δ.

The simplest method to calculate the FDs from the boundary list is by means of uniform resampling (figure 7.40). The first step is to construct a polygon which connects all points on the boundary list. The perimeter of this polygon is:

$$\hat{P} = (N_1 + N_2 \sqrt{2})\Delta \qquad (7.52)$$

where N_1 is the number of vertical and horizontal links in the list, and N_2 is the number of diagonal links. The second step is to divide the perimeter into a number of equally sized intervals. The positions of the intervals paced along the polygon are the new co-ordinates of the resampled contour.

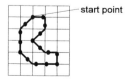

Figure 7.40 Uniform resampling of a contour associated with a boundary list

Figure 7.41a is a graphical representation of the boundary points obtained from figure 7.39b. Resampling this boundary to 64 equidistant samples (shown as dots in figure 7.41c) gives the representation in figure 7.41b.

Once a uniform sampled contour is available, the FDs can easily be calculated with the DFT (discrete Fourier transform):

$$Z_k = \frac{1}{N}\sum_{n=0}^{N-1} z_n \exp\left(\frac{-j2\pi nk}{N}\right) \qquad (7.53)$$

where z_n are the resampled boundary points, and N the number of points. The implementation is most efficient if N is a power of 2. In that case the FFT can be applied to find the FDs, see section 4.2.6. Figure 7.41d depicts the amplitude and phase of the first 32 FDs of the contour.

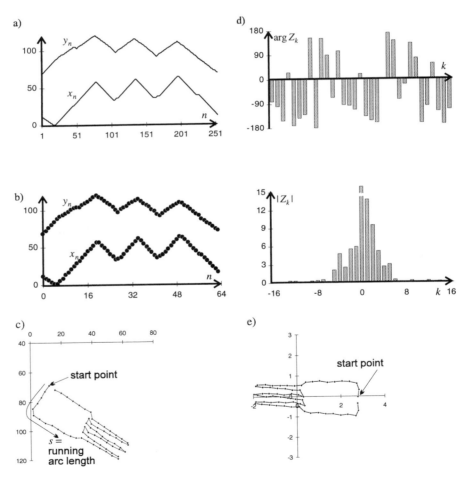

Figure 7.41 Fourier descriptors of the region shown in figure 7.39b
a) Boundary list (row and column) obtained by contour tracing (254 boundary points)
b-c) Uniform sampled contour (number of samples: 64)
d) Amplitude and phase spectrum derived from b). Only 32 FDs are shown
e) Contour reconstructed after normalisation

The objective of *normalisation* is to isolate the information in the FDs pertaining to size, position, orientation, start point and shape. Normalising the position is achieved by setting z_0 to zero. The normalisation with respect to size, orientation, and start point is achieved by scaling and rotation of the contour such that the amplitude of the first harmonic z_1 becomes unit and its phase zero. In order to normalise the starting point it is necessary to involve another harmonic. Normalisation of the start point is accomplished by shifting it such that the phase of the FD with largest amplitude (z_0 and z_1 excluded), say the L-th FD, becomes zero. The number of solutions satisfying the two phase constraints (i.e. $\arg z_1 = 0$ and $\arg z_L = 0$) is $|L-1|$. Only if $L = 2$, the normalisation of the orientation and start point is unambiguous.

We may try to select a unique solution by using a so-called *ambiguity-resolving function* that depends on several other FDs. Wallace and Wintz [1980] propose to use the following algorithm:

Algorithm 7.7: FD normalisation
> *Input:* Fourier descriptors z_k.
> 1. Set $z_0 = 0$ (position normalisation; z_0 is the centre of gravity of the contour).
> 2. Divide each of the z_k by $|z_1|$ (size normalisation; $|z_1|$ is a measure of size).
> 3. Find the FD z_L with largest amplitude (z_1 excluded).
> 4. Set $\theta_1 = \arg z_1$ and $\theta_2 = \arg z_L$
> 5. Multiply each of the z_k by $\exp[j((k-L)\theta_1 + (1-k)\theta_2)/(L-1)]$.
> 6. If $L = 2$, the normalisation is unique: exit.
> 7. Compute the ambiguity resolving function A and save it (A to be defined later).
> 8. If step 7 has been executed for all $|L-1|$ normalisations, go to 9. Otherwise, multiply each of the z_k by $\exp[j2\pi(k-1)/(L-1)]$; go to step 7.
> 9. Select the normalisation which maximises A; exit.
>
> *Output:* a set of normalised Fourier descriptors in z_k.

From the $|L-1|$ solutions the algorithm selects the one with maximum A. The ambiguity resolving function is defined by:

$$A = \sum_{k=0}^{N-1} \Re e(z_k)|\Re e(z_k)| \qquad (7.54)$$

Application of the normalisation procedure yields a set of *normalised Fourier descriptors*. The example in figure 7.42 indicates that the ambiguity of the class "resistor" is $|L-1|=2$, whereas the class "capacitor" is non-ambiguous. A normalised contour (figure 7.41e) can be found by application of the inverse DFT to the normalised FDs.

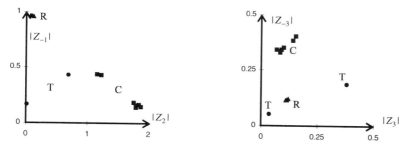

Figure 7.42. Scatter diagrams of the amplitude of some normalised FDs

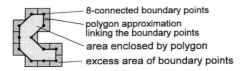

Figure 7.43 Area estimation

The normalisation procedure of Wallace and Wintz also produces parameters related to the position z_0 and the size $|Z_1|$. In fact, $|Z_1|$ is the radius of the best fitting circle. The orientation and start point associated with each normalisation are found as:

$$\beta = \frac{\theta_1 - \theta_2 + 2\pi m}{L-1} \quad \text{with:} \quad m = 0, \cdots, |L-1|-1 \qquad (7.55)$$

$$\alpha = -\theta_1 - \beta$$

where each instance of m corresponds to one normalisation.

Area and perimeter measurements

In accordance with equation (7.39), let $M_{0,0}$ denote the number of pixels of a region. Then $M_{0,0}\Delta^2$ is a direct estimate of the area of that region. Another approach to area measurement is the determination of the area enclosed by the polygon that links the boundary points, see figure 7.43. This actually raises the issue of accuracy [Kulpa 1983]. If the 8-connected boundary points are located on the true boundary of the region, $M_{0,0}\Delta^2$ is an overestimate. The excess area is almost proportional to the number of boundary points. On the other hand, if the boundary list represents an inner boundary, the area of the polygon is an underestimate.

The number of diagonal and non-diagonal links (as in equation 7.52) is a biased estimate of the perimeter. At orientations of $22.5°$ the bias is maximum (about 8%) due to the "ripple effect". The bias averaged over all orientations is about 5%. Therefore, an unbiased estimate of the perimeter is:

$$\hat{P} = 0.95(N_1 + N_2\sqrt{2})\Delta \qquad (7.56)$$

N_1 and N_2 are the number of non-diagonal and diagonal links, respectively.

Fourier descriptors are also useful to measure the area and perimeter of the region (equation 7.51). There are at least two factors that cause estimation errors due to sampling. The first one is the round-off error proportional to the sampling period Δ. The second one is the non-uniform sampling which causes errors in the estimation of the running arc length. These errors do not depend on the Δ. In the frequency domain, cancellation of the high frequency components will reduce the ripple effect:

$$Area = \pi \sum_{k=-K_{max}}^{K_{max}} k|Z_k|^2 \qquad P^2 = 4\pi^2 \sum_{k=-K_{max}}^{K_{max}} k^2|Z_k|^2 \qquad (7.57)$$

Provided that the sampling period is sufficiently small, a reasonable choice of the bandwidth is $K_{max} \approx 20$ [Heijden 1986].

Curvature

In the continuous domain, the *curvature* of a curve is the change of tangent angle per unit running arc length. If $\{x(s), y(s)\}$ is a parametric representation of a curve, and s is the running arc length (i.e. $dx^2(s) + dy^2(s) = ds^2$), the tangent angle is:

$$\Phi(s) = \tan^{-1}\left(\frac{y_s(s)}{x_s(s)}\right) \quad \text{with:} \quad x_s(s) = \frac{dx(s)}{ds} \quad y_s(s) = \frac{dy(s)}{ds} \quad (7.58)$$

The curvature, defined as $k(s) = \Phi_s(s)$, can be expressed in explicit form as:

$$k(s) = x_s(s) y_{ss}(s) - x_{ss}(s) y_s(s) \quad (7.59)$$

The estimation of the curvature of a curve, given only a finite boundary list, is similar to estimation of derivatives of a digital image. The derivatives of the parametric curves can be found by convolving the curves with the derivative of a Gaussian:

$$x_s(s) = \int_\xi h_s(s-\xi) x(\xi) d\xi \qquad h_s(s) = \frac{\partial}{\partial s}\left[\frac{1}{\sigma\sqrt{2\pi}} \exp\left(\frac{-s^2}{2\sigma^2}\right)\right]$$

$$x_{ss}(s) = \int_\xi h_{ss}(s-\xi) x(\xi) d\xi \qquad h_{ss}(s) = \frac{\partial^2}{\partial s^2}\left[\frac{1}{\sigma\sqrt{2\pi}} \exp\left(\frac{-s^2}{2\sigma^2}\right)\right] \quad (7.60)$$

etc.

The procedure can be regarded as Gaussian low-pass filtering combined with differentiation. A digital implementation is carried out efficiently in the Fourier domain. First the Fourier descriptors are determined (uniform resampling + FFT). Then, each FD is multiplied by the transfer function, e.g. $F\{h_s(s)\}$. Finally, inverse Fourier transformation gives the discrete representation of the derivative.

The parameter σ controls the amount of low-pass filtering. As such it determines the scale on which the operation takes place. As an example, figure 7.44a shows a contour that has been low-pass filtered at three different scales. With the availability of derivatives of the contour, the estimation of curvature according to (7.59) is straightforward. Some results are shown in figure 7.44b.

The interpretation of curvature is as follows. A straight line segment has zero curvature. The arc of a circle with radius r has a curvature $|k(s)| = 1/r$. Corners are seen as Dirac-functions which - due to the low-pass filtering - are replaced with Gaussians. Contour segments with positive curvature are convex. Segments with negative curvature are concave.

Contour segmentation

In many applications, the shape of a region is too complex to be analysed directly. In these cases it may be fruitful to decompose the region boundary into contour segments. The purpose of this is to break down the problem of region description into a number of subproblems defined at a lower hierarchical level. In fact, the situation is similar to image segmentation where the decomposition of the image plane into regions serves the same purpose.

Methods to segment a contour are comparable with image segmentation methods. One approach is to search contour segments in which a local property of

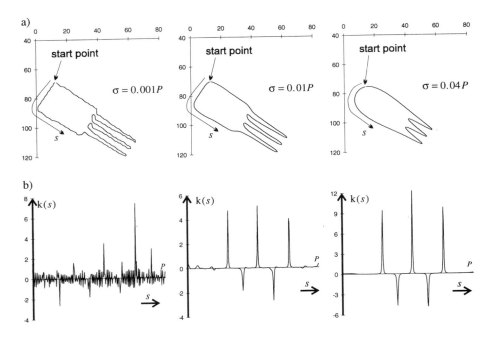

Figure 7.44 Curvature of a contour at three different scales
a) Low-pass filtered contours
b) Corresponding curvatures

the contour is uniform. The alternative approach is to find locations of discontinuities, e.g. corner detection.

In the latter approach, a sudden change of the tangent angle measured along the running arc length is an important feature of a corner. Similar to Canny's edge detector changes can be detected and located by convolving the tangent angle with the first derivative of a Gaussian. In fact, the output of the convolution happens to be the curvature estimated according to the procedure mentioned above. Locations on the running arc length where the curvature has a local maximum or a local

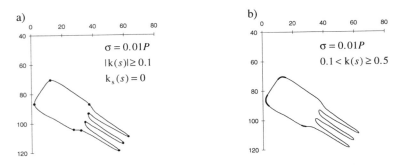

Figure 7.45 Contour segmentation based on curvature
a) Corner detection
b) Extraction of circle segments

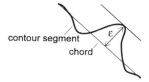

Figure 7.46. Distance between chord and contour segment

minimum are candidate corners. If the magnitude of the curvature exceeds a threshold, a candidate is accepted as corner, otherwise it is rejected. Figure 7.45a gives an example.

The curvature can also be helpful to determine segments in which some local property is required to be homogeneous. For instance, circle segments with fixed radius (within a tolerance interval) are characterised by a part of the running arc length the curvature of which is constant (within certain limits). Figure 7.45b gives an example. Here, multi-thresholding is applied to find contour segments that satisfies the curvature constraint. After that, small segments are suppressed, since they are probably stem from corners.

"Split-and-merge"-like techniques applies as well. The next algorithm implements a pure "split" scheme. It determines the vertices of a polygon that approximates the contour. The algorithm splits a segment into two parts whenever the chord connecting the end points of the segment does not sufficiently approximate the segment. For instance, whenever the distance between chord and segment exceeds a threshold (figure 7.46).

Algorithm 7.8 Polygon approximation of a closed contour

Input: a boundary list with start point $a = (n_0, m_0)$
1. Initiate: create an empty list "*links*" consisting of ordered pairs of boundary points, and a list "*open*".
2. Determine the point b on the boundary list, farthest from a.
3. Put (a,b) and (b,a) on *open*.
4. Remove the top element (c,d) from *open*.
5. Determine the point e on the boundary list between (c,d) which is farthest from the chord connecting c and d.
6. If the distance ε between e and the chord exceeds a threshold:
    6.1. Append (c,e) and (e,d) to *open*.
   else:
    6.2. Append (c,d) to *links*.
7. If *open* is not empty: go to 4.
8. Examine the two chords in *links* containing the start point a. if necessary: merge the two chords by replacing the two chords by one.
9. Exit.

output: the list "*links*" consisting of neighbouring vertices of the polygon.

Figure 7.47 shows some results. Further developments of this type of algorithm can be found in [Pavlidis 1977; Haralick and Shapiro 1992].

Properties of contour segments

Contour segments can be described with a number of parameters: start point, end point, length, average orientation, etc. Methods to find some of these parameters include the Hough transform and curve fitting.

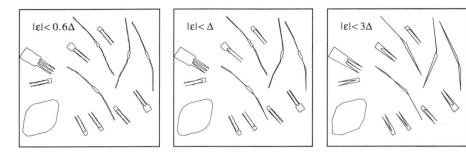

Figure 7.47 Application of polygon approximation

Parameters related to the curvature are: *average curvature* and *bending energy*:

$$\overline{k} = \frac{1}{L}\int_{s=0}^{L} k(s)ds \qquad \overline{k^2} = \frac{1}{L}\int_{s=0}^{L} k^2(s)ds \qquad (7.61)$$

where L is the length of the segment, and s the running arc length from start point to end point.

7.2.3 Relational description

In the preceding section each region was considered individually. In addition to that, relations between regions are often important clues to recognise an object. For instance in figure 7.48a, the regions are difficult to interpret without using the context in which they are placed.

The relations considered in this section are binary. It involves pairs of regions. Although in some applications useful, ternary and higher order relations are not discussed in this book.

Region adjacency graph

The *region adjacency graph* (RAG) is a graph that indicates which regions are direct neighbors. Nodes of the graph correspond to regions (or connected

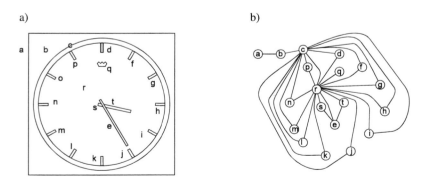

Figure 7.48 The region adjacency graph
a) A region map
b) Corresponding RAG

components). Two nodes are connected by an arc if the two corresponding regions are adjacent. The correspondence between a node and a region is established most conveniently by assigning a unique label to both of them. Sometimes the area outside the image plane - the *frame* - is also assigned a label. Figure 7.48b shows the RAG of figure 7.48a.

Some of the definitions in digital morphology (section 5.4.1) are borrowed from graph theory. For instance, a *path* in a graph is a sequence of nodes such that any two nodes adjacent in the sequence are connected in the graph. Two nodes are *connected* if a path exists between the nodes. A graph is *connected* if any pair of nodes are connected. A *component* of a graph is a connected subgraph whose nodes are not connected to any other subgraph.

Other definitions are as follows. The *degree* of a node is the number of nodes connected to it. The *node connectivity* of a graph is the minimum number of nodes which must be removed to create different components. In figure 7.48b, the node connectivity is 1 since removal of node b, c, or r would create components. These nodes are called *cutnodes*. The *arc connectivity* is the minimum number of arcs that must be removed to create different components. A *bridge* is an arc the removal of which would create new component(s). In figure 7.48b, the arc b-c is a bridge.

In terms of region relations, a node with degree 1 is a hole, i.e. a region (the frame not included) completely surrounded by another region. A cutnode corresponds to a region that surrounds a group of other regions.

Relations

We start with some relations between two regions A and B that are defined as ratios. With that, most relations can be made fairly invariant to changes of position, scale and orientation.

1. Adjacency $a(A,B)$:
   This is a measure which indicates to what extent region A is adjacent to region B. $a(A,B)=0$ expresses that there is no adjacency. If $a(A,B)=1$, then A is completely surrounded by B without surrounding another region. It can be defined as the fraction of the boundary of A which adjoins region B. Note that $a(A,B) \neq a(B,A)$.
2. Relative size $s(A,B)$:
   The size of region A related to that of region B can be expressed as the ratio between both areas

$$s(A,B) = \frac{\text{area}(A)}{\text{area}(B)} \qquad (7.62)$$

3. Normalised distance $d_{relative}(A,B)$:
   The distance between two regions can be defined in various ways. In morphology the distance is defined as the length of the shortest path connecting region A and B. Another possibility is to use the distance between the two centres of gravity. A relative measure, invariant to scale, is obtained by normalising the distance with respect to a scale parameter, for instance:

$$d_{relative}(A,B) = \frac{d(A,B)}{\sqrt{\bar{r}_A \bar{r}_B}} \qquad (7.63)$$

where $\bar{r}_A$ is the average radius of A (e.g. the amplitude $|z_1|$ of the first Fourier descriptor).

It is also possible to determine the distance taken in a specific direction (e.g. horizontal, vertical). One way to do that is to measure the distance between the centres of gravity projected onto the axis with the desired direction.

4. Congruence $c(A, B)$:
This is a measure of similarity of shape. If normalised Fourier descriptors are used to characterise shape, then the Euclidean distance between two contours can be used:

$$c(A, B) = \sqrt{\sum_{k=-K_{max}}^{K_{max}} |Z_k(A) - Z_k(B)|^2} \qquad (7.64)$$

5. Contrast $C(A, B)$:
This parameter characterises the relative change of average irradiance within the regions:

$$C(A, B) = \frac{2(E_A - E_B)}{E_A + E_B} \qquad (7.65)$$

where E_A is the average irradiance within region A. In the same way relative changes between texture parameters measured in the two regions can be defined.

Propositions are Boolean functions that establish binary relations between two regions. The relations given above are defined such that it is not too difficult to find reasonable thresholds with which propositions are formed. For instance: the proposition: $A\_is\_near\_B$ can be tested by checking whether $d_{relative}(A, B)$ is smaller than 0.5. Other examples of propositions are:

$A\_is\_adjacent\_to\_B$ $\qquad\qquad$ $A\_is\_larger\_than\_B$
$A\_is\_above\_B$ $\qquad\qquad$ $A\_is\_below\_B$
$A\_is\_left\_from\_B$ $\qquad\qquad$ $A\_is\_right\_from\_B$
$A\_surrounds\_B$ $\qquad\qquad$ $A\_is\_identical\_to\_B$

The representation of the relations found in a segmented image is most conveniently done by augmenting the RAG with labels attached to the arcs. In the same way region properties may be attached to the nodes so as to arrive at an attributed graph (figure 7.2). A possible implementation is the *node adjacency list* (figure 7.49).

7.3 OBJECT RECOGNITION

Regions, region parameters and regional relations are characteristics of an image. The imaged scene also possesses a number of characteristics: objects, object properties, relations between objects, and so on. The characteristic of the image are *intrinsic*; those of the scene *external*. Some of external characteristics are known before image acquisition has taken place. It belongs to the prior knowledge. Of course, a relationship must exist between the external characteristics and the internal characteristics. Without that, regions and related parameters do not add to the knowledge we have about the scene. In the preceding chapters some mathematical and physical models have already been passed: an illumination

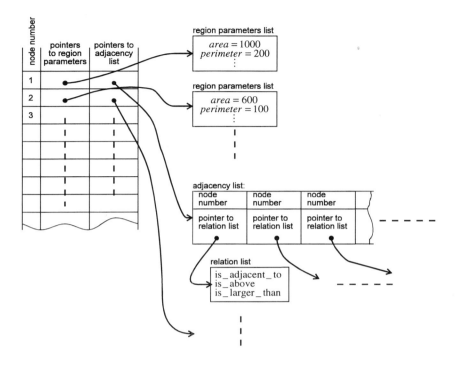

Figure 7.49 The labeled RAG implemented with adjacency lists

model, a reflection model, a projection model, image models, etc. In the next subsection, an *object model* will be added to this list. This model represents the relevant part of the prior knowledge of the scene.

Figure 7.50 gives an overview of a typical approach to 3-dimensional object recognition and scene description. The method is based on 2-dimensional, perspective projection imagery. The dimensionality of the measured data is one less than that of the measurand(s). Therefore, without specific prior knowledge the interpretation of the image is ambiguous. In particular, the knowledge about the 3-dimensional geometry of the objects is crucial. Together with the knowledge about the image formation system, the 3-dimensional geometry imposes object-specific properties to the 2-dimensional data structures. Matching techniques are techniques in which the 2-dimensional properties are checked to see whether they apply to the measured data at hand. With that, matching may give us a 3-dimensional interpretation.

Matching techniques can be applied at various levels. Perhaps the most rudimentary form is image correlation (section 5.3), i.e. template matching. The limitations of matching at a low level have already been mentioned. At the highest level, the interpretation of the (labeled) RAG involves *graph matching* and related techniques. Here, a sequence of image operations has to be invoked in order to arrive at such a description. The difficulty is that errors made in an early stage (e.g. segmentation errors) may give rise to aberrations of the RAG which are not foreseen. Without reusing the raw image data it is hard to recover such errors.

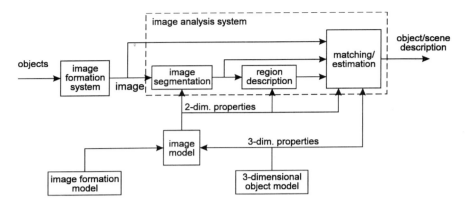

Figure 7.50 3-dimensional object recognition and scene description

The term *bottom-up control* is reserved for a strategy in which the raw image data is successively transformed to higher levels. After that a complete description has been formed the object model is used to arrive at an interpretation. *Top-down control* starts with an expectation that the image contains a particular object. From this expectation, predictions at lower levels are formed with which the actuality of the expectation is verified (hypothesis testing). Template matching is an example.

If the scene consists of many objects, and the configuration of objects is hardly constrained, neither bottom-up nor top-down control is appropriate. Systems with mixed control strategies, for instance *top-down control with feedback*, are better suited. The design of such systems is still the topic of many research projects. A unifying, flexible, and versatile approach has not been found as yet.

Another topic of research is the development of *model based* systems. In most vision systems, the structure of the models and the various procedures are intertwined with each other. In contrast, a model based system uses models that are stored explicitly in a knowledge base which can be consulted whenever needed. The advantage is that only a small part of the system needs to be changed in order to make it appropriate for another application.

7.3.1 Object models

Perhaps the simplest model occurs when the objects can be thought of as being concentrated in an object plane that is aligned perpendicular to the optical axis. Such a situation occurs in OCR and document analysis systems. Another application is in the quality control of material surfaces (metallic plates, textiles, etc.). The object model states that the plane is partitioned into different regions. Each region has its own radiometric properties. Many quality inspection problems are solved if regions with deviating radiometric properties can be detected and located. It requires a segmentation method that is carefully selected. Other applications (e.g. the analysis of engineering line drawings) need a relational object model in order to give the right interpretation.

Even if the objects are inherently 3-dimensional, a 2-dimensional object model may suffice provided that the application is such that the "frontal face" of the object is the crucial part.

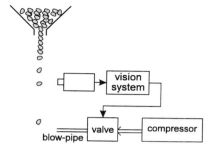

Figure 7.51 Schematic overview of a quality control system for small sized objects

Example 7.3 Inspection of nuts (continued)

Figure 7.51 shows a mechanical design for the inspection of small sized objects such as nuts. The transport mechanism is such that individual particles accelerate. This causes a separation distance between the particles. The next step is to illuminate the objects and to acquire an image. The latter can be done with a line-scan camera. A vision system decides whether the object is acceptable. If not, a electro-mechanical actuator removes the particle from the stream.

The attributed digraph in figure 7.52 is a relational object model describing the existence of nuts. Such a graph is called a *semantic network*. In this example, we discriminate four classes of objects: the "background", "healthy nuts", "nuts with burst shell", and "empty shells". Each class corresponds to a node in the graph. Nodes and arcs are labeled so as to represent properties of objects and relations between objects. For instance, it can be seen that the class "nut with burst shell" consists of parts with different properties. Note that the model reflects the property that particles are separated: they are always surrounded by background.

The model given in figure 7.52 is a simplification. It does not involve a stipulation of the region properties. In practice, the radiometric properties include spectral features. Furthermore, the frontal view of an empty shell does not always reveal "irregular" boundaries since the fracture may be located at the reverse side. Therefore, it may be necessary to use two or more multiple-view images of the same objects. Mirrors can be exploited in order to acquire these images concurrently.

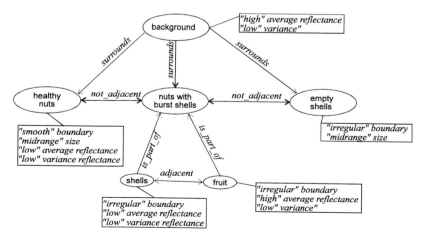

Figure 7.52 Object model represented by a semantic network

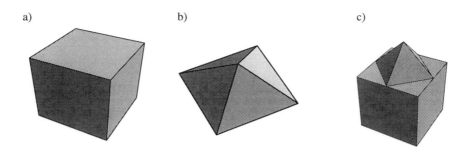

Figure 7.53 Modeling 3-dimensional objects with polyhedrons
a-b) Polyhedrons
c)   Combination of a and b

If the 3-dimensional nature of objects cannot be ignored, more complex models must be invoked. Examples are the objects which are built from *polyhedrons* (i.e. 3-dimensional solid cells the surfaces of which consist of planar faces connected by 3-dimensional edges; see figure 7.53).

Another example of a 3-dimensional model is the *generalised cylinder* [Nevatia and Binford 1977]. Here, a 3-dimensional curve, called the *spine* forms the skeleton of the object. With each point on the spine a planar cross section is associated which is orthogonal to the tangent of the spine taken at that particular point. The shape of the cross section is a parametric function of the position on the spine (i.e. of the running arc length of the spine). The complete set of cross sections makes up the volume of the object. Examples are given in figure 7.54.

If the polyhedrons and generalised cylinders are kept simple, i.e. defined such that a description with a few parameters suffices, they can be regarded as primitives. Translation, rotation, scaling, unions and/or intersections are the basic operations with which a number of these primitives can be combined into more complex objects. Syntactic rules restrict the permissible combinations. As a result, the geometric properties of primitives, and the relations between them, are constrained. This brings us back to an object model represented by semantic networks.

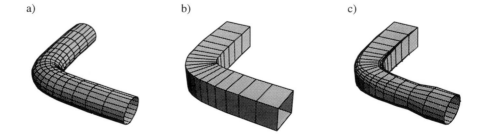

Figure 7.54 Modelling 3-dimensional objects with generalised cylinders
a) With rotationally symmetric cross section and curved spine
b) With rectangular cross section
c) With varying cross section

7.3.2 From object model to image model

Once an object model and an image formation model have been established, it becomes desirable to combine them in an image model. This model gives the relevant aspects of the 2-dimensional data considered at various hierarchical levels. Constraints, relations and properties of the 3-dimensional objects are reflected in a 2-dimensional version.

Even in the case that the objects are planar-like, an image model is needed. The reason is that an object model does not take into account the degradation during image formation and image acquisition. For instance, over- and under-segmentation may be due to poor image quality. It is an aspect that is not covered by the object model.

Example 7.3 Inspection of nuts (continued)

With a properly designed image formation and acquisition system, the semantic network from figure 7.52 readily transforms into a 2-dimensional image model. At the level of pixels the model states that regions associated with objects (nuts, empty shells) distinguish themselves from the background in that their radiance is substantially less than that of the background. Therefore, segmentation of the image can be based simply on multi-thresholding or edge detection. Problems may arise with parts of a "nut with burst shell" since the fruit of a nut can have a large reflectance making it indistinguishable from the background.

At the level of regions, we must anticipate errors which are made during segmentation. If we apply edge detection operating at a high resolution scale (this is necessary in order to preserve the "irregularity" of boundaries) we may expect that small erroneous regions may appear, see figure 7.30. Furthermore, we can also safely ignore "objects-of-interest" that touch the frame of the image (if a nut is only partly visible, we assume that it will become completely visible in the next frame). With these assumptions, the semantic network transforms into the graph shown in figure 7.55.

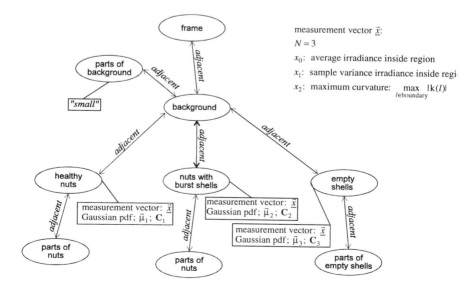

Figure 7.55 Object model transformed into an image model

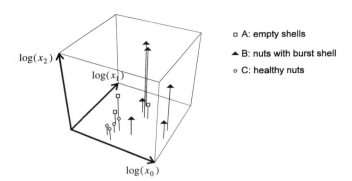

Figure 7.56 Scatter diagram of the measurement vectors of the objects in figure 7.30a

The regions that correspond to objects are characterised by a 3-dimensional measurement vector $\tilde{x}$. The vector embodies the region properties. The feature "irregular shape" is defined with the curvature of the contour. A scatter diagram of a few samples (figure 7.56) shows that each vector element contribute to the quality of the classification. The selected measurements are not well suited to discriminate between "empty shells" and "nuts with burst shell". However, since both classes are to be removed from the stream, this should not disturb us too much. There is one empty shell that cannot be distinguished from healthy nuts. The reason is that its fracture is on the reverse side. As mentioned above, multi-view acquisition should solve this problem.

Models of objects with a clear 3-dimensional nature impose constraints in the 2-dimensional image domain. An example, applicable to polyhedrons, is the model based on *aspect graphs* [Koenderink and van Doorn 1979]. An *aspect* of a polyhedron is the set of visible faces. One can imagine a sphere circumscribing the object. If we move our point of view along the surface of the sphere, then we can divide the surface into several partitions where the same faces are visible. Each partition corresponds to an appearance or aspect of the object. If the faces of an aspect are considered as nodes, and if nodes whose faces are adjacent in the image plane are connected by arcs, then the result is called an aspect graph. Figure 7.57 depicts the aspect graphs of the images of the objects in figure 7.53.

With fixed distance from camera to object the number of aspect graphs of a polyhedron is limited. Figures 7.58a and b show an object with all its aspect graphs. If the number of polyhedron types in a scene is limited, then recognition of an object can be done by storing all possible aspect graphs in a database, and by searching the aspect graph that matches the intrinsic RAG of an image best.

Figure 7.57 Aspect graphs of the images in figure 7.53

In practise, all kinds of problems must be faced in order to arrive at tractable solutions:

- Because the number of aspect graphs rapidly grows as the number of faces increases, the shape of the polyhedrons must not be too complex.
- The aspect graph of a polyhedron is certainly not unique. Other polyhedrons may have the same aspect graph.
- If combinations of polyhedrons are to be recognised, the intrinsic RAG of the image becomes a concatenation of aspect graphs. An object in the foreground may occlude an object in the background. Of course, this will affect the intrinsic RAG.
- Noise in the image acquisition system is likely to mutilate the intrinsic RAG.

The recognition can be made more robust. This requires an extension of the aspect graphs. For instance, attributes can be appended to the nodes so as to add information to the image of the faces (see figure 7.58c). Also, the interrelations between nodes can be made more expressive. For instance, it may be known in advance that one face is always on top. This could be expressed in a relation like "is_above".

Another source of information comes from the radiometry. If the illumination is conditioned, the orientation of a face imposes certain radiometric constraints in the image domain. In fact, the orientation of a face can be estimated from radiometric properties of a region of the image (shape from shading [Horn and Brooks 1989; Korsten 1989]). Shadow can also be utilised to improve the robustness. For that purpose, the aspect graph in figure 7.58d must be extended with an extra node. The position of the shadow depends on the position of the illuminator. This could be embedded in the model with a constraint like: "shadow_must_be_right_from_object". Additional constraints are in the shape of regions: the shape of a shadow region depends on the shape of the object.

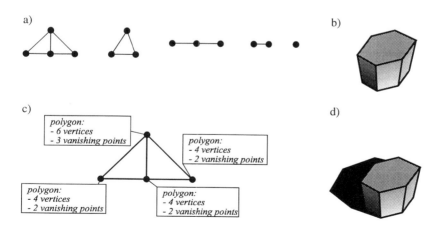

Figure 7.58 Aspect graph of a simply shaped object
a-b) The aspect graphs of the object given in b
c)   Attributed aspect graph.
d)   The aspect graph in c should be extended by an extra node so as to encompass the shadow region

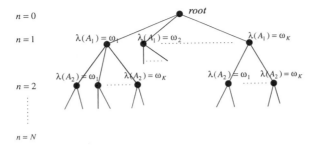

Figure 7.59 Search tree for region labeling

7.3.3 Matching techniques

An acquired image can be given an interpretation by setting up a correspondence between the intrinsic characteristics of the image and the external characteristics of the scene. Often, such a correspondence is established by assigning labels (symbols, classes) from the object model to the regions found in the image. More formally, suppose that the different types of objects mentioned in the object model are denoted by the symbol ω_k, and suppose that there are K of such types, i.e. the set of *external labels* is $\Omega = \{\omega_1, \cdots, \omega_K\}$. Furthermore, suppose that the regions are denoted by $A_1, \cdots, A_N$. The number of found regions is N. Then, an *interpretation* or a *labeling* is a set of N labels taken from Ω and assigned to the N regions: $\Lambda = \{\lambda(A_n) | n = 1, \cdots, N\}$ with $\lambda(A_n) \in \Omega$.

Clearly, the number of different labelings of an image is K^N. The problem is to select a labeling which represents a right semantic description of the scene. The description is assumed to be semantically correct when the intrinsic characteristics match the image model. For instance, suppose that the image model states that certain propositions must hold true in order to permit a certain labeling. Then, that particular labeling is assumed to be incorrect if these propositions are not confirmed by the intrinsic properties of the image.

A brute-force strategy to find a right description of the scene is to check all possible labelings, and to select the one with best match. One way to do that is to set up a search tree consisting of $N+1$ levels. Except for the root, each level corresponds to a region (figure 7.59). The tree is constructed such that the nodes between the root and level n contain all K^n possible labelings of the regions A_1 up to A_n. At level N all possible labels are covered. The brute-force method uses a graph traversal strategy to visit all nodes of the tree. For each labeling (i.e. each leaf) a measure of match is calculated, and the labeling with maximum measure of match is selected.

Unfortunately, the computational complexity of the brute-force strategy is on the order of K^N. This makes it impractical. The number of calculations can be considerably decreased by "branch-and-bound"-like techniques similar to the one given in section 6.5.1 and algorithm 7.2 in section 7.1.3. As an example, suppose that region properties and relations in the image model consist of propositions that have to be confirmed by the observed data in order to have a correct interpretation. Then, the following algorithm is a branch-and-bound algorithm:

IMAGE ANALYSIS

Algorithm 7.9: Region labeling

Input:
- a propositional image model.
- an attributed RAG of an observed image.

1. Create an empty list - called OPEN - in which nodes from the search tree can be stored.
2. Move *root* (= root of the search tree) to OPEN.
3. If OPEN is empty: print "No labeling found"; exit.
4. Remove the last entry - called *node* - from the list OPEN.
5. Check consistency of labeling according to *node*.
6. If *node* is not consistent, go to 3.
7. If *node* is a leaf, go to 11.
8. Append all children-nodes of *node* to OPEN.
9. Optional: Reorder the entries in the list OPEN.
10. Go to 3.
11. Store labeling according to *node* in Λ; exit

Output: a correct labeling stored in Λ (if one exists).

The algorithm stops as soon as a correct interpretation has been found. This interpretation is not necessarily unique. The computational savings come from step 6. Although these savings are considerable, the algorithm is still impractical (i.e. still on the order of K^N). Step 9 is optional. If it is omitted, the algorithm traverses the search tree according to *depth first* strategy. It goes directly to a leaf. If that leaf is not a correct labelling, the shortest path is chosen to find a next leaf. If step 9 is not omitted, a suitably selected sort criterion will speed up the algorithm just as in heuristic search (algorithm 7.1).

Reasoning

There is a formal theory to express facts and rules, and to combine them in order to infer new facts [Nilsson 1980]. This theory, called *predicate calculus*, can be used to represent propositional models and data structures. For that purpose, it uses *predicates, functions, variables,* and *constants*. A statement, for instance, that all transistors are either black or metallic could be expressed as:

$$(\forall x)[\text{TRANSISTOR}(x) \Rightarrow \text{BLACK}(x) \vee \text{METALLIC}(x)] \tag{7.66}$$

x is a variable; TRANSISTOR(·), BLACK(·), and METALLIC(·) are functions. The functions are connected by *logical connectives*:

$\forall x$	for all x
$\Rightarrow$	implies
$\vee$	or
$\exists x$	there exists an x
$\wedge$	and
$\neg$	not

The sentence above reads as: "for all x, if x is a transistor, then x is either black or metallic.

Relations are represented by predicates. For instance, the predicate ADJACENT_TO(region_x, frame) is true whenever region region_x is adjacent to frame. region_x is a variable; frame is a constant. From existing facts new facts can be derived using certain inference rules. Suppose that: $\neg\text{BLACK}(x) \wedge \neg\text{METALLIC}(x)$ holds true, then in combination with rule (7.66) we can infer that: $\neg\text{TRANSISTOR}(x)$. The inference rule applied here is called *Modus Tollens*.

Example 7.3 Inspection of nuts (continued)

The following logical statements follow from the object and image model in figures 7.52 and 7.55:

$$(\forall x)[\text{ADJACENT\_TO}(x,\text{frame})$$
$$\Rightarrow \text{BACKGROUND}(x)]$$
$$(\forall x)[(\exists y)[\text{BACKGROUND}(y) \wedge \text{SURROUNDS}(y,x)] \wedge \text{SMALL}(x)$$
$$\Rightarrow \text{BACKGROUND}(x)] \qquad (7.67)$$
$$(\forall x)[\neg \text{SMALL}(x) \wedge (\exists y)[\text{BACKGROUND}(y) \wedge \text{ADJACENT\_TO}(y,x)]$$
$$\Rightarrow \text{PART\_OF\_OBJECT}(x)]$$
$$(\forall x)[(\exists y)[\text{PART\_OF\_OBJECT}(y) \wedge \text{SURROUNDS}(y,x)]$$
$$\Rightarrow \text{PART\_OF\_OBJECT}(x)]$$

The first statement says that all regions adjacent to "frame" are background regions. The following line states that a small region that is surrounded by background is also background. The last two statements define the existence of regions that are part of an object (i.e. "nut", "nut with burst shell", or "empty shell"). Such regions are not small. They are either adjacent to the background, or adjacent to other object parts.

With the given statements it is easy to find a procedure that labels all "part of an object" regions in a RAG of a segmented image. First, all regions adjacent to the frame are labeled as background regions. Then, small holes in the background regions are also labeled as background. The remaining regions must be object parts.

The procedure is easily implemented in languages like Prolog and Lisp. Alternatively, an implementation using "if...then..." statements can be exploited:

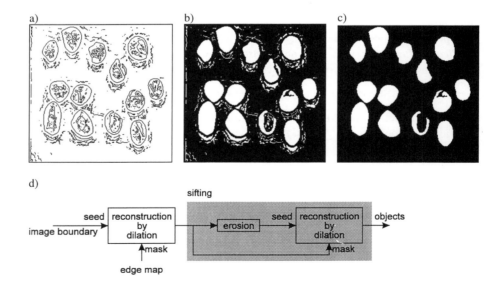

Figure 7.60 Detection of objects
a) Map with detected edges
b) Regions adjacent to the frame (shown in black)
c) Map with detected objects
d) Computational structure

if *(region is adjacent to frame)* then:
    region is background
if *(region is small and region is surrounded by background)* then:
    region is background
if *(region is not small and region is adjacent to background)* then:
    region is part of an object
if *(region is surrounded by object)* then:
    region is part of an object

If the image model is not too complex - such as in the current example - an implementation without explicitly building the RAG is feasible. Suppose that a map with detected edges (figure 7.60a) is available. Then the "surrounds" and "adjacent to" predicates, and the region property "small" can be implemented with morphological operations.

The first "if...then..." statement can be realised by propagating the border of the image area into the edge map (figure 7.60b). Small holes in the background are detected and removed with the sifting operation mentioned in example 5.4. The remaining regions are objects (figure 7.60c). The structure of the morphological operations is depicted in figure 7.60d.

Relaxation

The principle of (discrete) relaxation is to successively rule out labelings that are inconsistent with the intrinsic properties of the observed image. The process starts with the set of all K^N different labelings. In a sequence of iterations this number is reduced by eliminating those labelings that are incompatible.

With each region A_n we associate a label set L_n, initially containing all labels from Ω. In the first pass, the algorithm checks the so-called *node consistency* of labels in L_n. Individual region properties are compared with regional constraints induced by the image model. For some regions, this will rule out some labels. A record of that can be kept by removing those labels from L_n that are incompatible. After the first pass L_n is a subset of Ω: $L_n \subseteq \Omega$.

The next passes are iterative. Each iteration checks the so-called *arc consistency* between any pair of regions. Suppose that A_n and A_m are regions with label sets L_n and L_m, respectively. The essence of arc consistency is that every label $\lambda_i \in L_n$ is consistent with some label $\lambda' \in L_m$. If not, label λ_i is to be removed from L_n. The iterations continue until no changes occur. After that, the label sets L_n are consistent.

Example 7.4 recognition of blocks

Figure 7.61a shows the image of a scene consisting of two blocks. The boundaries of the regions found by image segmentation are overlaid. The purpose is to find all blocks in the scene. For that purpose, objects are modeled as polyhedrons with six rectangular faces. Adjacent faces are orthogonal. The object model is translated into an image model given in figure 7.61c. In this model regions are approximated by polygons (figure 7.61b). Vertices of polygons are called *T-corners* if the number of different regions in a small neighbourhood is at least three.

As an example, consider the (fictitious) segmented image in figure 7.61d. Here, we have four regions. A check on node consistency reveals that region A_4 is incompatible since the number of vertices exceeds six. Furthermore, arc consistency requires that walls are below a roof. This rules out the labels *wall_l* and *wall_r* for region A_1. With a similar argument it follows that certain labels are ruled out for regions A_2 and A_3. The final result is shown in figure 7.61f. In this simplified example, the final result is obtained after one iteration. In practise, a few iterations are needed in order to have full arc consistency.

a)
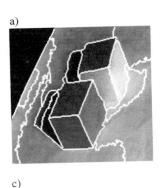

b)
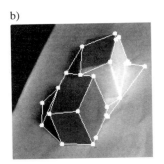

c)
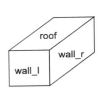

node consistency:
$4 \leq wall\_l.nr\_vertices \leq 6$ $wall\_l.nr\_T\_corners \geq 2$
$4 \leq wall\_r.nr\_vertices \leq 6$ $wall\_r.nr\_T\_corners \geq 2$
$4 \leq roof.nr\_vertices \leq 6$ $roof.nr\_T\_corners \geq 2$

arc consistency:
$wall\_l$ ADJACENT $wall\_r$ $wall\_l$ LEFT_FROM $wall\_r$
$wall\_l$ ADJACENT $roof$ $roof$ ABOVE $wall\_r$
$wall\_r$ ADJACENT $roof$ $roof$ ABOVE $wall\_l$

d)
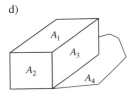

e)

	roof	wall_l	wall_r
A_1	1	1	1
A_2	1	1	1
A_3	1	1	1
A_4	0	0	0

f)

	roof	wall_l	wall_r
A_1	1	0	0
A_2	0	1	0
A_3	0	0	1
A_4	0	0	0

Figure 7.61 Discrete relaxation. (Reproduced by permission of K. Schutte, published by K. Schutte, Enschede)
a) Image of a scene consisting of blocks
b) Polygon approximation of regions. Inconsistent regions have been removed
c) Image model
d) Found regions
e) Consistency matrix after node consistency check. A "1" denotes consistency
f) Consistency matrix after arc consistency check

In the example above, the resulting label sets define an unambiguous labeling. This is certainly not guaranteed. Often, relaxation ends up in label sets containing a large number of labels. Therefore, it needs further processing in order to come to a true labeling. Nevertheless, relaxation is useful since it reduces the solution space to a fraction of K^N whereas the computational complexity of relaxation is on the order of $N^2 K^3$.

Relaxation can also be placed in a probabilistic framework. In that case, the first pass is an assignment of probabilities. Based on individual region properties, each region is given an initial probability that it carries a certain label. Then, the algorithm steps into an iterative scheme. From the relations between a region and neighboring regions, a so-called *support function* is formed that expresses the support from the neighbors that the region carries a label. This support is evaluated for each region. It is used to update the label probabilities. After that, a new iteration begins. The process terminates when no significant changes occur in the probabilities. Unfortunately, it is not always guaranteed that the iterations converge.

A thorough treatment of discrete relaxation can be found in [Henderson 1990]. A sound foundation for probabilistic relaxation has been given by Kittler et al [1989].

Graph matching

Suppose that the image model of an object is given as a graph $g_m = (V_m, E_m)$ where V_m is a set of vertices and $E_m \subseteq V_m \times V_m$ a set of arcs. For example, g_m could be an aspect graph of a polyhedron (section 7.3.2). In addition, suppose that $g_d = (V_d, E_d)$ is a graph representing the data derived from an observed image of a scene. We want to know whether the scene contains the object described by g_m.

The problem boils down to the question whether g_d is, or contains, g_m. More formally, two graphs g_d and g_m are said to be *isomorphic* if a bijective mapping $f: g_d \to g_m$ exists such that $(f(x), f(y)) \in E_m \Leftrightarrow (x, y) \in E_d$. A graph g_m is an *isomorphic subgraph* of g_d if there is a subgraph $g'_d \subseteq g_d$ that is isomorphic to g_m. Figure 7.62a gives an example of two graphs for which one is an isomorphic subgraph of the other.

With a small adaptation algorithm 7.9 (page 281) can be used to detect isomorphic subgraphs. The algorithm must construct a search tree such as the one in figure 7.62b. The construction of the tree takes place in step 8. In step 5 the current node of the search tree is checked to see whether the associated subgraph is still isomorphic. If not (step 6), the corresponding branch in the search tree is traced back so as to try another subgraph.

Unfortunately, the complexity of the algorithm is on the order of N^K, where N is the number of vertices in g_d, and K the number of vertices in g_m. As mentioned above, the search space can be reduced considerably by application of discrete relaxation. Bunke [1993] describes an implementation in which arc consistency is applied at each node of the search tree.

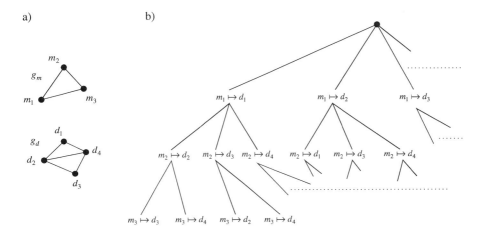

Figure 7.62 Subgraph isomorphism
a) g_m is an isomorhic subgraph of g_d
b) A search tree to find the subgraph(s) of g_d that are isomorphic to g_m

Usually, the data graph is an attributed graph, for instance, the RAG augmented with properties and relations. If these attributes are propositional, it is straightforward to extent the consistency check in step 5 of the isomorphic subgraph algorithm so as to encompass the additional requirements.

Suppose that the properties and relations of regions are not propositional. In that case, we cannot prune so easily the search tree because similarity is no more a discrete event. The obvious way to proceed is to quantify the similarity between two graphs. One method to do that is to assign cost to a mismatch. Here, we must take into account node similarity and arc similarity (i.e. relational similarity). If during a certain phase of the algorithm a number of M nodes of the model graph g_m is associated with nodes of the data graph g_d (say $d_i = f(m_i)$ $i=1,\cdots,M$), then the mismatch between m_i and d_i may be quantified by a cost function $C_{node}(m_i,d_i)$. Likewise, the mismatch between an arc (m_i,m_j) and its associated arc (d_i,d_j) may be quantified by a cost function $C_{arc}(m_i,m_j,d_i,d_j)$.

In addition to this, we may also want to assign cost to nodes in the model graph and the data graph that have not been incorporated in the association. For that purpose, we introduce a cost $C_{missing\_model\_node}$ and $C_{missing\_data\_node}$ for each node in g_m and g_d, respectively, that are not involved in the association.

Adding up the cost we have the following measure of mismatch:

$$C_{subgraph\_match} = \sum_{i=1}^{M} C_{node}(m_i,d_i) + \sum_{i=1}^{M}\sum_{\substack{j=1 \\ j \neq i}}^{M} C_{arc}(m_i,m_j,d_i,d_j) + \qquad (7.68)$$

$$+ N_m C_{missing\_model\_node} + N_d C_{missing\_data\_node}$$

where N_m and N_d are the number of missing nodes in the model graph and data graph, respectively.

As an example, suppose that the model prescribes that a certain node, say m_i should not be elongated. Then $C_{node}(m_i,d_i)$ could be defined as: $elongation(d_i)-1$, where $elongation(\cdot)$ is the ratio between the length of the maximum chord and the minimum chord of the region associated with d_i. The cost of an arc mismatch can be defined similarly. For instance, if the model prescribes that node m_i must surround m_j, the cost of a mismatch may be quantified as: $C_{arc}(m_i,m_j,d_i,d_j) = a(d_i,d_j)$ where $a(.,.)$ is the adjacency relation defined in section 7.2.3.

With the availability of a measure of match, the region labeling algorithm 7.9 on page 281 is applicable again provided that some minor modifications are accomplished:

Algorithm 7.10 Minimum cost graph matching

  Input:
  - an attributed model graph.
  - an attributed data graph derived from an observed image.
  1. Create an empty list - called OPEN - in which nodes from the search tree can be stored.
  2. Move *root* (= root of the search tree) to OPEN.
  3. Remove the top entry - called *node* - from the list OPEN.
  4. If *node* is a leaf, go to 8.
  5. Append all children-nodes of *node* to OPEN.

6. Sort the entries in the list OPEN according to decreasing cost of the subgraph match associated with the nodes in OPEN.
7. Go to 3.
8. Store labeling according to *node* in Λ; exit

Output: a labeling in Λ.

The algorithm is an implementation of the *A algorithm* described by Nilsson [1980]. It searches a minimum-cost path from the root to a leaf. The cost functions $C_{missing\_model\_node}$ and $C_{missing\_data\_node}$ play an important role. If they are set low, the algorithm is likely to find a low-cost, acceptable labeling. If they are set high, the algorithm becomes much faster, but also less reliable.

In practise we will not be able to obtain data graphs that contain exact replica of the model graphs. There are many factors that cause imperfections in our data graph. Noise in the input images may lead to over- or under-segmentation. Other disturbing factors are due to neighboring objects, for instance, *occlusion*. The distortions in the image may lead to missing or spurious nodes in our data graph.

One way to cope with distorted data graphs is to include a range of imperfect versions of the model graph in a database and to match all of them with the data graph. A more elegant method is to adapt the graph matching algorithm in order to be able to find *double subgraph isomorphisms*, i.e. isomorphism between subgraphs of g_m and subgraphs of g_d. An appropriate algorithm (clique finding in an association graph) can be found in [Ambler et al 1975; Ballard and Brown 1982]. Other possibilities are the algorithms for so-called *inexact graph matching* or *error tolerant graph matching*. These algorithms generalises the minimum cost graph matching algorithm. Nodes of g_d may be labeled as "spurious". Excessive nodes from g_m can be handled by addition of "missing" nodes in g_d. Therefore, the transformations needed to make g_d isomorphic to g_m are "insertion" and "deletion". Note that the insertion or deletion of nodes also implies that arcs are inserted or deleted. By the assignment of cost to inserted or deleted nodes and arcs a minimum cost solution is within reach. Details can be found in [Bunke 1993].

Another practical problem not discussed so far is that a scene often consists of multiple objects. A data graph containing multiple model graphs will be very large. This makes it impractical for graph matching. A solution for that is *grouping* [Huertas and Nevatia 1988; Price and Huertas 1992], i.e. selecting subgraphs from the data graph such that each subgraph is likely to contain exactly one model graph. *Pivoting* is a technique that can be used for finding the subgraphs. A pivot is a characteristic node from the model graph, that is easy to recognise in the data graph. If such a pivot is found in the data graph, a group is formed, for instance, by selecting the pivot, the neighbors of the pivot, and the neighbors of these neighbors.

Example 7.4 recognition of blocks (continued)

The model of the objects in figure 7.63a is given in figure 7.61. From this model it follows that a region that is closely approximated by a polygon with four vertices and two vanishing points is likely to be part of a block. Hence, a pivot could be defined as a region with the properties just mentioned.

Figure 7.63b gives an interpretation of the image. The object model and the graph matching method is robust enough to cope with the occlusion. Details of the implementation are given by Schutte [1994].

a) 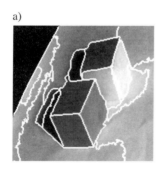 b)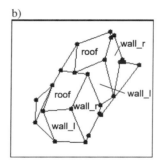

Figure 7.63 Recognition of block shaped objects (reproduced by permission of K. Schutte, published by K. Schutte, Enschede)
a) The region whose boundary segments are most "straight" is selected as a pivot
b) A labeling of the image found by graph matching

7.4 ESTIMATION OF 3-DIMENSIONAL BODY PARAMETERS

This section concerns the estimation of the parameters of an object in the scene. We assume that the geometric structure of the object is available either as prior knowledge or as the result from object recognition techniques. The missing knowledge is in the form of unknown parameters that describe shape, size, position, orientation, and/or radiometric properties of the object.

As an example, assume that the scene consists of a simple block shaped object (figure 7.64). The geometry of the measurement set-up is most conveniently given in terms of three co-ordinate systems. The object is described in a suitably chosen system of (homogeneous) *body co-ordinates* $\vec{\xi}^t = [\xi \quad \eta \quad \zeta \quad 1]$. The system is defined relative to a reference system $\vec{X}^t = [X \quad Y \quad Z \quad 1]$ called *world co-ordinates*. The position and orientation of the body with respect to world co-ordinates is conveniently described with three translation and three rotation parameters; see section 2.2.3.

Camera co-ordinates $\vec{x}^t = [x \quad y \quad z \quad 1]$ are chosen such that the z-axis coincides with the optical axis. The $x-y$ plane is at a certain distance from the pinhole (e.g. the focal distance). The transformation from body co-ordinates to camera co-ordinates is given by (see eq. 2.7):

$$\vec{x} = \mathbf{PM}_{cw}\mathbf{M}_{wb}\vec{\xi} \qquad (7.69)$$

The matrix $\mathbf{M}_{wb}$ consists of successive rotations around the X, Y, and Z axes. The rotations are followed by a translation. Similar parameters exists for $\mathbf{M}_{cw}$. The pinhole projection involves two parameters: the focal distance in $\mathbf{P}$ and the diaphragm (i.e. aperture size). The whole transformation is given by 14 parameters.

In addition to that, there is also a transformation from camera co-ordinates to the row and column indices of the CCD. The parameters needed to define this transformation include: sampling period, aspect ratio, translation, eventually a rotation (or even an affine transformation if the normal of the CCD surface is not properly aligned).

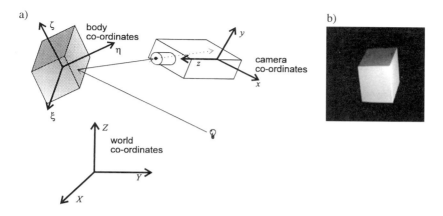

Figure 7.64 Estimation of the parameters of a block shaped object
a) Geometry of object and image formation system
b) Observed image

If the radiometry of the image is used to estimate body parameters, we have to take into account the illumination. In the simplest case, this can be modeled with an isotropic point source. Four parameters suffices to describe that (position and radiant flux).

Besides the position and orientation parameters in $\mathbf{M}_{wb}$ the body parameters include size and shape parameters (three in case of a block shaped object), and parameters describing the reflectance distribution. If the surface is mat, the reflectance is Lambertian, and the reflectance distribution may be modeled by a single parameter. If this does not suffice, a more complex reflection model is needed. For instance, the Torrance-Sparrow model mentioned in section 2.1.2.

As a whole, the measurement set up in figure 7.64a is described with at least 26 parameters. With a single observation (such as in figure 7.64b) not all of these parameters can be estimated. Radiant flux, aperture size, and reflectance coefficient are strictly coupled. Therefore, one cannot estimate these parameters simultaneously. The same holds true for parameters defining the matrices $\mathbf{M}_{cw}$ and $\mathbf{M}_{wb}$.

Our measurement system needs *calibration*, i.e. the determination of the transformation characteristics. If the goal is to measure the *pose* (position and orientation) of an object, then it would be desirable that all other parameters (the *system parameters*) are known in advance. A calibration procedure can provide us with that knowledge. Calibration objects, the parameters of which are known, enable us to determine the transformation characteristics accurately. For example, the geometric transformation parameters can be measured with the object shown in figure 7.65. The position of the markers on this object are well known. They are chosen such that the geometric parameters of the image formation system can be estimated.

Suppose that the parameters that we want to measure are denoted by $\vec{\alpha}$ and the (known) system parameters by $\vec{\beta}$. Let $f_{n,m}$ be the observed image. Apart from random elements (e.g. noise), the image formation and acquisition model (Chapters 2 and 4) give the dependency of the image data on $\vec{\alpha}$ and $\vec{\beta}$. We state this

290 IMAGE BASED MEASUREMENT SYSTEMS

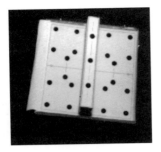

Figure 7.65 A calibration object (reproduced by permission of A.J. de Graaf, Hengelo)

dependency more explicitly by writing $\vec{f} = \vec{h}(\vec{\alpha})$. It would have been more precise to write $\vec{f} = \vec{h}(\vec{\alpha},\vec{\beta})$, but for the sake of brevity the former notation is preferred. $\vec{f}$ is a measurement vector containing all pixels from the image. $\vec{h}(.)$ represents the imaging process. In fact, with given $\vec{\alpha}$ and $\vec{\beta}$ computer graphics techniques enable us to generate an artificial image of our object by evaluating $\vec{h}(\vec{\alpha})$. The techniques include *ray tracing* or *rendering* [Foley et al 1989].

If the noise in the image is considered to be additive and object independent, we conceive the observed image as:

$$\vec{f} = \vec{h}(\vec{\alpha}) + \underline{\vec{n}} \qquad (7.70)$$

This model brings us to the discussion in section 6.2.5. According to (6.65) and (6.67) the least square error criterion gives an estimate of the parameter vector $\vec{\alpha}$ given by:

$$\hat{\vec{\alpha}}_{LSE}(\vec{f}) = \underset{\vec{\alpha}}{\operatorname{argmin}}\{\|\vec{f} - \vec{h}(\vec{\alpha})\|_2^2\} \qquad (7.71)$$

Furthermore, if the image formation would have been linear $\vec{f} = \mathbf{H}\vec{\alpha} + \underline{\vec{n}}$, then the solution of (7.71) would have been (6.68):

$$\hat{\vec{\alpha}}_{LSE}(\vec{f}) = (\mathbf{H}^t\mathbf{H})^{-1}\mathbf{H}^t\vec{f} \qquad (7.72)$$

In reality, image formation is certainly a non-linear process. However, we may expand $\vec{h}(\vec{\alpha})$ in a Taylor series and ignore second order and higher terms so as to arrive at a linear approximation:

$$\vec{f} = \vec{h}(\vec{\alpha}_0) + \mathbf{H}(\vec{\alpha}_0)\partial\vec{\alpha} + \underline{\vec{n}}' \quad \text{with:} \quad h_{i,j}(\vec{\alpha}_0) = \frac{\partial h_i(\vec{\alpha}_0)}{\partial \alpha_j} \quad \text{and:} \quad \vec{\alpha} = \vec{\alpha}_0 + \partial\vec{\alpha} \qquad (7.73)$$

The matrix $\mathbf{H}(\vec{\alpha}_0)$ is the Jacobian matrix of the vector function $\vec{h}(\vec{\alpha})$. Higher order terms of the Taylor series expansion are thought to be absorbed in the random vector $\underline{n}'$. Rewriting (7.73) gives us the LSE solution for $\partial\vec{\alpha}$:

$$\partial\vec{\alpha} = (\mathbf{H}^t(\vec{\alpha}_0)\mathbf{H}(\vec{\alpha}_0))^{-1}\mathbf{H}^t(\vec{\alpha}_0)(\vec{f} - \vec{h}(\vec{\alpha}_0)) \qquad (7.74)$$

With the result in (7.74) an iterative solution method, similar to Newton's method of root finding, is within reach. Suppose that in the k-th iteration of an estimation algorithm we have an estimation of $\vec{\alpha}$ denoted by $\hat{\vec{\alpha}}_k$. Then, a correction $\Delta\hat{\vec{\alpha}}_k$ of this estimate follows from (7.74). The correction gives us the estimate in the next iteration:

$$\Delta\hat{\vec{\alpha}}_k = \left(\mathbf{H}'(\hat{\vec{\alpha}}_k)\mathbf{H}(\hat{\vec{\alpha}}_k)\right)^{-1}\mathbf{H}'(\hat{\vec{\alpha}}_k)\left(\vec{f}-\vec{h}(\hat{\vec{\alpha}}_k)\right)$$
$$\hat{\vec{\alpha}}_{k+1} = \hat{\vec{\alpha}}_k + \Delta\hat{\vec{\alpha}}_k \quad (7.75)$$

The process is depicted in figure 7.66. In this process, $\vec{f}-\vec{h}(\hat{\vec{\alpha}}_k)$ are the residuals in the k-th iteration.

The process is successful if it converges to the true parameter vector (or at least close to it). This success depends on several factors: the initial guess $\hat{\vec{\alpha}}_0$, the number of parameters, the stability, the sensitivity to errors in the Jacobian matrix, etc. If the initial guess is too far from the correct parameter vector, the process will end up in a local minimum of the criterion function. Low-pass filtering of the residuals will increase the range of convergence, because this will smooth the criterion function.

Development of analytic expressions to compute the transposed Jacobian matrix is difficult. The alternative is to compute it by numerical differentiation, e.g.:

$$h_{i,j}(\hat{\vec{\alpha}}_k) = \frac{h_i(\hat{\vec{\alpha}}_k;\Delta\alpha_j) - h_i(\hat{\vec{\alpha}}_k)}{\Delta\alpha_j} \quad (7.76)$$

where $h_i(\hat{\vec{\alpha}}_k;\Delta\alpha_j)$ is the predicted grey level with parameter $\hat{\alpha}_{k_j}$ changed to $\hat{\alpha}_{k_j}+\Delta\alpha_j$. The step $\Delta\alpha_j$ must be selected carefully since aliasing may cause large errors.

Figure 7.67 illustrates the estimation process. In this example, two position parameters and one orientation parameter are simultaneously estimated. The problem is equivalent to pose estimation (six parameters) in which the object is constrained to be lying on a surface (such as in vehicle guidance). If more than three parameters are to be estimated, the process is likely to find a local minimum. The use of multiple view images (i.e. stereoscopic vision) may largely solve this problem [Graaf et al 1990].

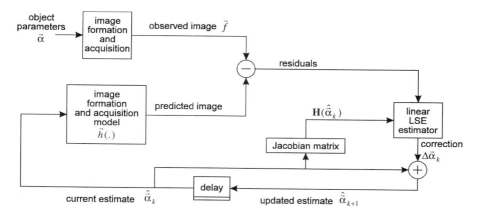

Figure 7.66 Linearised estimation of body parameters

292   IMAGE BASED MEASUREMENT SYSTEMS

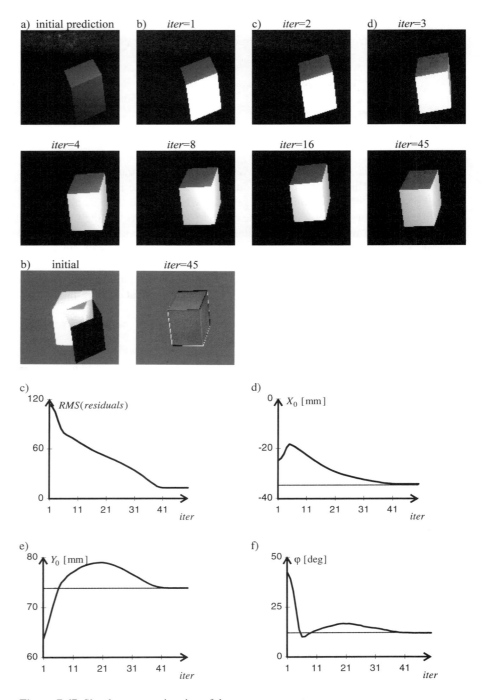

Figure 7.67 Simultaneous estimation of three pose parameters
a) Predicted images
b) Residuals
c)   *RMS* of the residuals
d-f) Estimated parameters

The estimation procedure discussed so far is also applicable in feature spaces. For instance, if the radiometry of the set up is hard to quantify, we may find it useful to transform the grey level image into a region map first. Estimation can be based on the minimisation of a distance measure between measured region map and predicted maps. As an alternative, region descriptors, such as moments and Fourier descriptors, may be derived from the region maps. These descriptors can also be used to define a cost function the minimisation of which yields the estimates of the object parameters.

REFERENCES

Ambler, A.P., Barrow, H.G., Brown, C.M., Burstall, R.M., and Popplestone, R.J., *A Versatile Computer-Controlled Assembly System*, Artificial Intelligence, Vol. 6, No. 2, 129-156, 1975.

Ballard, D.H., Brown, C.M., *Computer vision,* Prentice-Hall, Englewood Cliffs, London, 1982.

Canny, J.F., *A Computational Approach to Edge Detection.* IEEE Tr PAMI, Vol. 8, 679-698, November 1986.

Bunke, H., *Structural and Syntactic Pattern Recognition*, in: *Handbook of Pattern Recognition & Computer Vision* (ed. by Chen, C.H., Pau, L.F. and Wang, P.S.P), World Scientific, Singapore, 1993.

Graaf, A.J. de, Korsten, M.J., Houkes, Z., *Estimation of Position and Orientation of Objects from Stereo Images*, Proc. Mustererkennung 12. DAGM-Symposium (ed. R.E. Grosskopf), Springer-Verlag, Berlin, September 1990.

Foley, J.D., van Dam, A., Feiner, S.K., Hughes, J.F., *Computer Graphics, Principles and Practice*, Addison-Wesley Publishing Company, Massachusets, 1989.

Haralick, R.M. and Shapiro, L.G., *Computer and Robot Vision, Vol. I*, Addison-Wesley, Massachusetts, 1992.

Henderson, T.C., *Discrete Relaxation Techniques,* Oxford University Press, New York, 1990.

Heijden, F. van der, *On the Discretization Error of Fourier Descriptors of Planar Closed Curves*, Eusipco-86 Signal Processing III: Theory and Applications (Ed. I.T. Young et al), North-Holland, Amsterdam, 1986.

Heijden, F. van der, *A Statistical Approach to Edge and Line Detection in Digital Images*, Ph D Thesis, Faculty of Electrical Engineering, University of Twente, Enschede, 1991.

Horn, B.K.P. and Brooks, M.J., *Shape from Shading*, MIT Press, Massachusetts, 1989.

Hu, M.K., *Visual Pattern Recognition by Moments Invariants*, IRE Tr Info. Theory, Vol. IT-8, 179-187, 1962.

Huertas, A. and Nevatia, R., *Detecting Buildings in Aerial Images*, Computer

Vision, Graphics and Image Processing 41, 1988.

Jaehne, B., *Digital Image Processing - Concepts, Algorithms and Scientific Applications*, Springer-Verlag, Berlin, 1991

Jolion, J.M. and Rosenfeld, A., *A Pyramid Framework for Early Vision: Multiresolutional Computer Vision*, Kluwer Academic Publishers, Dordrecht, 1994.

Kittler, J. and Hancock, E.R., *Combining Evidence in Probabilistic Relaxation*, Int. J. of Pattern Recognition and Artificial Intelligence, Vol. 3, 412-424, January 1989.

Koenderink, J.J. and van Doorn, A.J., *The Internal Representation of Solid Shape with Respect to Vision*, Biological Cybernetics, Vol. 32, 211-216, 1979.

Korsten, M.J., *Three-Dimensional Body Parameter Estimation from Digital Images*, Ph D Thesis, Faculty of Electrical Engineering, University of Twente, Enschede, 1989.

Kreyszig, E.: *Introductory Mathematical Statistics - Principles and methods*, J. Wiley & Sons Inc, New York, 1970.

Kulpa, Z., *More about Areas and Perimeters of Quantized Objects*, Computer Graphics and Image Processing, Vol. 22, 268-276, 1983.

Martelli, A., *Edge Detection using Heuristic Search Methods*, Computer Graphics and Image Processing, Vol. 1, No. 2, 123-134, August 1972.

Nevatia, R. and Binford, T.O., *Description and Recognition of Curved Objects*, Artificial Intelligence, Vol. 8, 77-98, 1977.

Nilsson, N.J., *Principles of Artificial Intelligence*, Palo Alto, Tioga, 1980.

Pavlidis, T., *Structural pattern recognition*, Springer-Verlag, Berlin, 1977.

Price, K. and Huertas, A., *Using Perceptual Grouping to Detect Objects in Aerial Scenes*, ISPRS, Proceedings 17th Congres, Vol 29-3, Washington, 1992.

Schutte, K., *Knowledge Based Recognition of Man -Made Objects*, Ph.D. Thesis, Faculty of Electrical Engineering, University of Twente, Enschede, 1994.

Wallace, T.P. and Wintz, P.A., *An Efficient Three-Dimensional Aircraft Recognition Algorithm Using Normalized Fourier Descriptors*, Computer Graphics and Image Processing, Vol. 13, 99-126, 1980.

Zhou, Y.T., Venkateswar, R. and Chellappa, R., *Edge Detection and Linear Feature Extraction using a 2-D Random Field Model*, IEEE Tr PAMI, Vol. 11, 84-95, Janaury 1989.

APPENDIX A

TOPICS SELECTED FROM LINEAR ALGEBRA AND MATRIX THEORY

This appendix summarises some concepts from linear algebra and matrix theory. The concepts are part of the mathematical background required in this book. Mathematical peculiarities not relevant in this context are omitted. Instead, at the end of the appendix references to a detailed treatment are given.

A.1   LINEAR SPACES

A *linear space* (or *vector space*) over a field F is a set R with elements (*vectors*) $\mathbf{f}, \mathbf{g}, \mathbf{h}, \cdots$ equipped with two operations:
- *addition* $(\mathbf{f}+\mathbf{g})$: $\qquad\qquad\qquad R \times R \to R$.
- *scalar multiplication* ($\alpha \mathbf{f}$ with $\alpha \in F$): $\quad F \times R \to R$.

Usually, the field F is the set $\mathbb{R}$ of real numbers, or the set $\mathbb{C}$ of complex numbers. The addition and the multiplication operation must satisfy the following axioms:

a) $\mathbf{f}+\mathbf{g}=\mathbf{g}+\mathbf{f}$
b) $(\mathbf{f}+\mathbf{g})+\mathbf{h}=\mathbf{f}+(\mathbf{g}+\mathbf{h})$
c) there exists a so-called *zero element* $\mathbf{0} \in R$ such that $\mathbf{f}+\mathbf{0}=\mathbf{f}$
d) associated with $\mathbf{f}$ there exists a so-called *negative element* $-\mathbf{f}$ such that $\mathbf{f}+(-\mathbf{f})=\mathbf{0}$
e) $\alpha(\mathbf{f}+\mathbf{g})=\alpha\mathbf{f}+\alpha\mathbf{g}$
f) $(\alpha+\beta)\mathbf{f}=\alpha\mathbf{f}+\beta\mathbf{f}$
g) $(\alpha\beta)\mathbf{f}=\alpha(\beta\mathbf{f})$
h) $1\mathbf{f}=\mathbf{f}$

A *linear subspace* S of a linear space R is a subset of R which itself is linear. A condition sufficient and necessary for a subset $S \subset R$ to be linear is that $\alpha\mathbf{f}+\beta\mathbf{g} \in S$ for all $\mathbf{f},\mathbf{g} \in S$ and for all $\alpha,\beta \in F$.

Examples:

$\mathbb{C}[a,b]$ is the set of all complex functions $f(x)$ continuous in the interval $[a,b]$. With the usual definition of addition and scalar multiplication this set is a linear space[1]. The set of polynomials of degree N:

$$f(x) = c_0 + c_1 x + c_2 x^2 + \cdots + c_N x^N \qquad \text{with} \quad c_n \in \mathbb{C}$$

is a linear subspace of $\mathbb{C}[a,b]$.

The set $\mathbb{R}^\infty$ consisting of an infinite, countable series of real numbers $\mathbf{f} = (f_0, f_1, \cdots)$ is a linear space provided that the addition and multiplication takes place element by element[1]. The subset of $\mathbb{R}^\infty$ that satisfies the convergence criterion:

$$\sum_{n=0}^{\infty} |f_n|^2 < \infty$$

is a linear subspace of $\mathbb{R}^\infty$.

The set $\mathbb{R}^N$ consisting of N real numbers $\mathbf{f} = (f_0, f_1, \cdots f_{N-1})$ is a linear space provided that the addition and multiplication takes place element by element[1]. Any linear hyperplane containing the null vector (zero element) is a linear subspace.

Any vector that can be written as:

$$\mathbf{f} = \sum_{i=0}^{n-1} \alpha_i \mathbf{f}_i \qquad \alpha_i \in F \tag{a.1}$$

is called a *linear combination* of $\mathbf{f}_0, \mathbf{f}_1, \cdots, \mathbf{f}_{n-1}$. The vectors $\mathbf{f}_0, \mathbf{f}_1, \cdots, \mathbf{f}_{m-1}$ are *linear dependent* if a set of numbers β_i exists, not all zero, for which the following equation holds:

$$\sum_{i=0}^{m-1} \beta_i \mathbf{f}_i = \mathbf{0} \tag{a.2}$$

If no such set exists, the vectors $\mathbf{f}_0, \mathbf{f}_1, \cdots, \mathbf{f}_{m-1}$ are said to be *linear independent*.

The *dimension* of a linear space R is defined as the non-negative integer number N for which there exists N independent linear vectors in R, while a set of $N+1$ vectors in R is linear dependent. If for each number N there exists N linear independent vectors in R, the dimension of R is ∞.

Example:

$\mathbb{C}[a,b]$ and $\mathbb{R}^\infty$ have dimension ∞. $\mathbb{R}^N$ has dimension N.

A *norm* $\|\mathbf{f}\|$ of a linear space is a mapping $R \to \mathbb{R}$ (i.e. a *real function* or a *functional*) such that:

a) $\|\mathbf{f}\| \geq 0$, where $\|\mathbf{f}\| = 0$ if and only if $\mathbf{f} = \mathbf{0}$

---

[1] Throughout this appendix the examples relate to vectors which are either real or complex. However, these examples can be converted easily from real to complex or vice verse. The set of all real functions continuous in the interval $[a,b]$ are denoted by $\mathbb{R}[a,b]$. The set of infinite and finite countable complex numbers is denoted by $\mathbb{C}^\infty$ and $\mathbb{C}^N$, respectively.

b) $\|\alpha \mathbf{f}\| = |\alpha| \, \|\mathbf{f}\|$
c) $\|\mathbf{f} + \mathbf{g}\| \leq \|\mathbf{f}\| + \|\mathbf{g}\|$

A linear space equipped with a norm is called a *normed linear space*.

Examples:

The following real functions satisfy the axioms of a norm:
In $\mathbb{C}[a,b]$:

$$\|f(x)\|_p = \left(\int_{x=a}^{b} |f(x)|^p \, dx \right)^{1/p} \qquad \text{with: } p \geq 1 \qquad (a.3a)$$

In $\mathbb{R}^\infty$:

$$\|\mathbf{f}\|_p = \left(\sum_{n=0}^{\infty} |f_n|^p \right)^{1/p} \qquad \text{with: } p \geq 1 \qquad (a.3b)$$

In $\mathbb{R}^N$:

$$\|\mathbf{f}\|_p = \left(\sum_{n=0}^{N-1} |f_n|^p \right)^{1/p} \qquad \text{with: } p \geq 1 \qquad (a.3c)$$

These norms are called the L_p-*norm* (continuous case, e.g. $\mathbb{C}[a,b]$) and the l_p-*norm* (discrete case, e.g. $\mathbb{R}^\infty$). Graphical representations of the norm are depicted in figure A.1. Special cases occur for particular choices of the parameter p. If $p = 1$, the norm is the sum of the absolute magnitudes, e.g.:

$$\|f(x)\|_1 = \int_{x=a}^{b} |f(x)| \, dx \qquad \text{and} \qquad \|\mathbf{f}\|_1 = \sum_n |f_n| \qquad (a.4a)$$

If $p = 2$, we have the *Euclidean norm*:

$$\|f(x)\|_2 = \sqrt{\int_{x=a}^{b} |f(x)|^2 \, dx} \qquad \text{and} \qquad \|\mathbf{f}\|_2 = \sqrt{\sum_n |f_n|^2} \qquad (a.4b)$$

In $\mathbb{R}^3$ and $\mathbb{R}^2$ the norm $\|\mathbf{f}\|_2$ is the length of the vector as defined in geometry.
If $p \to \infty$, the L_p- and l_p-norm is the maximum of the absolute differences, e.g.:

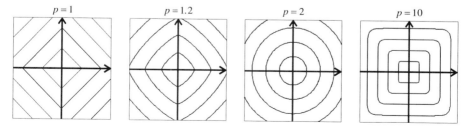

Figure A.1 "Circles" in $\mathbb{R}^2$ equipped with the l_p-norm

$$\|f(x)\|_\infty = \max_{x\in[a,b]} |f(x)| \qquad \text{and} \qquad \|\mathbf{f}\|_\infty = \max_n |f_n| \qquad (a.4c)$$

A *Euclidean space* (also called *inner product space*) R is a linear space for which the *inner product* is defined. The inner product $(\mathbf{f},\mathbf{g})$ between two vectors $\mathbf{f},\mathbf{g} \in R$ over a field F is a mapping $R \times R \to F$ that satisfies the following axioms:

a) $(\mathbf{f}+\mathbf{g},\mathbf{h}) = (\mathbf{f},\mathbf{h}) + (\mathbf{g},\mathbf{h})$
b) $(\alpha\mathbf{f},\mathbf{g}) = \alpha(\mathbf{f},\mathbf{g})$
c) $(\mathbf{g},\mathbf{f}) = \overline{(\mathbf{f},\mathbf{g})}$
d) $(\mathbf{f},\mathbf{f}) \geq 0$, real
e) $(\mathbf{f},\mathbf{f}) = 0 \Rightarrow \mathbf{f} = \mathbf{0}$

In c) the number $\overline{(\mathbf{f},\mathbf{g})}$ is the complex conjugated of $(\mathbf{f},\mathbf{g})$. Of course, this makes sense only if F equals $\mathbb{C}$. If F is the set of real numbers, then $(\mathbf{g},\mathbf{f}) = (\mathbf{f},\mathbf{g})$.

Examples:

$\mathbb{C}[a,b]$ is a Euclidean space if the inner product is defined as:

$$(f(x),g(x)) = \int_{x=a}^{b} f(x)\overline{g(x)}dx$$

$\mathbb{R}^\infty$ is a Euclidean space if with $\mathbf{f} = (f_0, f_1, \cdots)$ and $\mathbf{g} = (g_0, g_1, \cdots)$:

$$(\mathbf{f},\mathbf{g}) = \sum_{n=0}^{\infty} f_i g_i$$

$\mathbb{R}^N$ is a Euclidean space if with $\mathbf{f} = (f_0, f_1, \cdots, f_{N-1})$ and $\mathbf{g} = (g_0, g_1, \cdots, g_{N-1})$:

$$(\mathbf{f},\mathbf{g}) = \sum_{n=0}^{N-1} f_i g_i$$

In accordance with (a.4b), any Euclidean space becomes a normed linear space as soon as it is equipped with the *Euclidean norm*:

$$\|\mathbf{f}\| = +\sqrt{(\mathbf{f},\mathbf{f})} \qquad (a.5)$$

Given this norm, any two vectors $\mathbf{f}$ and $\mathbf{g}$ satisfy the *Schwarz inequality*:

$$|(\mathbf{f},\mathbf{g})| \leq \|\mathbf{f}\|\|\mathbf{g}\| \qquad (a.6)$$

Two vectors $\mathbf{f}$ and $\mathbf{g}$ are said to be *orthogonal* (notation: $\mathbf{f} \perp \mathbf{g}$) whenever $(\mathbf{f},\mathbf{g}) = 0$. If two vectors $\mathbf{f}$ and $\mathbf{g}$ are orthogonal then (*Pythagoras*):

$$\|\mathbf{f}+\mathbf{g}\|^2 = \|\mathbf{f}\|^2 + \|\mathbf{g}\|^2 \qquad (a.7)$$

Given a vector $\mathbf{g} \neq \mathbf{0}$. An arbitrary vector $\mathbf{f}$ can be decomposed into a component that coincides with $\mathbf{g}$ and an orthogonal component: $\mathbf{f} = \alpha \mathbf{g} + \mathbf{h}$ with $\mathbf{g} \perp \mathbf{h}$. The scalar α follows from:

$$\alpha = \frac{(\mathbf{f},\mathbf{g})}{(\mathbf{g},\mathbf{g})} = \frac{(\mathbf{f},\mathbf{g})}{\|\mathbf{g}\|^2} \qquad (a.8)$$

The term $\alpha\mathbf{g}$ is called the *projection* of $\mathbf{f}$ on $\mathbf{g}$.
The *angle* φ between two vectors $\mathbf{f}$ and $\mathbf{g}$ is defined such that it satisfies:

$$\cos(\varphi) = \frac{(\mathbf{f},\mathbf{g})}{\|\mathbf{f}\|\|\mathbf{g}\|} \qquad (a.9)$$

A.2 METRIC SPACES

A *metric space* R is a set equipped with a *distance measure* $\rho(\mathbf{f},\mathbf{g})$ that maps any couple of elements $\mathbf{f},\mathbf{g} \in R$ into a non-negative real number: $R \times R \to \mathbb{R}^+$. The distance measure must satisfy the following axioms:
a) $\rho(\mathbf{f},\mathbf{g}) = 0$ if and only if $\mathbf{f} = \mathbf{g}$.
b) $\rho(\mathbf{f},\mathbf{g}) = \rho(\mathbf{g},\mathbf{f})$
c) $\rho(\mathbf{f},\mathbf{h}) \leq \rho(\mathbf{f},\mathbf{g}) + \rho(\mathbf{g},\mathbf{h})$

Examples:

All normed linear spaces become a metric space, if we set:

$$\rho(\mathbf{f},\mathbf{g}) = \|\mathbf{f} - \mathbf{g}\| \qquad (a.10)$$

Consequently, the following mappings satisfy the axioms of a distance measure:
In $\mathbb{C}[a,b]$:

$$\rho(f(x),g(x)) = \left(\int_{x=a}^{b} |f(x) - g(x)|^p \, dx \right)^{1/p} \qquad \text{with: } p \geq 1 \qquad (a.11a)$$

In $\mathbb{R}^\infty$:

$$\rho(\mathbf{f},\mathbf{g}) = \left(\sum_{n=0}^{\infty} |f_n - g_n|^p \right)^{1/p} \qquad \text{with: } p \geq 1 \qquad (a.11b)$$

In $\mathbb{R}^N$:

$$\rho(\mathbf{f},\mathbf{g}) = \left(\sum_{n=0}^{N-1} |f_n - g_n|^p \right)^{1/p} \qquad \text{with: } p \geq 1 \qquad (a.11c)$$

A theorem related to this measure is *Minkowski's inequality*. In $\mathbb{R}^N$ and $\mathbb{R}^\infty$ this equality states that[2]:

$$\left(\sum_n |f_n + g_n|^p \right)^{1/p} \leq \left(\sum_n |f_n|^p \right)^{1/p} + \left(\sum_n |g_n|^p \right)^{1/p} \qquad (a.12)$$

---

[2] In $\mathbb{C}[a,b]$ the inequality is similar.

Special cases occur for particular choices of the parameter p. If $p=1$, the distance measure equals the sum of absolute differences between the various coefficients, e.g.:

$$\rho(f(x),g(x)) = \int_{x=a}^{b} |f(x)-g(x)|dx \quad \text{and} \quad \rho(\mathbf{f},\mathbf{g}) = \sum_{n} |f_n - g_n| \qquad (a.13a)$$

This measure is called the *magnitude distance* or the *city-block distance*. If $p=2$, we have the ordinary *Euclidean distance measure*. If $p \to \infty$ the maximum of the absolute differences between the various coefficients are determined, e.g.:

$$\rho(f(x),g(x)) = \max_{x\in[a,b]} |f(x)-g(x)| \quad \text{and} \quad \rho(\mathbf{f},\mathbf{g}) = \max_{n} |f_n - g_n| \qquad (a.13b)$$

This measure is the *maximum distance* or the *chessboard distance*.

In section A.4 we will define a so-called *self-adjoint operator* **A**. If the eigenvalues of **A** are non-negative, then it can be shown (section A.8) that the following function satisfies the axioms of a distance measure:

$$\rho(\mathbf{f},\mathbf{g}) = +\sqrt{(\mathbf{f}-\mathbf{g},\mathbf{Af}-\mathbf{Ag})} \qquad (a.14)$$

This measure finds its application in, for instance, pattern classification (the Mahalanobis distance).

Another distance measure is:

$$\rho(\mathbf{f},\mathbf{g}) = \begin{cases} 1 & \text{if } \mathbf{f}=\mathbf{g} \\ 0 & \text{if } \mathbf{f} \neq \mathbf{g} \end{cases} \qquad (a.15)$$

An application of this measure is in Bayesian estimation and classification theory where it is used to express a particular cost function. Note that in contrast with the preceding examples this measure cannot be derived from a norm. For every metric derived from a norm we have $\rho(\alpha\mathbf{f},\mathbf{0}) = |\alpha|\rho(\mathbf{f},\mathbf{0})$. However, this equality does not hold for (a.15).

A.3 ORTHONORMAL SYSTEMS AND FOURIER SERIES

In a Euclidean space R with the norm given by (a.5), a subset $S \subset R$ is an *orthogonal system* if every couple $\mathbf{b}_i, \mathbf{b}_j$ in S is orthogonal; i.e. $(\mathbf{b}_i, \mathbf{b}_j) = 0$ whenever $i \neq j$. If in addition each vector in S has unit length, i.e. $\|\mathbf{b}_i\| = 1$, then S is called an *orthonormal system*.

Examples:

In $\mathbb{C}[a,b]$ the following harmonic functions form an orthonormal system:

$$b_n(x) = \frac{1}{\sqrt{b-a}} \exp\left(\frac{2\pi j n x}{b-a}\right) \quad \text{with:} \quad j=\sqrt{-1} \qquad (a.16a)$$

In $\mathbb{C}^N$ the following vectors are an orthonormal system:

$$\mathbf{b}_n = \left[\frac{1}{\sqrt{N}} \quad \frac{1}{\sqrt{N}}\exp\left(\frac{2\pi j n}{N}\right) \quad \frac{1}{\sqrt{N}}\exp\left(\frac{2\pi j 2n}{N}\right) \quad \cdots \quad \frac{1}{\sqrt{N}}\exp\left(\frac{2\pi j (N-1)n}{N}\right)\right] \qquad (a.16b)$$

Let $S = \{\mathbf{b}_0 \ \mathbf{b}_1 \ \cdots \ \mathbf{b}_{N-1}\}$ be an orthonormal system in a Euclidean space R, and let $\mathbf{f}$ be an arbitrary vector in R. Then the *Fourier coefficients* of $\mathbf{f}$ with respect to S are the inner products:

$$\phi_k = (\mathbf{f}, \mathbf{b}_k) \qquad k = 0, 1, \cdots, N-1 \tag{a.17}$$

Furthermore, the series:

$$\sum_{k=0}^{N-1} \phi_k \mathbf{b}_k \tag{a.18}$$

is called the *Fourier series* of $\mathbf{f}$ with respect to the system S. Suppose one wishes to approximate $\mathbf{f}$ by a suitably chosen linear combination of the system S. The best approximation (according to the norm in R) is given by (a.18). This follows from the following inequality, that can be proven quite easily:

$$\left\| \mathbf{f} - \sum_{k=0}^{N-1} \phi_k \mathbf{b}_k \right\| \leq \left\| \mathbf{f} - \sum_{k=0}^{N-1} \beta_k \mathbf{b}_k \right\| \qquad \text{for arbitrary } \beta_k \tag{a.19}$$

The approximation improves as the number N of vectors increases. This follows readily from *Bessel's inequality*:

$$\sum_{k=0}^{N-1} |\phi_k|^2 \leq \|\mathbf{f}\|^2 \tag{a.20}$$

Let $S = \{\mathbf{b}_0 \ \mathbf{b}_1 \ \cdots\}$ be an orthonormal system in a Euclidean space R. The number of vectors in S may be infinite. Suppose that no vector $\tilde{\mathbf{b}}$ exists for which the system S augmented by $\tilde{\mathbf{b}}$, i.e. $\tilde{S} = \{\tilde{\mathbf{b}} \ \mathbf{b}_0 \ \mathbf{b}_1 \ \cdots\}$, is also an orthonormal system. Then, S is called an *orthonormal basis*. In that case, the smallest linear subspace containing S is the whole space R.

The number of vectors in an orthonormal basis S may be finite (as in $\mathbb{R}^N$ and $\mathbb{C}^N$), countable infinite (as in $\mathbb{R}^\infty$, $\mathbb{C}^\infty$, $\mathbb{R}[a,b]$, and $\mathbb{C}[a,b]$ with $-\infty < a < b < \infty$), or uncountable infinite (as in $\mathbb{R}[-\infty, \infty]$, and $\mathbb{C}[-\infty, \infty]$).

Examples:

In $\mathbb{C}[a,b]$ with $-\infty < a < b < \infty$ an orthonormal basis is given by the harmonic functions in (a.16a). The number of such functions is countable infinite. The Fourier series defined in (a.18) essentially corresponds to the Fourier series expansion of a periodic signal. In $\mathbb{C}[-\infty, \infty]$ this expansion evolves into the Fourier integral[3].

In $\mathbb{C}^N$ an orthonormal basis is given by the vectors in (a.16b). The Fourier series in (a.18) is equivalent to the discrete Fourier transform.

The examples given above are certainly not the only orthonormal bases. In fact, even in $\mathbb{R}^N$ (with $N > 1$) infinitely many orthonormal bases exist.

---

[3] In fact, $\mathbb{C}[-\infty, \infty]$ must satisfy some conditions in order to assure the existence of the Fourier expansion.

If S is an orthonormal basis, and $\mathbf{f}$ an arbitrary vector with Fourier coefficients ϕ_k with respect to S, then the following theorems hold:

$$\mathbf{f} = \sum_k \phi_k \mathbf{b}_k \tag{a.21}$$

$$(\mathbf{f},\mathbf{g}) = \sum_k (\mathbf{f},\mathbf{b}_k)(\mathbf{g},\mathbf{b}_k) \qquad \text{(Parseval)} \tag{a.22}$$

$$\|\mathbf{f}\|^2 = \sum_k |\phi_k|^2 \qquad \text{(Parseval/Pythagoras)} \tag{a.23}$$

Equation (a.21) corresponds to the inverse transform in Fourier analysis. Equation (a.23) follows directly from (a.22). The equation shows that in case of an orthonormal basis Bessel's inequality transforms to an equality.

A.4 LINEAR OPERATORS

Given two normed linear spaces R_1 and R_2. A mapping A of a subset of R_1 into R_2 is called an *operator* from R_1 to R_2. The subset of R_1 (possibly R_1 itself) for which the operator A is defined is called the *domain* D_A of A. The *range* R_A is the set $\{\mathbf{g}|\, \mathbf{g} = A\mathbf{f}, \mathbf{f} \in D_A\}$. In the sequel we will assume that D_A is a linear (sub)space.

An operator is *linear* if for all vectors $\mathbf{f},\mathbf{g} \in D_A$ and for all α and β:

$$A(\alpha\mathbf{f} + \beta\mathbf{g}) = \alpha A\mathbf{f} + \beta A\mathbf{g} \tag{a.24}$$

Examples:
Any orthogonal transform is a linear operation. In $\mathbb{R}^N$ and $\mathbb{C}^N$ any matrix-vector multiplication is a linear operation. In $\mathbb{R}^\infty$ and $\mathbb{C}^\infty$ left sided shifts, right sided shifts, and any linear combination of them (i.e. discrete convolution) are linear operations. In $\mathbb{C}[a,b]$ convolution integrals and differential operators are linear operators.

Some special linear operators are:
- The *null operator* 0 assigns the null vector to each vector: $0\mathbf{f} = \mathbf{0}$.
- The *identity operator* I carries each vector into itself: $I\mathbf{f} = \mathbf{f}$.

An operator A is *invertible* if for each $\mathbf{g} \in R_A$ the equation $\mathbf{g} = A\mathbf{f}$ has a unique solution in D_A. The operator A^{-1} that uniquely assigns this solution $\mathbf{f}$ to $\mathbf{g}$ is called the *inverse operator* of A:

$$\mathbf{g} = A\mathbf{f} \Leftrightarrow \mathbf{f} = A^{-1}\mathbf{g} \tag{a.25}$$

The following properties are shown easily:
- $A^{-1}A = I$
- $AA^{-1} = I$
- The inverse of a linear operator - if it exists - is linear.

Suppose that in a linear space R two orthonormal bases $S_a = \{\mathbf{a}_0 \ \mathbf{a}_1 \ \cdots\}$ and $S_b = \{\mathbf{b}_0 \ \mathbf{b}_1 \ \cdots\}$ are defined. According to (a.21) each vector $\mathbf{f} \in R$ has two representations:

$$\mathbf{f} = \sum_k \alpha_k \mathbf{a}_k \qquad \text{with:} \qquad \alpha_k = (\mathbf{f}, \mathbf{a}_k)$$

$$\mathbf{f} = \sum_k \beta_k \mathbf{b}_k \qquad \text{with:} \qquad \beta_k = (\mathbf{f}, \mathbf{b}_k)$$

Since both Fourier series represent the same vector we conclude that:

$$\mathbf{f} = \sum_k \alpha_k \mathbf{a}_k = \sum_k \beta_k \mathbf{b}_k$$

The relationship between the Fourier coefficients α_k and β_k can be made explicit by calculation of the inner product:

$$(\mathbf{f}, \mathbf{b}_n) = \sum_k \alpha_k (\mathbf{a}_k, \mathbf{b}_n) = \sum_k \beta_k (\mathbf{b}_k, \mathbf{b}_n) = \beta_n \qquad (a.26)$$

The Fourier coefficients α_k and β_k can be arranged as vectors $\vec{\alpha} = (\alpha_0, \alpha_1, \cdots)$ and $\vec{\beta} = (\beta_0, \beta_1, \cdots)$ in $\mathbb{R}^N$ or $\mathbb{C}^N$ (if the dimension of R is finite), or in $\mathbb{R}^\infty$ and $\mathbb{C}^\infty$ (if the dimension of R is infinite). In one of these spaces equation (a.26) defines a linear operator U:

$$\vec{\beta} = U\vec{\alpha} \qquad (a.27a)$$

The inner product in (a.26) could equally well be accomplished with respect to a vector $\mathbf{a}_n$. This reveals that an operator U^* exists for which:

$$\vec{\alpha} = U^* \vec{\beta} \qquad (a.27b)$$

Clearly, from (a.25) and (a.27):

$$U^* = U^{-1} \qquad (a.28)$$

Suppose we have two vectors $\mathbf{f}_1$ and $\mathbf{f}_2$ represented in S_a by $\vec{\alpha}_1$ and $\vec{\alpha}_2$, and in S_b by $\vec{\beta}_1$ and $\vec{\beta}_2$. Since the inner product $(\mathbf{f}_1, \mathbf{f}_2)$ must be independent on the representation, we conclude that $(\mathbf{f}_1, \mathbf{f}_2) = (\vec{\alpha}_1, \vec{\alpha}_2) = (\vec{\beta}_1, \vec{\beta}_2)$. Therefore:

$$(\vec{\alpha}_1, U^{-1} \vec{\beta}_2) = (U \vec{\alpha}_1, \vec{\beta}_2) \qquad (a.29)$$

Each operator that satisfies (a.29) is called a *unitary* operator. A corollary of (a.29) is that any unitary operator preserves the Euclidean norm.

The *adjoint* A^* of an operator A is an operator that satisfies:

$$(A\mathbf{f}, \mathbf{g}) = (\mathbf{f}, A^* \mathbf{g}) \qquad (a.30)$$

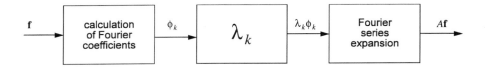

Figure A.2 Eigenvalue decomposition of a self-adjoint operation

From this definition, and from (a.29), it follows that an operator U for which its adjoint U^* equals its inverse U^{-1} is unitary. This is in accordance with the notation used in (a.28). An operator A is called *self-adjoint*, if $A^* = A$.

Suppose that A is a linear operator in a space R. A vector $\mathbf{e}_k$ that satisfies:

$$A\mathbf{e}_k = \lambda_k \mathbf{e}_k \qquad \mathbf{e}_k \neq \mathbf{0} \qquad (a.31)$$

with λ_k a real or complex number is called an *eigenvector* of A. The number λ_k is the *eigenvalue*. The eigenvectors and eigenvalues of an operator may be found be solving the equation $(A - \lambda_k I)\mathbf{e}_k = \mathbf{0}$.

Operators that are self-adjoint have - under mild conditions - some nice properties related to their eigenvectors and eigenvalues. The properties relevant in our case are:

1. All eigenvalues are real.
2. To each eigenvalue at least one eigenvalue is associated. However, an eigenvalue may have multiple eigenvectors. These eigenvectors span a linear subspace.
3. There is an orthonormal basis $V = \{\mathbf{v}_0 \ \mathbf{v}_1 \ \cdots\}$ formed by normalised eigenvectors $\mathbf{e}_k / \|\mathbf{e}_k\|$. Due to possible multiplicities of eigenvalues (see above) this basis may not be unique.

A corollary of the properties is that any vector $\mathbf{f} \in R$ can be represented by a Fourier series with respect to V, and that in this representation the operation becomes simply a linear combination, that is:

$$\mathbf{f} = \sum_k \phi_k \mathbf{v}_k \qquad \text{with:} \qquad \phi_k = (\mathbf{f}, \mathbf{v}_k) \qquad (a.32a)$$

$$A\mathbf{f} = \sum_k \lambda_k \phi_k \mathbf{v}_k \qquad (a.32b)$$

The connotation of this decomposition of the operation is depicted in figure A.2. The set of eigenvalues is called the *spectrum* of the operator.

A.5 VECTORS AND MATRICES

The remaining sections in this appendix focus the attention on linear spaces with finite dimension, i.e. $\mathbb{R}^N$ and $\mathbb{C}^N$. Vectors in these spaces are denoted by an arrow, e.g. $\vec{f}$, $\vec{g}$. Furthermore, the elements in a vector are either arranged vertically (a column-vector) or horizontally (a row-vector). For example:

$$\vec{f} = \begin{bmatrix} f_0 \\ f_1 \\ \vdots \\ f_{N-1} \end{bmatrix} \qquad \text{or:} \qquad \vec{f}^t = \begin{bmatrix} f_0 & f_1 & \cdots & f_{N-1} \end{bmatrix} \qquad (a.33)$$

The superscript t is used to convert column-vectors to row-vectors. Vector addition and scalar multiplication are defined as in section A.1.

A *matrix* $\mathbf{H}$ with dimension $N \times M$ is an arrangement of NM numbers $h_{n,m}$ (the elements) on an orthogonal grid of N rows and M columns:

$$\mathbf{H} = \begin{bmatrix} h_{0,0} & h_{0,1} & \cdots & h_{0,M-1} \\ h_{1,0} & h_{1,1} & \cdots & h_{1,M-1} \\ h_{0,0} & \vdots & & \vdots \\ \vdots & \vdots & & \vdots \\ h_{N-1,0} & \cdots & \cdots & h_{N-1,M-1} \end{bmatrix} \qquad (a.34)$$

Often, the elements are real or complex. Vectors can be regarded as $N \times 1$-matrices (column-vectors) or $1 \times M$-matrices (row-vectors). A matrix can be regarded as an horizontal arrangement of M column-vectors with dimension N, for example:

$$\mathbf{H} = \begin{bmatrix} \vec{h}_0 & \vec{h}_1 & \cdots & \vec{h}_{M-1} \end{bmatrix} \qquad (a.35)$$

Of course, a matrix can also be regarded as a vertical arrangement of N row-vectors.

The *scalar-matrix multiplication* $\alpha \mathbf{H}$ replaces each element in $\mathbf{H}$ with $\alpha h_{n,m}$. The *matrix-addition* $\mathbf{H} = \mathbf{A} + \mathbf{B}$ is only defined if the two matrices $\mathbf{A}$ and $\mathbf{B}$ have equal size $N \times M$. The result $\mathbf{H}$ is an $N \times M$-matrix with elements $h_{n,m} = a_{n,m} + b_{n,m}$. These two operations satisfy the axioms of a linear space (section A.1). Therefore, the set of $N \times M$-matrices is another example of a linear space.

The *matrix-matrix product* $\mathbf{H} = \mathbf{AB}$ is defined only when the number of columns of $\mathbf{A}$ equals the number of rows of $\mathbf{B}$. Suppose that $\mathbf{A}$ is an $N \times P$-matrix, and that $\mathbf{B}$ is a $P \times N$-matrix, then the product $\mathbf{H} = \mathbf{AB}$ is an $N \times M$-matrix with elements:

$$h_{n,m} = \sum_{k=0}^{P-1} a_{n,p} b_{p,m} \qquad (a.36)$$

Since a vector can be regarded as an $N \times 1$-matrix, this also defines the *matrix-vector product* $\vec{g} = \mathbf{H}\vec{f}$ with $\vec{f}$ an M-dimensional column-vector, $\mathbf{H}$ an $N \times M$-matrix, and $\vec{g}$ an N-dimensional column-vector. In accordance with these definitions, the inner product between two real N-dimensional vectors given in section A.1 can be written as:

$$\sum_{k=0}^{N-1} f_k g_k = \vec{f}^t \vec{g} \qquad (a.37)$$

It is easy to show that a matrix-vector product $\vec{g} = \mathbf{H}\vec{f}$ defines a linear operator from $\mathbb{R}^N$ into $\mathbb{R}^M$ and $\mathbb{C}^N$ into $\mathbb{C}^M$. Therefore, all definitions and properties related to linear operators also apply to matrices.

Some special matrices are:
- The *null-matrix* $\mathbf{0}$. This is a matrix fully filled with zero. It corresponds to the null-operator: $\mathbf{0}\vec{f} = \vec{0}$.
- The *unit matrix* $\mathbf{I}$. This matrix is square ($N = M$), fully filled with zero, except for the diagonal elements which are unit:

$$\mathbf{I} = \begin{bmatrix} 1 & & 0 \\ & \ddots & \\ 0 & & 1 \end{bmatrix}$$

This matrix corresponds to the unit operator: $\mathbf{I}\vec{f} = \vec{f}$.
- A *diagonal matrix* Λ is a square matrix, fully filled with zero, except for its diagonal elements $\lambda_{n,n}$:

$$\Lambda = \begin{bmatrix} \lambda_{0,0} & & 0 \\ & \ddots & \\ 0 & & \lambda_{N-1,N-1} \end{bmatrix}$$

Often, diagonal matrices are denoted by upper case Greek symbols.
- The *transposed matrix* $\mathbf{H}^t$ of an $N \times M$-matrix $\mathbf{H}$ is an $M \times N$-matrix, its elements are given by: $h^t_{m,n} = h_{n,m}$.
- A *symmetric matrix* is a square matrix for which $\mathbf{H}^t = \mathbf{H}$.
- The *conjugated* of a matrix $\mathbf{H}$ is a matrix $\overline{\mathbf{H}}$ the elements of which are the complex conjugated of the one of $\mathbf{H}$.
- With the inner product given by (a.37), the *adjoint* of a matrix $\mathbf{H}$ is a matrix $\mathbf{H}^*$ which is the conjugated and the transposed of $\mathbf{H}$, that is: $\mathbf{H}^* = \overline{\mathbf{H}}^t$. A matrix $\mathbf{H}$ is *self-adjoint* or *Hermitian* if $\mathbf{H}^* = \mathbf{H}$. This is the case only if $\mathbf{H}$ is square and $h_{n,m} = \overline{h}_{m,n}$.
- The *inverse* of a square matrix $\mathbf{H}$ is the matrix $\mathbf{H}^{-1}$ that satisfies $\mathbf{H}^{-1}\mathbf{H} = \mathbf{I}$. If it exists, it is unique. In that case the matrix $\mathbf{H}$ is called *regular*. If $\mathbf{H}^{-1}$ doesn't exist, $\mathbf{H}$ is called *singular*.
- A *unitary matrix* $\mathbf{U}$ is a square matrix that satisfies: $\mathbf{U}^{-1} = \mathbf{U}^*$. A real unitary matrix is called *orthonormal*. These matrices satisfy $\mathbf{U}^{-1} = \mathbf{U}^t$.
- A square matrix $\mathbf{H}$ is *Toeplitz* if its elements satisfy: $h_{n,m} = g_{(n-m)}$ in which g_n is a sequence of $2N-1$ numbers.
- A square matrix $\mathbf{H}$ is *circulant* if its elements satisfy: $h_{n,m} = g_{(n-m)\%N}$. Here,

- $(n-m)\%N$ is the remainder of $(n-m)/N$.
- A matrix **H** is *separable* if it can be written as the product of two vectors: $\mathbf{H} = \bar{f}\bar{g}^t$.

Some properties with respect to the matrices mentioned above:

$$(\mathbf{H}^{-1}\mathbf{H})^{-1} = \mathbf{H} \tag{a.38}$$

$$(\mathbf{AB})^{-1} = \mathbf{B}^{-1}\mathbf{A}^{-1} \tag{a.39}$$

$$(\mathbf{H}^*)^* = \mathbf{H} \tag{a.40}$$

$$(\mathbf{AB})^* = \mathbf{B}^*\mathbf{A}^* \tag{a.41}$$

$$(\mathbf{H}^{-1})^* = (\mathbf{H}^*)^{-1} \tag{a.42}$$

Examples:

Defined in a finite interval, the discrete convolution between a sequence f_k and g_k:

$$g_n = \sum_{k=0}^{N-1} h_{n-k} f_k \qquad \text{with:} \qquad n = 0, 1, \cdots, N-1 \tag{a.43}$$

can be written economically as a matrix-vector product $\bar{g} = \mathbf{H}\bar{f}$. The matrix **H** is a Toeplitz matrix:

$$\mathbf{H} = \begin{bmatrix} h_0 & h_{-1} & h_{-2} & \cdots & h_{1-N} \\ h_1 & h_0 & h_{-1} & & h_{2-N} \\ h_2 & h_1 & \ddots & \ddots & \vdots \\ \vdots & & \ddots & \ddots & h_{-1} \\ h_{N-1} & h_{N-2} & \cdots & h_1 & h_0 \end{bmatrix} \tag{a.44}$$

If this "finite interval"-convolution is replaced with a *circulant* (wrap-around) discrete convolution, then (a.43) becomes:

$$g_n = \sum_{k=0}^{N-1} h_{(n-k)\%N} f_k \qquad \text{with:} \qquad n = 0, 1, \cdots, N-1 \tag{a.45}$$

In that case, the matrix-vector relation $\bar{g} = \mathbf{H}\bar{f}$ still holds. However, the matrix **H** is now a circulant matrix:

$$\mathbf{H} = \begin{bmatrix} h_0 & h_{N-1} & h_{N-2} & \cdots & h_{N-1} \\ h_1 & h_0 & h_{N-1} & & h_{N-2} \\ h_2 & h_1 & \ddots & \ddots & \vdots \\ \vdots & & \ddots & \ddots & h_{N-1} \\ h_{N-1} & h_{N-2} & \cdots & h_1 & h_0 \end{bmatrix} \tag{a.46}$$

Figure A.3 Discrete circulant convolution accomplished in the Fourier domain

The *Fourier matrix* $\mathbf{W}$ is a unitary $N \times N$-matrix with elements given by:

$$w_{n,m} = \frac{1}{\sqrt{N}} \exp\left(\frac{-2\pi j n m}{N}\right) \quad \text{with:} \quad j = \sqrt{-1} \qquad (a.47)$$

The (row-)vectors in this matrix are complex conjugated of the basisvectors given in (a.16b). It can be shown that the circulant convolution in (a.45) can be transformed into an element-by-element multiplication provided that the vector $\vec{g}$ is represented by the orthonormal basis of (a.16b). In this representation the circulant convolution $\vec{g} = \mathbf{H}\vec{f}$ becomes; see (a.27b):

$$\mathbf{W}^*\vec{g} = \mathbf{W}^*\mathbf{H}\vec{f} \qquad (a.48)$$

Writing $\mathbf{W} = \lfloor \vec{w}_0 \ \vec{w}_1 \ \cdots \ \vec{w}_{N-1} \rfloor$ and carefully examining $\mathbf{H}\vec{w}_k$ reveals that the basisvectors $\vec{w}_k$ are the eigenvectors of the circulant matrix $\mathbf{H}$. Therefore, we may write $\mathbf{H}\vec{w}_k = \lambda_k \vec{w}_k$ with $k = 0,1,\cdots,N-1$. The numbers λ_k are the eigenvalues of $\mathbf{H}$. If these eigenvalues are arranged in a diagonal matrix:

$$\Lambda = \begin{bmatrix} \lambda_0 & & 0 \\ & \ddots & \\ 0 & & \lambda_{N-1} \end{bmatrix} \qquad (a.49)$$

the N equations $\mathbf{H}\vec{w}_k = \lambda_k \vec{w}_k$ can be written economically as: $\mathbf{HW} = \mathbf{W}\Lambda$. Right sided multiplication of this equation by $\mathbf{W}^{-1}$ yields: $\mathbf{H} = \mathbf{W}\Lambda\mathbf{W}^{-1}$. Substitution in (a.48) gives:

$$\mathbf{W}^*\vec{g} = \mathbf{W}^*\mathbf{W}\Lambda\mathbf{W}^{-1}\vec{f} = \Lambda\mathbf{W}^{-1}\vec{f} \qquad (a.50)$$

Note that the multiplication Λ with the vector $\mathbf{W}^{-1}\vec{f}$ is an element-by-element multiplication because the matrix Λ is diagonal. The final result is obtained if we perform a left sided multiplication in (a.50) by $\mathbf{W}$:

$$\vec{g} = \mathbf{W}\Lambda\mathbf{W}^{-1}\vec{f} \qquad (a.51)$$

The interpretation of this result is depicted in figure A.3.

A.6 TRACE AND DETERMINANT

The *trace* $\mathrm{tr}(\mathbf{H})$ of a square matrix $\mathbf{H}$ is the sum of its diagonal elements:

$$\mathrm{tr}(\mathbf{H}) = \sum_{k=0}^{N-1} h_{k,k} \qquad (a.52)$$

Properties related to the trace are (**A** and **B** are two $N \times N$-matrices, $\vec{f}$ and $\vec{g}$ are two N-dimensional vectors):

$$\text{tr}(\mathbf{AB}) = \text{tr}(\mathbf{BA}) \tag{a.53}$$

$$(\vec{f}, \vec{g}) = \text{tr}(\vec{f}\vec{g}^*) \tag{a.54}$$

The *determinant* $|\mathbf{H}|$ of a square matrix **H** is recursively defined with its co-matrices. The co-matrix $\mathbf{H}_{n,m}$ is an $(N-1) \times (N-1)$-matrix that is derived from **H** by exclusion of the n-th row and the m-th column. The following equations define the determinant:

$$\begin{aligned}\text{If } N = 1: & \quad |\mathbf{H}| = h_{0,0} \\ \text{If } N > 1: & \quad |\mathbf{H}| = \sum_{m=0}^{N-1} (-1)^{1+m} h_{0,m} |\mathbf{H}_{0,m}|\end{aligned} \tag{a.55}$$

Some properties related to the determinant:

$$|\mathbf{AB}| = |\mathbf{A}||\mathbf{B}| \tag{a.56}$$

$$|\mathbf{A}^{-1}| = \frac{1}{|\mathbf{A}|} \tag{a.57}$$

$$|\mathbf{A}^t| = |\mathbf{A}| \tag{a.58}$$

$$\mathbf{U} \text{ is unitary:} \quad \Rightarrow \quad |\mathbf{U}| = \pm 1 \tag{a.59}$$

$$\Lambda \text{ is diagonal:} \quad \Rightarrow \quad |\Lambda| = \prod_{n=0}^{N-1} \lambda_{n,n} \tag{a.60}$$

The *rank* of a matrix is the maximum number of column-vectors (or row-vectors) that are linearly independent. The rank of a regular $N \times N$-matrix is always N. In that case, the determinant is always non-zero. The reverse holds true too. The rank of a singular $N \times N$-matrix is always less than N, and the determinant is zero.

A.7 DIFFERENTIATION OF VECTORS AND MATRICES

Suppose $f(\vec{x})$ is a real or complex function of the N-dimensional vector $\vec{x}$. Then, the first derivative of $f(\vec{x})$ with respect to $\vec{x}$ is an N-dimensional vector function (the *gradient*):

$$\frac{\partial f(\vec{x})}{\partial \vec{x}} \quad \text{with elements:} \quad \frac{\partial f(\vec{x})}{\partial x_n} \tag{a.61}$$

If $f(\vec{x}) = \vec{a}'\vec{x}$ (i.e. the inner product between $\vec{x}$ and a vector $\vec{a}$), then:

$$\frac{\partial[\vec{a}'\vec{x}]}{\partial \vec{x}} = \vec{a} \qquad (a.62)$$

Likewise, if $f(\vec{x}) = \vec{x}'\mathbf{H}\vec{x}$ (i.e. a quadratic form defined by the matrix $\mathbf{H}$), then:

$$\frac{\partial[\vec{x}'\mathbf{H}\vec{x}]}{\partial \vec{x}} = 2\mathbf{H}\vec{x} \qquad (a.63)$$

The second derivative $f(\vec{x})$ with respect to $\vec{x}$ is an $N \times N$-matrix called the *Hessian* matrix:

$$\mathbf{H}(\vec{x}) = \frac{\partial^2 f(\vec{x})}{\partial \vec{x}^2} \quad \text{with elements:} \quad h_{n,m} = \frac{\partial^2 f(\vec{x})}{\partial x_n \partial x_m} \qquad (a.64)$$

The determinant of this matrix is called the *Hessian*.

The *Jacobian matrix* of an N-dimensional vector function $\vec{f}(\vec{x})$ (that is, $f()\colon \mathbb{R}^M \to \mathbb{R}^N$) is defined as an $M \times N$-matrix:

$$\mathbf{H}(\vec{x}) = \frac{\partial \vec{f}(\vec{x})}{\partial \vec{x}} \quad \text{with elements:} \quad h_{n,m} = \frac{\partial f_n(\vec{x})}{\partial x_m} \qquad (a.65)$$

The determinant of this matrix is called the *Jacobian*.

The differentiation of a function of a matrix, e.g. $f(\mathbf{H})$, with respect to this matrix is defined similar to the differentiation in (a.61). The result is a matrix:

$$\frac{\partial f(\mathbf{H})}{\partial \mathbf{H}} \quad \text{with elements:} \quad \frac{\partial f(\mathbf{H})}{\partial h_{n,m}} \qquad (a.66)$$

Suppose $\mathbf{A}$, $\mathbf{B}$, and $\mathbf{C}$ are square matrices of equal size. Then, some properties related to the derivatives of the trace and the determinant are:

$$\frac{\partial}{\partial \mathbf{A}} \operatorname{tr}(\mathbf{A}) = \mathbf{I} \qquad (a.67)$$

$$\frac{\partial}{\partial \mathbf{A}} \operatorname{tr}(\mathbf{BAC}) = \mathbf{B}'\mathbf{C}' \qquad (a.68)$$

$$\frac{\partial}{\partial \mathbf{A}} \operatorname{tr}(\mathbf{ABA}') = 2\mathbf{AB} \qquad (a.69)$$

$$\frac{\partial}{\partial \mathbf{A}} |\mathbf{BAC}| = |\mathbf{BAC}|(\mathbf{A}^{-1})' \qquad (a.70)$$

A.8 DIAGONALISATION OF SELF-ADJOINT MATRICES

Recall from section A.5 that - on the condition that the inner product defined in (a.37) is adopted - an $N \times N$-matrix $\mathbf{H}$ is called self-adjoint or Hermitian if $\mathbf{H}^* = \mathbf{H}$. From the discussion on self-adjoint operators in section A.4 it is clear that associated with $\mathbf{H}$ there exists an orthonormal basis $V = \{\vec{v}_0 \ \vec{v}_1 \ \cdots \ \vec{v}_{N-1}\}$ which we arrange in a unitary matrix $\mathbf{V} = [\vec{v}_0 \ \vec{v}_1 \ \cdots \ \vec{v}_{N-1}]$. Each vector $\vec{v}_k$ is an eigenvector with corresponding (real) eigenvalue λ_k. These eigenvalues are arranged as the diagonal elements in a diagonal matrix Λ.

The operation $\mathbf{H}\vec{f}$ can be written as (see a.32 and a.37):

$$\mathbf{H}\vec{f} = \sum_{k=0}^{N-1} \lambda_k (\vec{v}_k, \vec{f}) \vec{v}_k = \sum_{k=0}^{N-1} \lambda_k \vec{v}_k \vec{f}^* \vec{v}_k = \sum_{k=0}^{N-1} \lambda_k \vec{v}_k \vec{v}_k^* \vec{f} \qquad (a.71)$$

Suppose that the rank of $\mathbf{H}$ equals R. Then, there are exactly R non-zero eigenvalues. Consequently, the number of terms in (a.71) can be replaced with R. From this, it follows that $\mathbf{H}$ is a composition of its eigenvectors according to:

$$\mathbf{H} = \sum_{k=0}^{R-1} \lambda_k \vec{v}_k \vec{v}_k^* \qquad (a.72)$$

The summation on the right hand side can written more economically as: $\mathbf{V}\Lambda\mathbf{V}^*$. Therefore:

$$\mathbf{H} = \mathbf{V}\Lambda\mathbf{V}^* \qquad (a.73)$$

The unitary matrix $\mathbf{V}^*$ transforms the domain of $\mathbf{H}$ such that the $\mathbf{H}$ becomes a diagonal matrix in this new domain. The matrix $\mathbf{V}$ accomplishes the inverse transform. In fact, (a.67) is the matrix-version of the decomposition shown in figure A.2.

If the rank R equals N, there are exactly N non-zero eigenvalues. In that case, the matrix Λ is invertible, and so is $\mathbf{H}$:

$$\mathbf{H}^{-1} = \mathbf{V}\Lambda^{-1}\mathbf{V}^* = \sum_{k=0}^{N-1} \frac{\vec{v}_k \vec{v}_k^*}{\lambda_k} \qquad (a.74)$$

It can be seen that the inverse $\mathbf{H}^{-1}$ is self-adjoint.

Example:

A self-adjoint matrix $\mathbf{H}$ is *positive definite* if its eigenvalues are all positive. In that case the expression $\rho(\vec{f}, \vec{g}) = \sqrt{(f-\vec{g})^* \mathbf{H}(f-\vec{g})}$ satisfies the conditions of a distance measure (section A.2). To show this it suffices to prove that $\sqrt{f^* \mathbf{H} f}$ satisfies the conditions of a norm (see equation a.10). These conditions are given in section A.1. We use the diagonal form of $\mathbf{H}$, see (a.74):

$$\vec{f}\mathbf{H}\vec{f}^* = \vec{f}^* \mathbf{V}\Lambda\mathbf{V}^* \vec{f} \qquad (a.75)$$

Since $\mathbf{V}$ is a unitary matrix, the vector $\mathbf{V}^*\vec{f}$ can be regarded as the representation of $\vec{f}$ with respect to the orthonormal basis defined by the vectors in $\mathbf{V}$. Let this

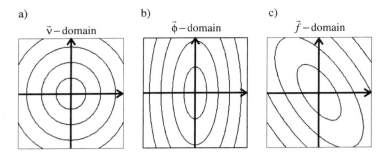

Figure A.4 Vectors in $\mathbb{R}^2$ that are equidistant with respect to the origin

representation be denoted by: $\vec{\phi} = \mathbf{V}^*\vec{f}$. The expression $\vec{f}^*\mathbf{H}\vec{f}$ equals:

$$\vec{f}^*\mathbf{H}\vec{f} = \vec{\phi}^*\Lambda\vec{\phi} = \sum_{k=0}^{N-1} \lambda_k |\phi_k|^2 \qquad (a.76)$$

Written in this form, it is easy to show that all conditions of a norm are met.

With the norm $\sqrt{f^*\mathbf{H}f}$ the set of points equidistant to the origin, i.e. the vectors that satisfy $f^*\mathbf{H}f = \text{constant}$, are ellipsoids. See figure A.4. This follows from (a.76):

$$\vec{f}^*\mathbf{H}\vec{f} = \text{constant} \quad \Leftrightarrow \quad \sum_{k=0}^{N-1} \lambda_k |\phi_k|^2 = \text{constant}$$

Hence, if we introduce a new vector $\vec{v}$ with elements defined as: $v_k = \phi_k/\sqrt{\lambda_k}$, we must require that:

$$\sum_{k=0}^{N-1} |v_k|^2 = \text{constant}$$

In the $\vec{v}$-domain the ordinary Euclidean norm applies. Therefore, the solution space in this domain is a sphere (figure A.4a). The operation defined by $v_k = \phi_k/\sqrt{\lambda_k}$ is merely a scaling of the axes by factors $\sqrt{\lambda_k}$. This transforms the sphere into an ellipsoid, the principal axes of which line up with the basisvectors $\vec{v}_k$ (figure A.4b). Finally, the unitary transform $\vec{f} = \mathbf{V}\vec{\phi}$ rotates the principal axes, but without affecting the shape of the ellipsoid (figure A.4c).

REFERENCES

Kolmogorov, A.N. and Fomin, S.V.: *Introductory Real Analysis*, Dover Publications, New York, 1970.

Bellman, R.E.: *Introduction to Matrix Analysis*, McGraw-Hill, New York, 1970.

APPENDIX B
PROBABILITY THEORY AND STOCHASTIC PROCESSES

This appendix summarises concepts from probability theory and stochastic processes. This summary only concerns those concepts that are part of the mathematical background required in this book. Mathematical peculiarities not relevant here are omitted. At the end of the appendix references to a detailed treatment are given.

B.1 PROBABILITY THEORY AND RANDOM VARIABLES

The axiomatic development of *probability* involves the definitions of three concepts. Taken together these concepts are called an *experiment* consisting of:
a) A set or space Ω consisting of outcomes ω. This set is called the *certain event*.
b) A Borel field A consisting of certain subsets of Ω. Each subset $\alpha \in A$ is called an *event*.
c) A real function $P(\alpha)$ defined on A. This function, called *probability*, satisfies the following axioms:
   I: $P(\alpha) \geq 0$
   II: $P(\Omega) = 1$
   III: If $\alpha, \beta \in A$ and $\alpha \cap \beta = \emptyset$ then $P(\alpha \cup \beta) = P(\alpha) + P(\beta)$

Example:
> The space of outcomes corresponding to the colours of a traffic-light is: $\Omega = \{\text{red, green, yellow}\}$. The Borel field A may consist of subsets like: $\emptyset$, red, green, yellow, red $\cup$ green, red $\cap$ green, red $\cup$ green $\cup$ yellow, $\cdots$. With that, $P(\text{green})$ is the probability that the light will be green. $P(\text{green} \cup \text{yellow})$ is the probability that the light will be green, or yellow, or both. $P(\text{green} \cap \text{yellow})$ is the probability that at the same time the light is green and yellow.

A *random variable* $\underline{x}(\omega)$ is a mapping of Ω onto a set of numbers, for instance: integer numbers, real numbers, complex numbers, etc. The *distribution function* $F_{\underline{x}}(x)$ is the probability of the event that corresponds to $\underline{x} \leq x$:

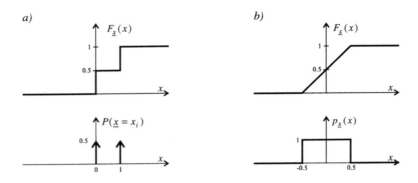

Figure B.1
a) Discrete distribution function with probabilities
b) Continuous distribution function with a uniform probability density

$$F_{\underline{x}}(x) = P(\underline{x} \leq x) \tag{b.1}$$

The random variable $\underline{x}$ is said to be *discrete* if a finite number (or infinite countable number) of events $x_1, x_2, \cdots$ exists for which:

$$P(\underline{x} = x_i) = P_i > 0 \quad \text{and} \quad \sum_{\text{all } i} P_i = 1 \tag{b.2}$$

The random variable is *continuous* if a function $p_{\underline{x}}(x)$ exists for which

$$F_{\underline{x}}(x) = \int_{\xi=-\infty}^{x} p_{\underline{x}}(\xi) d\xi \tag{b.3}$$

This function is called the *probability density* of $\underline{x}$. The discrete case can be included in the continuous case by permitting $p_{\underline{x}}(x)$ to contain Dirac-functions of the type $P_i \delta(x - x_i)$.

Examples:
We consider the experiment consisting of tossing a (fair) coin. The outcomes are {head, tail}. The random variable $\underline{x}$ is defined according to:

head → $\underline{x} = 0$
tail → $\underline{x} = 1$

This random variable is discrete. Its distribution function and probabilities are depicted in figure B.1a.
An example of a continuous random variable is the round-off error which occurs when a real number is replaced with its nearest integer number. This error is uniformly distributed between -0.5 and 0.5. The distribution function and associated probability density is shown in figure B.1.b. Note that, since $F_{\underline{x}}(x)$ is a non-decreasing function of x, and $F_{\underline{x}}(\infty) = 1$, the density must be a non-negative function with $\int p_{\underline{x}}(x) dx = 1$.

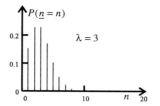 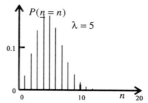 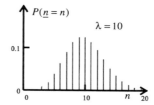

Figure B.2 Poisson distributions

The *moment of order n* of a random variable is defined as:

$$E\{\underline{x}^n\} = \int_{x=-\infty}^{\infty} x^n p_{\underline{x}}(x) dx \qquad (b.4)$$

This definition is such that $E\{\underline{x}^0\}$ is always a unit. Another notation of $E\{\underline{x}^n\}$ is $\overline{\underline{x}^n}$.

The first order moment is called the *expectation*. This quantity is often denoted by μ_x or (if confusion is not to be expected) briefly μ. The *central moments of order n* are defined as:

$$E\{(\underline{x}-\mu)^n\} = \int_{x=-\infty}^{\infty} (x-\mu)^n p_{\underline{x}}(x) dx \qquad (b.5)$$

The first central moment is always zero. The second central moment is called *variance*:

$$Var\{\underline{x}\} = E\{(\underline{x}-\mu)^2\} = E\{\underline{x}^2\} - \mu^2 \qquad (b.6)$$

The *(standard) deviation* of $\underline{x}$ denoted by σ_x, or briefly σ, is the square root of the variance:

$$\sigma_x = \sqrt{Var\{\underline{x}\}} \qquad (b.7)$$

Examples:

Radiant energy is carried by a discrete number of photons. In daylight situations, the average number of photons per unit area is on the order of $10^{12}/s \cdot mm^2$. However, in fact this number is a discrete random variable $\underline{n}$ which obeys the so-called *Poisson distribution* with parameter λ:

$$P(\underline{n} = n) = \frac{\lambda^n \exp(-\lambda)}{n!} \qquad (b.8)$$

Examples of this distribution are shown in figure B.2. The expectation $E\{\underline{n}\}$ and the variance $Var\{\underline{n}\}$ of $\underline{n}$ are both equal to λ. Therefore, the relative deviation $\sigma_n/E\{\underline{n}\}$ is $1/\sqrt{\lambda}$. An image sensor with an integrating area of $100\mu m^2$, an integration time of $25ms$, and an illuminance of $250lx$ receives about $\lambda = 10^6$ photons. Hence, the relative deviation is about 0.1%. In most imaging sensors this deviation is negligible compared with other noise sources. However, if an image intensifier is used, such as in some X-ray equipment, each photon may give rise to about 10.000 new photons. In that case the relative deviation is about 10%. This so-called *quantum mottle* is quite perceptible.

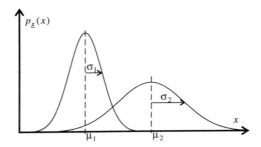

Figure B.3 Gaussian probability densities

A well-known example of a continuous random variable is the one with a *Gaussian* (or *normal*) probability density:

$$p_{\underline{x}}(x) = \frac{1}{\sigma\sqrt{2\pi}} \exp\left(\frac{-(x-\mu)^2}{2\sigma^2}\right) \qquad (b.9)$$

The parameters μ and σ^2 are the expectation and the variance, respectively. Two examples of the probability density with different μ and σ^2 are shown in figure B.3.

Gaussian random variables occur whenever the underlying process is caused by the outcomes of many independent experiments the associated random variables of which add up linearly (the *central limit theorem*). An example is thermal noise in an electrical current. The current is proportional to the sum of the velocities of the individual electrons. Another example is the Poisson distributed random variable mentioned above. The envelop of the Poisson distribution approximates the Gaussian distribution as λ tends to infinity. As illustrated in figure B.3, the approximation looks quite reasonable already when $\lambda > 10$.

B.2 BIVARIATE RANDOM VARIABLES

In this section we consider an experiment with which two random variables $\underline{x}$ and $\underline{y}$ are associated. The *joint distribution function* is the probability that $\underline{x} \le x$ and $\underline{y} \le y$, i.e.:

$$F_{\underline{x},\underline{y}}(x,y) = P(\underline{x} \le x, \underline{y} \le y) \qquad (b.10)$$

The function $p_{\underline{x},\underline{y}}(x,y)$ for which:

$$F_{\underline{x},\underline{y}}(x,y) = \int_{\xi=-\infty}^{x}\int_{\eta=-\infty}^{y} p_{\underline{x},\underline{y}}(x,y)\,d\eta\,d\xi \qquad (b.11)$$

is called the *joint probability density*. Strictly speaking, definition (b.11) holds true only when $F_{\underline{x},\underline{y}}(x,y)$ is continuous. However, by permitting $p_{\underline{x},\underline{y}}(x,y)$ to contain Dirac-functions the definition also applies to the discrete case.

From definitions (b.10) and (b.11) it is clear that the *marginal distribution* $F_{\underline{x}}(x)$ and the *marginal density* $p_{\underline{x}}(x)$ are given by:

$$F_{\underline{x}}(x) = F_{\underline{x},\underline{y}}(x,\infty) \tag{b.12a}$$

$$p_{\underline{x}}(x) = \int_{y=-\infty}^{\infty} p_{\underline{x},\underline{y}}(x,y)dy \tag{b.12b}$$

Two random variables $\underline{x}$ and $\underline{y}$ are *independent* if:

$$F_{\underline{x},\underline{y}}(x,y) = F_{\underline{x}}(x)F_{\underline{y}}(y) \tag{b.13}$$

This is equivalent to:

$$p_{\underline{x},\underline{y}}(x,y) = p_{\underline{x}}(x)p_{\underline{y}}(y) \tag{b.14}$$

Suppose that $h(\cdot,\cdot)$ is a function $\mathbb{R}\times\mathbb{R}\to\mathbb{R}$. Then $h(\underline{x},\underline{y})$ is a random variable. The expectation of $h(\underline{x},\underline{y})$ equals:

$$\mathrm{E}\{h(\underline{x},\underline{y})\} = \int_{x=-\infty}^{\infty} \int_{y=-\infty}^{\infty} h(\underline{x},\underline{y}) p_{\underline{x},\underline{y}}(x,y) dy dx \tag{b.15}$$

The *joint moments* m_{ij} of two random variables $\underline{x}$ and $\underline{y}$ are defined as the expectations of the functions $\underline{x}^i \underline{y}^j$:

$$m_{ij} = \mathrm{E}\{\underline{x}^i \underline{y}^j\} \tag{b.16}$$

The quantity $i+j$ is called the *order* of m_{ij}. It can easily be verified that: $m_{00}=1$, $m_{10} = \mathrm{E}\{\underline{x}\} = \mu_x$, and $m_{01} = \mathrm{E}\{\underline{y}\} = \mu_y$.

The *joint central moments* μ_{ij} of order $i+j$ are defined as:

$$\mu_{ij} = \mathrm{E}\{(\underline{x}-\mu_x)^i (\underline{y}-\mu_y)^j\} \tag{b.17}$$

Clearly, $\mu_{20} = \mathrm{Var}\{\underline{x}\}$ and $\mu_{02} = \mathrm{Var}\{\underline{y}\}$. Furthermore, the parameter μ_{11} is called the *covariance* (sometimes denoted by $\mathrm{Cov}\{\underline{x},\underline{y}\}$). This parameter can be written as: $\mu_{11} = m_{11} - m_{10}m_{01}$. Two random variables are called *uncorrelated* if their covariance is zero. Two independent random variables are always uncorrelated. The reverse is not necessarily true.

Example:

Two random variables $\underline{x}$ and $\underline{y}$ are Gaussian if their joint probability density is:

$$p_{\underline{x},\underline{y}}(x,y) = \frac{1}{2\pi\sigma_x\sigma_y\sqrt{1-r^2}}\exp\left(\frac{-1}{2(1-r^2)}\left(\frac{(x-\mu_x)^2}{\sigma_x^2} - \frac{2r(x-\mu_x)(y-\mu_y)}{\sigma_x\sigma_y} + \frac{(y-\mu_y)^2}{\sigma_y^2}\right)\right) \tag{b.18}$$

The parameters μ_x and μ_y are the expectations of $\underline{x}$ and $\underline{y}$, respectively. The parameters σ_x and σ_y are the standard deviations. The parameter r is called the *correlation coefficient*, defined as:

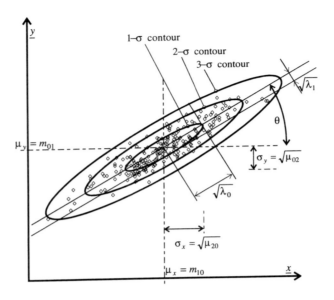

Figure B.4 Scatter diagram of two Gaussian random variables

$$r = \frac{\mu_{11}}{\sqrt{\mu_{20}\mu_{02}}} = \frac{\text{Cov}\{\underline{x},\underline{y}\}}{\sigma_x \sigma_y} \tag{b.19}$$

Figure B.4 shows a scatter diagram with 121 realisations of two Gaussian random variables. In this figure, a geometrical interpretation of (b.18) is also given. The solutions of x and y which satisfy:

$$p_{\underline{x},\underline{y}}(x,y) = p_{\underline{x},\underline{y}}(\mu_x,\mu_y)\exp\left(\frac{-1}{2}\right)$$

(i.e. the 1-σ contour) turn out to be an ellipse. The centre of this ellipse coincides with the expectation. The eccentricity, size, and orientation of the 1-σ contour describe the scattering of the samples around this centre. Their projections on the principal axes are denoted by: $\sqrt{\lambda_0}$ and $\sqrt{\lambda_1}$. The angle between the principal axis associated to λ_0 and the x-axis is θ. With these conventions, the variances of $\underline{x}$ and $\underline{y}$, and the correlation coefficient r, are:

$$\sigma_x^2 = \lambda_0 \cos^2\theta + \lambda_1 \sin^2\theta \tag{b.20a}$$

$$\sigma_y^2 = \lambda_1 \cos^2\theta + \lambda_0 \sin^2\theta \tag{b.20b}$$

$$r = \frac{(\lambda_0 - \lambda_1)\sin\theta \cos\theta}{\sigma_x \sigma_y} \tag{b.20c}$$

Consequently, $r = 0$ whenever $\lambda_0 = \lambda_1$ (the ellipse degenerates into a circle), or θ is a multiple of $\pi/2$ (the ellipse lines up with the x-axis). In both cases the random variables are independent. The conclusion is that two Gaussian random variables which are *uncorrelated* are also *independent*.

The situation $r=1$ or $r=-1$ occurs only if $\lambda_0 = 0$ or $\lambda_1 = 0$. The ellipse degenerates into a straight line. Therefore, if two Gaussian random variables are completely *correlated*, then this implies that two constants a and b can be found for which $\underline{y} = a\underline{x} + b$.

The expectation and variance of the random variable defined by $\underline{z} = a\underline{x} + b\underline{y}$ are:

$$E\{\underline{z}\} = aE\{\underline{x}\} + bE\{\underline{y}\} \tag{b.21}$$

$$\sigma_z^2 = a^2\sigma_x^2 + b^2\sigma_y^2 + 2ab\,\text{Cov}\{\underline{x},\underline{y}\} \tag{b.22}$$

The first equation shows that the expectation is a linear operation (regardless of the joint distribution function). The variance is not a linear operation. However, (b.22) shows that if two random variables are uncorrelated, then $\sigma_{x+y}^2 = \sigma_x^2 + \sigma_y^2$.

Another important aspect of two random variables is the concept of conditional probabilities and moments. The *conditional distribution* $F_{\underline{x}|\underline{y}}(x|y)$ is the probability that $\underline{x} \leq x$ given that $\underline{y} \leq y$. Written symbolically:

$$F_{\underline{x}|\underline{y}}(x|y) = P(\underline{x} \leq x | \underline{y} \leq y) \tag{b.23}$$

The so-called *conditional probability density* associated with $F_{\underline{x}|\underline{y}}(x|y)$ is denoted by $p_{\underline{x}|\underline{y}}(x|y)$. Its definition is similar to (b.3). An important property of conditional probability densities is (*Bayes' theorem for conditional probabilities*):

$$p_{\underline{x}|\underline{y}}(x|y)p_{\underline{y}}(y) = p_{\underline{x},\underline{y}}(x,y) = p_{\underline{y}|\underline{x}}(y|x)p_{\underline{x}}(x) \tag{b.24}$$

The *conditional moments* are defined as:

$$E\{\underline{x}^n | \underline{y} = y\} = \int_{x=-\infty}^{\infty} x^n p_{\underline{x}|\underline{y}}(x|y)dx \tag{b.25}$$

Conditional expectation and *conditional variance* are sometimes denoted by $\mu_{x|y}$ and $\sigma_{x|y}^2$, respectively.

B.3 RANDOM VECTORS

In this section, we discuss a finite sequence of N random variables: $\underline{x}_0, \underline{x}_1, \cdots, \underline{x}_{N-1}$. We assume that these variables are arranged in an N-dimensional random vector $\vec{\underline{x}}$. The *joint distribution function* $F_{\vec{\underline{x}}}(\vec{x})$ is defined as:

$$F_{\vec{\underline{x}}}(\vec{x}) = P(\underline{x}_0 \leq x_0, \underline{x}_1 \leq x_1, \cdots, \underline{x}_{N-1} \leq x_{N-1}) \tag{b.26}$$

The *probability density* $p_{\vec{\underline{x}}}(\vec{x})$ of the vector $\vec{\underline{x}}$ is the function that satisfies:

$$F_{\vec{\underline{x}}}(\vec{x}) = \int_{\vec{\xi}=-\infty}^{\vec{x}} p_{\vec{\underline{x}}}(\vec{\xi})d\vec{\xi} \tag{b.27}$$

with:
$$\int_{\underline{\xi}=-\infty}^{\bar{x}} p_{\bar{x}}(\underline{\xi})d\underline{\xi} = \int_{\xi_0=-\infty}^{x_0}\int_{\xi_1=-\infty}^{x_1} \cdots \int_{\xi_{N-1}=-\infty}^{x_{N-1}} p_{\bar{x}}(\underline{\xi})d\xi_0 d\xi_1 \cdots d\xi_{N-1}$$

The *expectation* of a function $g(\bar{x})$: $\mathbb{R}^N \to \mathbb{R}$ is:

$$E\{g(\bar{x})\} = \int_{\bar{x}=-\infty}^{\infty} g(\bar{x})p_{\bar{x}}(\bar{x})d\bar{x} \qquad (b.28)$$

Note that similar definitions apply to vector-to-vector mappings ($\mathbb{R}^N \to \mathbb{R}^N$) and vector-to-matrix mappings ($\mathbb{R}^N \to \mathbb{R}^N \times \mathbb{R}^N$). Particularly, the *expectation vector* $\bar{\mu}_x = E\{\bar{x}\}$ and the *covariance matrix* $\mathbf{C}_x = E\{\bar{x}\bar{x}^t\}$ are frequently used.

Example:

A Gaussian random vector has a probability density given by:

$$p_{\bar{x}}(\bar{x}) = \frac{1}{\sqrt{(2\pi)^N |\mathbf{C}_x|}} \exp\left(\frac{-(\bar{x}-\bar{\mu}_x)^t \mathbf{C}_x^{-1}(\bar{x}-\bar{\mu}_x)}{2}\right) \qquad (b.29)$$

The parameters $\bar{\mu}_x$ (expectation vector) and $\mathbf{C}_x$ (covariance matrix) fully define the probability density.

A random vector is called *uncorrelated* if its covariance matrix is a diagonal matrix. A random vector is *independent* if its probability density is the product of the probability densities of its element:

$$p_{\bar{x}}(\bar{x}) = \prod_{i=0}^{N-1} p_{x_i}(x_i) \qquad (b.30)$$

An independent random vector is uncorrelated. The reverse holds true in some specific cases, e.g. for all Gaussian random vectors.

The *conditional probability density* of two random vectors $\bar{x}$ and $\bar{y}$ is defined as:

$$p_{\bar{x}|\bar{y}}(\bar{x}|\bar{y}) = \frac{p_{\bar{x},\bar{y}}(\bar{x},\bar{y})}{p_{\bar{y}}(\bar{y})} \qquad (b.31)$$

The definitions of *conditional expectation vector* and *conditional covariance matrix* are similar.

Suppose the random vector $\bar{y}$ results from a linear (matrix-)operation $\bar{y} = \mathbf{A}\bar{x}$. The input vector of this operator $\bar{x}$ has expectation vector $\bar{\mu}_x$ and covariance matrix $\mathbf{C}_x$, respectively. Then, the expectation of the output vector and its covariance matrix are:

$$\begin{aligned}\bar{\mu}_y &= \mathbf{A}\bar{\mu}_x \\ \mathbf{C}_y &= \mathbf{A}\mathbf{C}_x\mathbf{A}^t\end{aligned} \qquad (b.32)$$

These relations hold true regardless of the type of distribution functions. In general, the distribution function of $\bar{y}$ may differ from the one of $\bar{x}$. For instance, if the elements from $\bar{x}$ are uniformly distributed, then the elements of $\bar{y}$ will not be

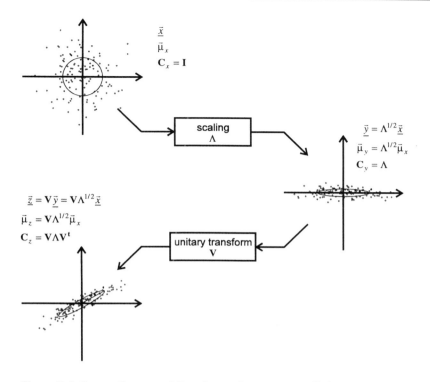

Figure B.5 Scatter diagrams of Gaussian random vectors applied to two linear operators

uniform except for trivial cases (e.g. when $A = I$). However, if $\underline{x}$ has a Gaussian distribution, then so has $\underline{y}$.

Example:
> Figure B.5 shows the scatter diagram of a Gaussian random vector $\underline{x}$ the covariance matrix of which is the identity matrix I. Such a vector is called *white*. Multiplication of $\underline{x}$ with a diagonal matrix $\Lambda^{1/2}$ yields a vector $\underline{y}$ with covariance matrix Λ. This vector is still uncorrelated. Application of a unitary matrix V to $\underline{y}$ yields the random vector $\underline{z}$, which is correlated.

B.3.1 Decorrelation

Suppose a random vector $\underline{z}$ with covariance matrix C_z is given. *Decorrelation* is a linear operation A which, applied to $\underline{z}$, will give a white random vector $\underline{x}$. The operation A can be found by diagonalisation of the matrix C_z. To see this it suffices to recognise that the matrix C_z is self-adjoint, i.e. $C_z = C_z^*$. According to section A.8 a unitary matrix V and a (real) diagonal matrix Λ must exist such that $C_z = V\Lambda V^*$. The matrix V consists of the eigenvectors of C_z. The matrix Λ contains the eigenvalues. Therefore, application of the unitary transform V^* yields a random vector $\underline{y} = V^*\underline{z}$ the covariance matrix of which is Λ. Furthermore, the operation $\Lambda^{-1/2}$ applied to $\underline{y}$ gives the white vector $\underline{x} = \Lambda^{-1/2}\underline{y}$. Hence, the decorrelation/whitening operation A equals $\Lambda^{-1/2}V^*$. Note that the operation $\Lambda^{-1/2}V^*$ is the inverse of the operations shown in figure B.5.

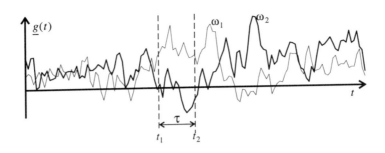

Figure B.6 Two realisations of a stochastic process

B.4 STOCHASTIC PROCESSES

Stochastic processes are mathematical models of randomly fluctuating phenomena. Usually, the independent variable t is the time. However, this is not always the case. In handling images, fluctuations may occur in space (one or more dimensions), in time, and/or in the wavelength domain. In this section, only one independent variable - denoted by t - will be used.

As in section B.1, the triad $\{\Omega, A, P(\cdot)\}$ defines an experiment. A *stochastic process* $\underline{g}(t,\omega)$ is a mapping of Ω onto a set $\{g(t,\omega)|\omega \in \Omega\}$ of functions of t. This set is called the *ensemble*. If ω is fixed, $g(t,\omega)$ is an ordinary function called a *realisation* of the process. With t fixed, $\underline{g}(t,\omega)$ is a random variable. Figure B.6 gives an example of a stochastic process corresponding to two outcomes, i.e. two realisations.

The *n-th order distribution function* is the probability:

$$F_{\underline{g}(t_1),\underline{g}(t_2),\cdots,\underline{g}(t_n)}(g_1,g_2,\cdots,g_n) = P\left(\underline{g}(t_1) \le g_1, \underline{g}(t_2) \le g_2, \cdots, \underline{g}(t_n) \le g_n\right) \quad \text{(b.33)}$$

A stochastic process is *statistically determined* if for each order n these distribution functions are known. Often we confine ourselves to the first order and second order statistics. The first and second order probability densities are functions such that:

$$F_{\underline{g}(t)}(g) = \int_{\xi=-\infty}^{g} p_{\underline{g}(t)}(\xi)d\xi \quad \text{(b.34a)}$$

$$F_{\underline{g}(t_1),\underline{g}(t_2)}(g_1,g_2) = \int_{\xi_1=-\infty}^{g_1}\int_{\xi_2=-\infty}^{g_2} p_{\underline{g}(t_1),\underline{g}(t_2)}(\xi_1,\xi_2)d\xi_1 d\xi_2 \quad \text{(b.34b)}$$

Some moments of a stochastic process are:
- The *expectation*:

$$\mu_{\underline{g}}(t) = \mathrm{E}\{\underline{g}(t)\} = \int_{\xi=-\infty}^{\infty} \xi p_{\underline{g}(t)}(\xi)d\xi \quad \text{(b.35)}$$

- The *autocorrelation function*:

$$R_{gg}(t_1,t_2) = \mathrm{E}\{\underline{g}(t_1)\underline{g}(t_2)\} = \int_{\xi_1=-\infty}^{\infty} \int_{\xi_2=-\infty}^{\infty} \xi_1\xi_2 p_{\underline{g}(t_1),\underline{g}(t_1)}(\xi_1,\xi_2)d\xi_2\xi_1 \quad\quad (b.36)$$

- The *autocovariance function*:

$$C_{gg}(t_1,t_2) = R_{gg}(t_1,t_2) - \mu_g(t_1)\mu_g(t_2) \quad\quad (b.37)$$

- The *variance*:

$$\sigma_g^2(t) = \mathrm{Var}\{\underline{g}(t)\} = C_{gg}(t,t) = \mathrm{E}\{\underline{g}^2(t)\} - \mu_g^2(t) = \int_{\xi=-\infty}^{\infty}\xi^2 p_{\underline{g}(t)}(\xi)d\xi - \mu_g^2(t) \quad\quad (b.38)$$

- The *signal power*:

$$\mathrm{E}\{\underline{g}^2(t)\} = \sigma_g^2(t) + \mu_g^2(t) = R_{gg}(t,t) \quad\quad (b.39)$$

B.4.1 Stationarity and power spectrum

A stochastic process is called *stationary* if the statistics of a process is not affected by a shift τ of the origin of t. This means that the probability density of any order n is such that:

$$p_{\underline{g}(t_1+\tau),\underline{g}(t_2+\tau),\cdots,\underline{g}(t_n+\tau)}(g_1,g_2,\cdots,g_n) = p_{\underline{g}(t_1),\underline{g}(t_2),\cdots,\underline{g}(t_n)}(g_1,g_2,\cdots,g_n) \quad\quad (b.40)$$

If a process is stationary, then the first order probability density does not depend on t. In that case the expectation and the variance are constants. The autocorrelation function and the autocovariance function depends only on the difference $\tau = t_2 - t_1$ (see figure B.6). Therefore, in case of stationarity the notation $R_{gg}(\tau)$ and $C_{gg}(\tau)$ suffices.

A stochastic process is not fully determined by its expectation and its autocorrelation function. Exceptions to this rule are the Gaussian processes, the probability densities of which are completely determined once the expectation and autocorrelation function are known. Often, the characterisation of a stochastic process is solely in terms of these two parameters. In that case, a definition of stationarity in terms of expectation and autocorrelation function will do. A process is called *wide sense stationary* if its expectation is constant, and its autocorrelation function depends on the difference $\tau = t_2 - t_1$ only.

The *power (density) spectrum* $S_{gg}(u)$ of a (wide sense) stationary stochastic process is the Fourier transform of the autocorrelation function:

$$S_{gg}(u) = F\{R_{gg}(\tau)\} = \int_{\tau=-\infty}^{\infty} R_{gg}(\tau)\exp(-2\pi j u\tau)d\tau$$

$$R_{gg}(\tau) = F^{-1}\{S_{gg}(u)\} = \int_{u=-\infty}^{\infty} S_{gg}(u)\exp(2\pi j u\tau)du \quad\quad (b.41)$$

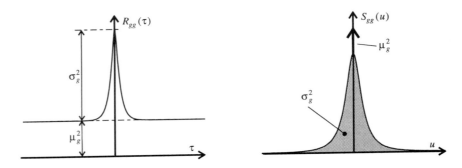

Figure B.7
The autocorrelation function and power spectrum of the stochastic process shown in figure B.6

From (b.39) and this definition it follows readily that:

$$E\{\underline{g}^2(t)\} = R_{gg}(0) = \int_{u=-\infty}^{\infty} S_{gg}(u) du \qquad (b.41)$$

which shows that the power spectrum is a density function that distributes the signal power $E\{\underline{g}^2(t)\}$ along the frequencies u. Figure B.7 gives an example. Note that the autocorrelation function is composed of a non-fluctuating part μ_g^2 and a fluctuating part σ_g^2. The power spectrum distributes the fluctuating part along the frequencies, while as the non-fluctuating part is seen as a Dirac-function with weight μ_g^2 at $u=0$.

A process for which the power spectrum is constant, i.e. $S_{gg}(u) = V$, is called *white noise*. Physically, white noise cannot be realised since it would require too much energy. However, in case of a signal source and a noise source, the noise is called white if the spectrum of the noise is constant for all frequency-bands for which the signal energy is non-zero.

B.4.2 Average, expectation and ergodicity

The expectation operator of a stochastic process relates to the ensemble of process. In contrast with that, the operation *"averaging"* relates to single realisations. Averaging takes place along the time axis. For instance, if $g(t,\omega_1)$ is the realisation associated to the outcome ω_1, and if this realisation is observed within an interval $-T < t < T$, then the average of this realisation is:

$$\mu_{g,T}(\omega_1) = \frac{1}{2T} \int_{t=-T}^{T} g(t,\omega_1) dt \qquad (b.42a)$$

Clearly, this integral assigns a number to each outcome. Therefore, it can be regarded as a new random variable:

$$\underline{\mu}_{g,T} = \frac{1}{2T} \int_{t=-T}^{T} \underline{g}(t) dt \qquad (b.42b)$$

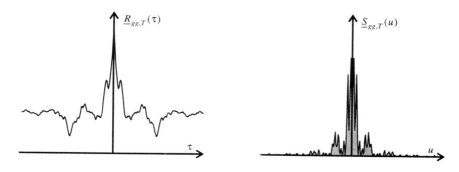

Figure B.8
Estimates of the autocorrelation function and power spectrum shown in figure B.7

Sometimes the definition of expectation is given as the limiting case of the time-average, i.e.:

$$\underline{\mu}_{g,\infty} = \lim_{T \to \infty} \underline{\mu}_{g,T}$$

However, it is easy to define a stochastic process such that the variance of $\underline{\mu}_{g,\infty}$ is non-zero. Therefore, as a definition $\underline{\mu}_{g,\infty}$ is not consistent. Nevertheless, a particular class of stochastic processes exists for which the equation $\underline{\mu}_{g,\infty} = \mu_g$ holds. These processes are called *ergodic* with respect to the expectation.

Example:
The time-average:

$$\underline{R}_{gg,\infty}(\tau) \qquad \text{with} \qquad \underline{R}_{gg,T}(\tau) = \frac{1}{2T} \int_{t=-T}^{T} \underline{g}(t)\underline{g}(t+\tau)dt \qquad (b.43)$$

is sometimes used as a definition of the autocorrelation function. This definition is consistent only when the stochastic process is ergodic with respect to the second moments. This is the case if the process is stationary, and if some conditions with respect to the fourth moments are met. The stochastic process shown in figure B.6 satisfies these conditions. A realisation of an estimate $\underline{R}_{gg,T}(\tau)$ is given in figure B.8.

The square-magnitude of the Fourier transform of a realisation, i.e.:

$$\underline{S}_{gg,T}(u) = \frac{1}{2T} \left| \int_{t=-T}^{T} \underline{g}(t)\exp(-2\pi j u t)dt \right|^2 \qquad (b.44)$$

is called the *periodogram*. An example is given in figure B.8. As an estimate of the true power spectrum a periodogram is less appropriate. It can be proven that the variance:

$$E\left\{ \left(\underline{S}_{gg,T}(u) - S_{gg}(u) \right)^2 \right\}$$

is non-zero, and independent on T. Hence, the periodogram is not ergodic with respect to the power spectrum.

B.4.3 Time-invariant linear systems

Suppose we have a wide sense stationary stochastic process $\underline{f}(t)$ that is applied as the input to a linear time-invariant system with impulse response $h(t)$. Since the system is linear and time-invariant, the output

$$\underline{g}(t) = h(t) * \underline{f}(t) = \int_{\xi=-\infty}^{\infty} h(\xi)\underline{f}(t-\xi)d\xi \qquad \text{(b.45)}$$

is also wide sense stationary. To show this it suffices to calculate the expectation and autocorrelation function of $\underline{g}(t)$.

The expectation of $\underline{g}(t)$ follows from:

$$E\{\underline{g}(t)\} = h(t) * E\{\underline{f}(t)\} = \int_{\xi=-\infty}^{\infty} h(\xi) E\{\underline{f}(t-\xi)\} d\xi \qquad \text{(b.46a)}$$

If $\underline{f}(t)$ is wide sense stationary, its expectation does not depend on time. Hence, we may write $E\{\underline{f}(t)\} = \mu_f$. With that:

$$E\{\underline{g}(t)\} = \mu_f \int_{\xi=-\infty}^{\infty} h(\xi) d\xi \qquad \text{(b.46b)}$$

Clearly, the expectation $E\{\underline{g}(t)\}$ does not depend on time either. This can be made explicit by writing $E\{\underline{g}(t)\} = \mu_g$. Furthermore, with the introduction of the transfer function of the linear system, i.e. $H(u) = F\{h(x)\}$, the expectation of the output becomes simply:

$$\mu_g = \mu_f H(0) \qquad \text{(b.46c)}$$

The autocorrelation function of $\underline{g}(t)$ follows from:

$$\begin{aligned} R_{gg}(t_1,t_2) &= E\{\underline{g}(t_1)\underline{g}(t_2)\} \\ &= E\{(\underline{f}(t_1) * h(t_1))(\underline{f}(t_2) * h(t_2))\} \\ &= h(t_1) * E\{\underline{f}(t_1)\underline{f}(t_2)\} * h(t_2) \\ &= h(t_1) * R_{ff}(t_1,t_2) * h(t_2) \\ &= \int_\xi \int_\eta h(\xi) R_{ff}(\xi-t_1, \eta-t_2) h(\eta) d\xi d\eta \end{aligned} \qquad \text{(b.47a)}$$

If $\underline{f}(t)$ is wide sense stationary, the autocorrelation function depends only on a differences in time. Therefore we may write $R_{ff}(\tau)$ with $\tau = t_2 - t_1$.

$$\begin{aligned} R_{gg}(t_1,t_2) &= \int_\xi \int_\eta h(\xi) R_{ff}(\eta - \tau - \xi) h(\eta) d\xi d\eta \\ &= h(\tau) * R_{ff}(\tau) * h(-\tau) \end{aligned} \qquad \text{(b.47b)}$$

The conclusion is that $R_{gg}(t_1,t_2)$ depends only on τ as well. Therefore, $R_{gg}(t_1,t_2)$ can be replaced with $R_{gg}(\tau)$, and with that it has been shown that $\underline{g}(t)$ is wide sense stationary. Note that (b.47b) is quite similar to (b.32).

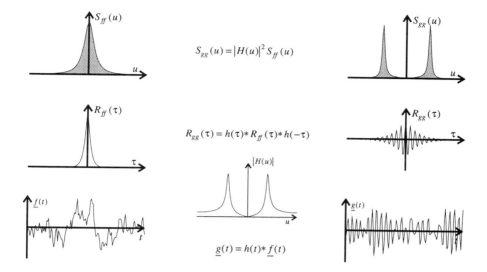

Figure B.9
Input-output relations of a linear, time-invariant system with stationary stochastic input.
In this example, the signal energy of the output is concentrated in a narrow frequency band

With equation (b.47b) it is easy to determine the power spectrum of $\underline{g}(t)$. Since $F\{h(-\tau)\} = H^*(u)$, it follows that:

$$S_{gg}(u) = H(u) S_{ff}(u) H^*(u) = |H(u)|^2 S_{ff}(u) \tag{b.48}$$

The signal power of the output is:

$$E\{\underline{g}^2(t)\} = \int_{u=-\infty}^{\infty} |H(u)|^2 S_{ff}(u) du \tag{b.49}$$

An example of a realisation of a stochastic process applied to a linear system is given in figure B.9. This figure also summarises the relations between input and output processes.

REFERENCES

Papoulis, A.: *Probability, Random Variables and Stochastic Processes*, McGraw-Hill, New York, 1965 (third edition: 1991).

Mortensen, R.E.: *Random Signals and Systems*, J. Wiley & Sons, New York, 1987.

Shanmugan, K. and Breipohl, A.M.: *Random Signals: Detection, Estimation and Data Analysis*, J. Wiley & Sons, New York, 1988.

BIBLIOGRAPHY

In addition to the articles and books cited in the text this bibliography provides references from which readers may continue theirs studies of the topics in the present book. No attempt has been made to provide a comprehensive guide to the very wide range of literature covering the fields of measurement science, computer vision, pattern classification and estimation.

- Alexandrov, V.V. and Gorsky, N.D., *Image representation and processing: a recursive approach,* Kluwer Academic Publishers, Dordrecht, 1993.
- Anzai, Y., *Pattern recognition and machine learning,* Academic Press, Boston, 1992.
- Batchelor, B.G., *Intelligent image processing in Prolog,* Springer, London, 1991.
- Beck, J.V. and Arnold, K.J., *Parameter estimation in engineering and science,* J Wiley & Sons, New York, 1977.
- Bow, S.T., *Pattern recognition and image preprocessing,* M Dekker, New York, 1992.
- Bretschi, J., *Automated inspection systems for industry: scope for intelligent measuring,* IFS (Publications), Bedford, 1981.
- Castleman, K.R., *Digital image processing,* Prentice-Hall, Englewood Cliffs, 1979.
- Catlin, D.E., *Estimation, control, and the discrete Kalman filter,* Springer-Verlag, New York, 1989.
- Duda, R.O. and Hart, P.E., *Pattern classification and scene analysis,* J. Wiley & Sons, New York, 1973.
- Eykhoff, P., *System identification: parameter and state estimation,* J Wiley & Sons, London, 1974.
- Fu, K.S., *Syntactic methods in pattern recognition,* Academic Press, New York, 1974.
- Fukunaga, K., *Introduction to statistical pattern recognition,* 2nd edition, Academic Press, Boston, 1990.
- Hall, E.L., *Computer image processing and recognition,* Academic Press, New York, 1979.

- Heijmans, H.J.A.M., *Morphological Image Operators,* Academic Press Inc., San Diego, 1994.
- Horn, B.K.P., *Robot vision,* MIT Press, Cambridge, Mass., McGraw-Hill, New York, 1986.
- Jaehne, B., *Spatio-temporal image processing: theory and scientific applications,* Springer-Verlag, Berlin, 1993.
- Kanatani, K., *Geometric Computation for Machine Vision,* Clarendon Press, Oxford, 1993.
- Kay, S.M., *Fundamentals of statistical signal processing: estimation theory,* Prentice-Hall, Englewood Cliffs, 1993.
- Kazakos, D. and Kazakos, P.O., *Detection and estimation,* Computer Science Press, New York, 1990.
- Kovalevsky, V.A., *Image pattern recognition,* Springer-Verlag, New York, 1980.
- Kubacek, L., *Foundations of estimation theory,* Elsevier, Amsterdam, 1988.
- Lewis, R., *Practical digital image processing,* Ellis Horwood, New York, 1990.
- Loughlin, C., *Sensors for industrial inspection,* Kluwer Academic Publishers, Dordrecht, 1993.
- Marr, D., *Vision,* Freeman, San Fransisco, 1982.
- Marion, A., *An introduction to image processing,* Chapman and Hall, London, 1991.
- Marshall, A.D. and Martin, R.R., *Computer vision, models and inspection,* World Scientific, Singapore, 1992.
- McLachlan, G.J., *Discriminant analysis and statistical pattern recognition,* J Wiley & Sons, New York, 1992.
- Naddler, M., and Smith, E.P., *Pattern Recognition Engineering,* J. Wiley & Sons, New York, 1993.
- Pao, Y.H., *Adaptive pattern recognition and neural networks,* Addison-Wesley, Reading, MA, 1989.
- Pavel, M., *Fundamentals of pattern recognition,* 2nd edition, M Dekker, New York, 1993.
- Pitas, I., *Digital image processing algorithms,* Prentice Hall, New York,. 1993.
- Serra, J., *Image Analysis and Mathematical Morphology,* Academic Press, New York, 1982.
- Silverman, B.W., *Density estimation for statistics and data analysis,* Chapman and Hall, London, 1986.
- Sonka, M., Hlavac, V. and Boyle, R., *Image processing, analysis and machine vision,* Chapman & Hall Computing, London, 1993.
- Staudte, R.G. and Sheather, S.J., *Robust estimation and testing,* J. Wiley & Sons, New York, 1990.
- Tarantola, A., *Inverse problem theory - methods for data fitting and model parameter estimation,* Elsevier, Amsterdam, 1987.
- Teuber, J., *Digital image processing,* Prentice Hall, New York, 1993.
- Therrien, C.W., *Decision, Estimation and Classification,* J. Wiley & Son, New York, 1989
- Tou, J.T. and Gonzalez, R.C., *Pattern recognition principles,* Addison-Wesley, Reading, MA, 1974.
- Wahl, M., *Digital image signal processing,* ArtechHouse, Boston, 1987.
- Zuech, N. and Miller R.K., *Machine vision,* Fairmont press, Lilburn, GA, 1987.

PERMISSION SOURCES NOTES

Figure 3.6c, chapter 3 page 44: reproduced by permission of TNO-FEL, The Hague.

Algorithm 6.1, chapter 6, page 178: reproduced by permission of Prentice Hall, Hemel Hempstead.

Algorithm 6.2, chapter 6, page 179: reproduced by permission of Prentice Hall, Hemel Hempstead.

Figure 7.11, chapter 7, page 221: reproduced by permission of P Brodatz, published by Dover Publications, Inc, New York.

Figure 7.61, chapter 7, page 284: reproduced by permission of K. Schutte, published by K. Schutte, Enschede.

Figure 7.63, chapter 7, page 288: reproduced by permission of K. Schutte, published by K. Schutte, Enschede.

Figure 7.65, chapter 7, page 290: reproduced by permission of A.J. de Graaf, Hengelo

INDEX

2-dimensional circulant convolution, 91
2-dimensional convolution, 22, 81
2-dimensional filtering, 40
2-dimensional Fourier transform, 28
2-dimensional look-up table, 217
3-dimensional object recognition, 273
4-connectedness, 110
4-neighborhood, 110
8-connectedness, 110
8-neighborhood, 110

A algorithm, 250, 287
absolute value cost function, 152
absolute value operation, 71
absorption, 2, 16
accuracy, 266
active imaging, 2
adaptive thresholding, 230
ADC, 65
adjacency, 110, 271
adjoint matrix, 306
adjoint, 89, 303
aliasing error, 58
aliasing, 55
ambiguity-resolving function, 265
amplitude discretisation, 65
amplitude spectrum, 28
analog-to-digital conversion, 49, 65
aperture function, 59

aperture, 21
applications, 4
AR process, 164
arc connectivity, 271
arc consistency, 283
arc, 207
area CCD, 50
area, 263, 266
aspect graph, 278
aspect, 278
associative law, 112
attributed graph, 207
autocorrelation function, 39, 62, 323
autocovariance function, 39, 323
autoregressive process, 164
average curvature, 270
average risk, 137
average squared distance, 187
average, 169
averaging operator, 82, 216
averaging, 324

background, 108
backpropagation neural network, 184
backward difference, 95
Bayes classification, 136
Bayes estimation, 152
Bayes' theorem, 319
bending energy, 270
Bessel function, 29
between-scatter matrix, 188
Bhattacharyya bounds, 191
Bhattacharyya distance, 199, 224
bias, 162
bilinear interpolation, 57
binary measurement vector, 171, 195
binomial distribution, 171, 193
bins, 173
bitmap, 253
blooming, 52
body co-ordinates, 288
body parameter estimation, 288
bottom-up control, 274
boundary extraction, 119
boundary list, 261
boundary segment, 250
boundary, 260
boxcar function, 242
branch, 207
branch-and-bound, 196, 280
breadth first, 250
bridge, 271
brightness, 77
Butterworth filter, 92

calibration, 289
camera co-ordinates, 17, 288
candela, 14
Canny operators, 240
CCD, 50
ceiling operation, 70
central limit theorem, 84, 316
central moment, 256, 315
centre of gravity, 256
centroid, 256
charge-coupled device, 50

charge-storage element, 50
Chernoff bound, 191
Chernoff distance, 191
chessboard distance, 109, 300
chromaticity, 77
circulant convolution, 81, 90, 307
circulant matrix, 306
city-block distance, 109, 300
classification, 134
clip operation, 70
clique finding, 287
closing, 114
cluster analysis, 167
co-occurrence matrix, 259
coherent light, 10
collimator, 32
color CCD, 51
color image, 216
color transform, 76
color, 10
commutative law, 112
commutativity, 23
compactness, 237, 258
complement, 109
component labeling, 253
component, 271
computer graphics, 3
computer vision, 3
condensing, 179
conditional covariance matrix, 154
conditional dilation, 123
conditional distribution, 319
conditional expectation, 319
conditional mean, 153
conditional probability density, 135, 151, 320
conditional risk, 136, 152
confusion matrix, 150
congruence, 272
connected component, 110
connectedness, 271
connection structure, 183
connectivity conversions, 121
constriction, 237
continuation, 228
continuous image representation, 49
contour segmentation, 267
contour tracing, 260
contour, 260
contrast, 272
convex deficiency, 123

convex hull, 123
convex, 123
convolution kernel, 81
convolution mask, 81
convolution, 22, 81
corner enhancement, 103
correction, 166
correlation coefficient, 39, 317
correlation, 23, 103, 226, 250, 228, 260
cost function, 136, 152, 249, 286
cost matrix, 136
cost, 136
covariance matrix, 320
covariance, 260, 317
cross correlation coefficient, 107
cross correlation, 40, 105, 226
cubic model, 35
cumulative histogram, 74
curvature, 267
curve fitting, 250
cutnode, 271
cutset, 234
cylinder-function, 29

dark current noise, 52
dark field, 32
decision boundary, 132
decision function, 135
decision making, 134
decomposition, 112
decorrelation, 321
degree, 271
depth first search, 197, 281
depth map, 21
design set, 167
detection quality, 45
detection, 104, 148, 240
determinant, 309
DFT, 64, 89, 264
diagonal fill, 121
diagonalisation, 311, 321
diameter, 258
difference, 95
differential operator, 94
diffraction, 22
diffuse illumination, 32
digital image, 2
digital morphology, 108
digraph, 208
dilation, 111

dimension, 296
dimensionality, 193
Dirac comb-function, 54
Dirac function, 23
directed graph, 208
directional derivative, 94
discontinuities, 43
discrete convolution, 80
discrete Fourier transform, 64, 89, 264
discrete image representation, 49
discrete Kalman filter, 166
discriminant functions, 179
distance measure, 109, 299
distance transform, 125
distribution function, 313
distributive law, 113
domain, 302
double subgraph isomorphism, 287
duality, 113
dyadic operation, 76
dynamic estimation, 163
dynamic range, 52
dynamic system, 151, 164

eccentricity, 257
edge based segmentation, 238
edge detection, 101, 238
edge linking, 247
edge map, 34
edge model, 44
edge spread function, 24
edge, 102, 207, 238
efficiency, 163
eigenvalue, 304
eigenvector, 304
electromagnetic spectrum, 9, 11
elongation, 258
energy, 105, 260
ensemble, 322
entropy, 260
epoch, 181
equal covariance matrices, 200
equal expectation vectors, 201
equidistant quantisation, 66
ergodic, 325
erosion, 111
error backpropagation, 183
error function, 146

error rate, 139, 186, 193
error tolerant graph
 matching, 287
Euclidean distance, 104
Euclidean space, 298
evaluation set, 193
even field, 51
event, 313
expectation vector, 320
expectation, 36, 315, 322
experiment, 313
extensivity, 113: 116
external characteristics, 272, 280
external labels, 280
extrapolation, 247

false alarm, 148, 227
fast Fourier transform, 65
fast transform, 89
FD normalisation, 265
FDs, 262
feature extraction, 193, 218, 226, 230
feature selection, 193: 195
feature vector, 198, 222
FFT, 65
filtering of stochastic processes, 41
first order statistics, 34
Fisher's linear discriminant, 190
fixed-pattern noise, 52
flat model, 35, 216, 236
floor operation, 70
flux, 12
foreground, 108
formal language, 211
forward difference, 96
Fourier analysis, 24
Fourier coefficients, 301
Fourier descriptors, 261
Fourier matrix, 308
Fourier series, 301
Fourier transform, 26, 81, 89
frame storage, 51
frame, 271
Freeman chain code, 261
front illumination, 31
functional, 296

gamma correction, 72
gamma, 73
Gauss differential operator, 99

Gauss function, 29
Gaussian probability density, 316
Gaussian filter, 82, 83, 216
Gaussian random vector, 140
generalised cylinder, 276
generalised linear discriminant functions, 182
geodesic distance, 126
geometric parameters, 255
geometry, 232
gradient vector, 94, 230
grammar, 211
graph matching, 273, 285
graph searching, 248
graph, 207, 271
grey level thresholding, 213
grey scale morphology, 127
grouping, 287

Hadamard transform, 92
harmonic function, 24
Hessian matrix, 310
heuristic search, 249, 281
hidden layer, 184
hierarchy, 208
high-emphasis filtering, 85
high-pass filtering, 85
histogram egalisation, 75
histogram manipulation, 73
histogram, 73
histogramming, 36, 73, 173
hit-and-miss transform, 114
homogeneity criterion, 236
homogeneous co-ordinates, 19, 288
homotopy preserving thinning, 120
homotopy, 120
Hough transform, 250
HSV model, 77
hue, 77
hypothesis testing, 236

idempotence, 116
IDFT, 65
if...then..." statements, 282
illuminance, 13
image acquisition, 49
image analysis, 207
image based measurement systems, 1
image coding, 4
image conditioning, 31

image enhancement, 4
image feature extraction, 4
image formation, 9
image model, 33, 277
image operation, 69
image restoration, 4, 85
image segmentation, 212
image, 2
impulse modulation, 53
incoherent light, 10
increasingness, 113, 116
independent random variables, 317
inertia, 260
inexact graph matching, 287
inner boundary point, 119
inner product space, 298
inner product, 86, 298
input layer, 184
interclass and intraclass distance, 186, 204, 217
interclass distance, 189
interference error, 58
interior point, 119
interlacing, 51
interpolation function, 57, 174
interpretation, 280
intraclass distance, 189
intrinsic characteristics, 272, 280
inverse discrete Fourier transform, 65
inverse matrix, 306
inverse operator, 302
irradiance, 13
isolated pixel, 118
isomorphic graphs, 285
isomorphic subgraph, 285
isotropic, 15

Jacobian matrix, 310, 290
joint distribution function, 316, 319
joint moments, 317

k-nearest neighbor rule, 177
k-NNR, 177
Kalman filter, 166
Kalman gain matrix, 166
kernel, 23, 81
kurtosis, 36

labeling, 280
labels, 33

Lambertian surfaces, 16
Laplacian, 95, 227, 230
leakage, 63
learning rate, 185
learning, 167
least squares estimation, 159
least squares fitting, 158
left direct matrix product, 88
length, 125
lens system, 22
light field, 32
likelihood ratio, 144
likelihood, 144, 158
line detection, 227
line element, 45
line model, 45
line spread function, 24
line-scan CCD, 50
linear combination, 296
linear decision function, 141
linear dependent, 296
linear discriminant functions, 180
linear feature extraction, 198, 222
linear image filtering, 81
linear mean square estimation, 162
linear model, 35
linear operator, 78, 302
linear (sub)space, 295
linearity, 22, 23
link, 207
LMSE-estimation, 162
local homogeneity, 260
local neighborhood, 79
localisation, 104, 240
log-likelihood ratio test, 145
log-likelihood ratio, 225, 231
logical connectives, 281
logical operation, 76
look-up table, 71
low-pass filtering, 82
LSE, 159, 290
luminance, 16, 77
luminous energy, 13
luminous exitance, 13
luminous flux, 13, 14
luminous intensity, 13
LUT, 72, 117

Mahalanobis distance, 141, 146
majority voting operation, 118

MAP estimation, 155
map, 34
marginal distribution, 316
Marr-Hildreth operation, 102, 244
mask, 123
matched filter, 48
matching techniques, 280
mathematical morphology, 107
matrix, 305
maximum a posterior estimation, 155
maximum chord, 258
maximum likelihood estimation, 158, 195
maximum posterior probability classification, 138
maximum probability, 260
mean squared error estimation, 153
measurement space, 132
measurement vector, 131, 212
median, 153
metric space, 299
minimum chord, 258
minimum cost graph matching, 286
minimum cost path, 248
minimum distance classificatiion, 142
minimum Mahalonobis distance classification, 141
minimum mean square error estimation, 153
minimum risk classification, 137
minimum risk estimation, 153
minimum risk, 152
minimum variance estimation, 153
minimum-cost path, 287
missed event, 148
ML-estimation, 158, 195
modality, 2
mode, 153
model based systems, 274
modulation transfer function, 31
moment invariants, 257
moments, 255, 315
monadic operation, 70
monochromatic light, 10

morphological operation, 107, 223, 247, 283
MSE, 153
multi-edit algorithm, 178
multinomial distribution, 171
multiple responses, 229, 230, 242

nearest neighbor classification, 175
nearest neighbor estimation, 175
nearest neighbor rule, 177
neighborhood operation, 80, 108
neighborhood, 80, 214, 224
neural network, 183
neurons, 183
NNR, 177
node adjacency list, 272
node connectivity, 271
node consistency, 283
node, 207
noise propagation, 86: 97
noise suppression, 99, 117
noise whitening, 161
noise, additive, 35
noise, Gaussian stochastic process, 35
noise, multiplicative, 35
non-local maximum suppression, 104, 226, 229
non-parametric learning, 168, 173
norm, 105, 296
normalisation, 75
normalised central moments, 257
normalised distance, 271
normalised Fourier descriptors, 265
normed linear space, 297
Nyquist criterion, 58
Nyquist frequency, 58

object model, 274
object recognition, 272
occlusion, 287
odd field, 51
off-specular reflection, 17
offset operation, 70
opening, 114
operator, 302
optical transfer function, 31
orientation, 255

orthogonal transform, 86
orthogonal, 298
orthogonality, 25
orthonormal basis, 301
orthonormal system, 300
outcomes, 313
outer boundary, 119
output layer, 184
over-segmentation, 234

parameter estimation, 132, 151
parametric curve, 261
parametric learning, 168, 195
parsing, 211
Parzen estimation, 173, 214
passive imaging, 2
path, 110, 271
pattern classification, 131, 207, 210, 212, 221
pattern, 131
perceptron, 181
performance measure, 186, 195
perimeter, 263, 266
perspective projection, 17, 19, 20, 221
phase spectrum, 28
photometric quantities, 13
photometry, 10
physical process, 131
pinhole camera model, 19
pivoting, 287
pixel classification, 212
pixel, 53, 109
pixel-to-pixel operation, 69
point operation, 69
point spread function, 20, 21
Poisson distribution, 315
polychromatic light, 10
polygon approximation, 269
polyhedrons, 276
pose estimation, 289
position, 255, 263
posterior probability density, 153
posterior probability, 136
power spectrum, 28, 41, 62, 260, 323
predicate calculus, 281
predicates, 281
prediction, 159: 166
preprocessing, 178
presampling filtering, 61
Prewitt operator, 240

Prewitt's compass operator, 240
primary colors, 76
primitives, 207
principal axes, 256
principal moments, 256
prior probability, 134, 151
probability density, 35, 314
probability, 313
projection function, 243
projection, 17, 19, 299
propagation, 123
proposition, 272
pruning, 121
PSF-matrix, 81

QPT, 233
quadratic cost function, 152
quadratic decision function, 141
quadratic discriminant functions, 182
quadratic model, 35
quantisation error, 67
quantum noise, 53
quartic picture tree, 233

radiance, 16
radiant energy, 9, 12, 13
radiant exitance, 13
radiant flux, 13
radiant intensity, 13
radiometric parameters, 258
radiometric quantities, 13
radiometry, 9, 232
RAG, 270
ramp, 239
random variable, 313
range imaging, 34
range, 302
random vectors, 319
ray tracing, 290
realisation, 322
rear illumination, 31
reasoning, 281
receiver operating characteristic, 148
reconstruction by dilation, 123
reconstruction error, 58
reconstruction filter, 57, 99
reconstruction, 56
rectangle-function, 28
reflectance distribution, 16
reflection models, 16

reflection, 2, 16
region adjacency graph, 270
region aggregation, 232
region based segmentation, 231
region filling, 123
region growing, 233
region labeling, 281
region map, 34, 253
region of interest, 214
region parameters, 254
region properties, 253
region, 33, 109, 212
rejection option, 149
rejection rate, 149
relational description, 270
relative size, 271
relaxation, 283
rendering, 290
reproducing density, 163
reset noise, 52
residuals, 159, 236
resolution error, 58
resolution, 99
RGB model, 77
risk, 136, 153
Roberts operator, 240
robustness, 163
ROC, 148
ROI, 214
roof edge, 239
rotation, 263
rotationally invariant, 95
rule based segmentation, 238
running arc length, 261

sampled stochastic process, 61
saturation, 77
scale operation, 70
scale, 99, 102, 243
scaling, 263
scanning, 49
scatter diagram, 38, 132
scatter matrix, 187
scattering, 2
scene, 1
Schwarz inequality, 298
scratch detection, 45
search tree, 280, 286
second directional derivative, 95
second order statistical parameters, 39
second order statistics, 37

seed pixel, 232
seed, 123
segment, 33, 212
self-adjoint matrix, 306
self-adjoint operator, 300, 304
semantic network, 275
semantics, 211
sensoring, 49
sensory system, 131
separability, 23
separable matrix, 307
separable PSF, 81
separable transform, 88
sequency ordering, 93
set theory, 107
shape based segmentation, 237
shape, 255
shift invariant, 21, 22, 80
shift variant, 69
sifting integral, 23
sifting, 124
sigmoid function, 184
signal power, 323
signal-to-noise ratio, 47, 241
sinc-function, 28
size, 255, 263
skeleton, 120
skewness, 36
sloped model, 35, 236
smear, 52
smearing, 63
SNR, 47, 98, 241
Sobel operator, 240
soft threshold-function, 184
solid angle, 9
space invariant, 70, 80
space variant, 69
spatial frequencies, 24
spatial sampling, 49, 53
spectral feature, 216
spectral irradiance, 13
spectral luminous efficacy, 14
spectral power density, 42
spectral radiant exitance, 13
spectral radiant flux, 13
spectral radiant intensity, 13
spectral responsivity, 10
spectrum, 216, 304
specular illumination, 32

specular reflection, 16
split-and-merge, 233
spot detection, 225
spots, 45, 225
SQE, 182
squared error performance, 182
square operation, 70
square root operation, 70
stability, 163
standard deviation, 29, 36, 315
start point, 263
state variable, 164
static system, 151
stationarity, 37, 221, 323
statistical pattern classification, 131, 207, 210, 212, 221
steepest descent procedure., 182
step edge, 239
step-like transition, 44
stereoscopic vision, 291
stochastic process, 322
structural pattern recognition, 207
structured light, 33
structuring element, 111
Student's t test, 237
superposition principle, 78
supervised learning, 167
support function, 284
symbolic image, 34
symmetric difference, 96
symmetries, 263
syntactic pattern recognition, 211
syntactic rule, 211
system parameters, 289

t-distribution, 237
table-look-up operation, 71
target, 183
template, 105
textural parameters, 258
texture energy, 260
texture feature, 219, 260
texture measure, 237
thermal noise, 53
thickening operation, 122
thickness, 124, 258

thinness, 228
thinning, 120
threshold operation, 70
threshold, 45
time-invariant linear system, 326
top surface, 127
top-down control, 274
Torrance-Sparrow model, 17
trace, 308
training set, 167, 193
transfer function, 81, 184
translation, 113, 263
transmission, 16
tristimulus values, 76

umbra, 127
unconditional probability density, 135
uncorrelated random variables, 317
under-segmentation, 235
unidirected graph, 208
uniform cost function, 138, 152, 155
uniform distribution, 314
uniform resampling, 263
uniformity predicate, 234
unitary matrix, 306
unitary operator, 303
unitary transform, 89
unsupervised learning, 167

variance, 36, 315, 323
vector space, 295
vertex, 207

wavelength, 9
white noise, 42, 324
whitening, 189, 201
window function, 63
Wishart distribution, 170
within-scatter matrix, 188
world co-ordinates, 17, 288

X-ray imaging, 22
XYZ-model, 77

zero crossing detection, 102, 244

WILEY SERIES IN MEASUREMENT SCIENCE AND TECHNOLOGY

Chief Editor

Peter H. Sydenham
University of South Australia

Instruments and Experiences: Papers on Measurement and Instrument Design
R. V. Jones

Handbook of Measurement Science, Volume 1
Edited by P. H. Sydenham

Handbook of Measurement Science, Volume 2
Edited by P. H. Sydenham

Handbook of Measurement Science, Volume 3
Edited by P. H. Sydenham and R. Thorn

Introduction to Measurement Science & Engineering
P. H. Sydenham, N. H. Hancock and R. Thorn

Temperature Measurement
L. Michalski, K. Eckersdorf and J. McGhee

Technology of Electrical Measurements
Edited by L. Schnell

Traceable Temperatures: An Introduction to Temperature Measurement and Calibration
J. V. Nicholas and D. R. White

Image Based Measurement Systems
F. van der Heijden

DESIGN AND MEASUREMENT IN ELECTRONIC ENGINEERING

Series Editors

D. V. Morgan
School of Electrical, Electronic and Systems Engineering
University of Wales College of Cardiff,
Cardiff, UK

H. L. Grubin
Scientific Research Associates Inc.
Glastonbury, Connecticut, USA

THYRISTOR DESIGN AND REALIZATION
P. D. Taylor

ELECTRONICS OF MEASURING SYSTEMS
Tran Tien Lang

DESIGN AND REALIZATION OF BIPOLAR TRANSISTORS
Peter Ashburn

ELECTRICAL CHARACTERIZATION OF GaAs MATERIALS AND DEVICES
David C. Look

COMPUTERISED INSTRUMENTATION
Tran Tien Lang

PRINCIPLES AND TECHNOLOGY OF MODFETs
VOLUMES 1 AND 2
H. Morkoç, H. Unlu, G. Ji

RELIABILITY OF GALLIUM ARSENIDE MMICs
Edited by Aris Christou

INTEGRATING QUALITY AND RELIABILITY INTO MICROELECTRONICS MANUFACTURING
Aris Christou

DESIGN OF SMALL ELECTRICAL MACHINES
E. S. Hamdi

IMAGE BASED MEASUREMENT SYSTEMS
F. van der Heijden